REMAKING BOSTON

REMAKING BOSTON

AN ENVIRONMENTAL HISTORY OF THE CITY AND ITS SURROUNDINGS

Edited by Anthony N. Penna
and Conrad Edick Wright

UNIVERSITY OF PITTSBURGH PRESS

Published by the University of Pittsburgh Press, Pittsburgh, Pa., 15260

Manufactured in the United States of America
Printed on acid-free paper
10 9 8 7 6 5 4 3 2 1

Library of Congress Cataloging-in-Publication Data

Remaking Boston : an environmental history of the city and its surroundings / edited by Anthony N. Penna and Conrad Edick Wright.
p. cm.
Includes index.
ISBN-13: 978-0-8229-4381-5 (cloth : alk. paper)
ISBN-10: 0-8229-4381-6 (cloth : alk. paper)
1. Boston Region (Mass.)—Environmental conditions—History. 2. Environmental policy—Massachusetts—Boston Region—History. 3. Environmental degradation—Massachusetts—Boston Region—History. 4. Nature—Effect of human beings on—Massachusetts—Boston Region—History. I. Penna, Anthony N. II. Wright, Conrad Edick.
GE155.M4R46 2009
333.7709744'61—dc22 2009029725

CONTENTS

ACKNOWLEDGMENTS

THIS collection developed out of a conference in May 2006 at the Massachusetts Historical Society, which in turn grew out of the Society's seminar series on environmental history. Since 2001, the Society has hosted the Boston Environmental History Seminar. The success of the series over the years encouraged its steering committee to organize the conference "Remaking Boston: The City and Environmental Change over the Years." Many of the contributions presented on that occasion seemed to warrant revision and publication, and the book in your hands is the result.

The volume's editors are grateful to past and present members of the steering committee for the Boston Environmental History Seminar as well as to the conference participants, past and present members of the Society's staff, and the staff of the University of Pittsburgh Press for all they have done to make this book possible. On the steering committee we owe thanks to Phyllis Andersen, Chris Bosso, Karl Haglund, Bill Meyer, Bill Newman, Steve Rudnick, and Nancy Seasholes. From the conference participants, in addition to those who are represented by chapters in this collection, we are in debt to Robert Allison, Martyn Bowden, Robert Clarke, Michael Conzen, Michael S. Dukakis, Jim Lambrechts, Colin Polsky, Barbara Gutmann Rosenkrantz, Joel A. Tarr, Alexander von Hoffman, Sam Bass Warner Jr., and Peter K. Weiskel. We are also grateful to the Lowell Institute, which underwrote the conference's keynote address by Brian Donahue. From the Society's staff, we owe special thanks to Anne E. Bentley, Chris Carberry, Peter Drummey, Dennis A. Fiori, William M. Fowler Jr., James Harrison, Nancy Heywood, Ondine E. Le Blanc, Cherylinne T. Pina, Melissa Pino, Erin Pipkin, Jean Powers, Jen Smith, Seth M. Vose III, and Donald Yacovone. And to the staff of the University of Pittsburgh Press, our thanks go to freelance editor Amy Smith Bell, Deborah Meade, Cynthia Miller, and Ann Walston as well as to the series editors, Joel A. Tarr and Martin V. Melosi.

MICHAEL RAWSON'S contribution, "What Lies Beneath: Science, Nature, and the Making of Boston Harbor," first appeared in the *Journal of Urban History* 35 (2009). We publish it in this book by arrangement with the *Journal of Urban History.*

REMAKING BOSTON

1

Introduction

BOSTON FROM PENINSULA TO METROPOLIS

Anthony N. Penna

IN 1630, English settlers established the town of Boston. Four years later, William Wood, a visitor to the frontier outpost, published a description of its physical features. His book, *New England's Prospect,* became for many readers an early example of boosterism, as it extolled the virtues of this bountiful new land. Wood's picture of Boston is especially revealing because the town that he described bore only a remote resemblance to the city since created by centuries of land making. In his vivid portrayal, Boston stood on a mostly square peninsula connected to the mainland at Roxbury by a very narrow neck. "Tramont," Wood's term, denominated the peninsula's most prominent feature, the three hills that looked down on the town.

Boston's home, the Shawmut Peninsula, was in the most precarious of locations, exposed to a deep-channeled harbor and the unpredictable Atlantic Ocean. An isthmus 120 feet wide at high tide supported the single road (the future Washington Street) that connected the town to the mainland. Although we now think of Boston as a major coastal city, a product of centuries of engagement in global trade, its modest origins identify it before the early nineteenth century as effectively an island. Tidal mudflats—the

Back Bay with the Charles River running through it—separated the town from the mainland. Despite some land making—namely, a wall to seal off the rising sea and some rocks, gravel, sand, and soil dumped for fill—as late as the early 1800s, the center of gravity of this small and constricted community remained its downtown, which consisted of Beacon Hill, the North End, and the West End.

With an expanding commercial global economy, symbolized by wharves that thrust toward the deep water of the harbor, land making became a necessity. Rope making was a thriving industry in a town where sailing ships dotted the harborscape, for example, and by the end of the eighteenth-century rope makers had to have new space to work. "Using hemp fibers and flammable tar," the geologist William A. Newman and the sociologist Wilfred E. Holton have written, "ropes were woven in a straight line while the rope makers walked backward in long wooden sheds, called ropewalks."[1] Fires, a constant threat in crowded cities built with flammable materials, visited Boston throughout its early history, and in 1796 one of its most serious blazes destroyed six ropewalks as well as ninety-six other buildings. Fear of recurring conflagrations inspired ropewalk owners to petition the town for permission to construct facilities over marshlands and mudflats west of Boston Common. Thus a long history of land making started with a municipal and commercial agreement.

Land making began on Charles Street with gravel from Mount Vernon, west of Beacon Hill, in 1803. The new land connected Charles Street to the new West Boston Bridge, linking Cambridge to Boston, and created new housing in the Charles Street and Beacon Street area. Additional land making widened the neck connecting Boston to Roxbury in the 1820s, and within thirty years new land extended along Beacon Street to the present location of Clarendon Street. By 1822, when incorporation legally changed the town into a city, major transformations were already under way. In response to a growing population, the marshes and mud flats became a sink for dumping. Recycling animal and human wastes were a principal attribute of the complex ecosystem here, and when the process overwhelmed human senses and sensitivities, efforts commenced to find relief. Land making served as a remedy, and it also responded to the need for urban space for a growing population. Capital growth, commercial and residential construction, and the din of busy city streets could not mask Boston's growing ecological footprint. A city with a manageable population of about 61,000 in 1830, it reached 177,000 within three decades. Boston's growing population of humans and draft animals metabolized increasing quantities of food to do work. Disposal of wastes continues to be an ongoing challenge for all cities, including Boston; as a coastal city, Boston's adjacent waterways served the function of ultimate sink.

With nineteenth-century industrialization and the construction of mill dams using tidal power to run factory machinery, the ability of ten-foot tidal surges to flush the flats each day disappeared. The process of harnessing waterpower began in colonial Boston with the damming of the cove separating the North End and the West End and creating the Mill Pond. The construction of the Mill Dam and the Cross Dam coupled with the creation of a Full Basin and a Receiving Basin sealed off the Back Bay from Boston Harbor in the 1820s. As a significant engineering achievement that allowed industry to run around the clock, these dams and basins made a noteworthy contribution to Boston's built environment. Commercial success, however, did not mask the effects of these engineering marvels on the natural rhythms of a coastal environment. Sewage and trash accumulated and solidified in the basins. Rather than moving out to sea with the ebb tide, waste became a constant reminder to Boston's growing population that manufacturing and wealth creation left as an unseemly by-product.

Turning these effluent-gathering mudflats into land upon which to build fashionable residences for the city elites solved two problems. It addressed the matter of waste management and stemmed the flight of the wealthy to the suburbs. This greatest of land-making projects lasted for almost a quarter of a century until its completion in 1885.[2] Today, Boston's Back Bay bears no resemblance to its premodern conditions and the memory of its history is lost to many of its residents and visitors. Victorian townhouses bordered Commonwealth Avenue, a two-hundred-foot-wide thoroughfare in the tradition of the grand boulevards of Napoleon III, with a wide, tree-lined walking path in the center. Its green promenade, characteristic of the nineteenth-century walking city, integrated nature into an urban-built environment. The avenue's green space "would later be connected to a string of green spaces stretching out to Franklin Park known as Boston's 'Emerald Necklace'; the Back Bay is thus an important link in the park system that starts with [the] Boston Common and the Public Garden."[3]

Although Commonwealth Avenue is a particularly distinctive feature of Boston's urban design, Back Bay today brings walkers into contact with an array of late-nineteenth- and early-twentieth-century stone and brick residences along many tree-lined streets. The magnet that attracts residents, professionals, and visitors is the vitality of the district. Two modern urban malls, Copley Place and the Prudential Center, fill daily with shoppers, while expensive boutiques and art galleries on Newbury Street cater to upper-income tourists. Boylston Street—a mixture of fashionable shops, chain stores, and office buildings—bustles with activity. A walker in the city encounters many of the Back Bay's landmarks: Trinity Church and the Boston Public Library in Copley Square, Symphony Hall, the Massachusetts Historical Society, the Charles River Esplanade, and many other centers of the city's cultural and educational life.[4]

Land making, although vital to Boston's physical expansion, represented only a fraction of the city's growth. An era of annexation turned many of the nation's nineteenth-century cities into metropoles. For Boston, annexation began rapidly with Roxbury in 1868, followed by Roslindale in 1873, and the following year Charlestown, Dorchester, Brighton, and West Roxbury. Failure to annex Brookline in 1911, however, brought an end to this activity. Expansion broadened Boston's tax base, gave it access to much needed potable water, and provided a protective umbrella of city services to the smaller residential communities.

At the time of this writing, Boston's population approached 560,000, after a modest decline of 5 percent during the first five years of the twenty-first century. It is a relatively young city, with a median age of thirty-one and with more than 33 percent of the population under forty. Women outnumber men in Boston 52 percent to 48 percent, while Caucasians represent 54 percent, African-Americans 25 percent, Asians 8 percent, and others 13 percent. The city covers 48.4 square miles of land, with an average elevation of 20 feet above sea level and 41.2 square miles of water. Physical changes in Boston's environment are a central theme throughout this book. Other continuing themes include the link of rural and urban spaces and how Bostonians made sense of their environment. Each of the collection's three parts—on the Boston Harbor, on the town and the countryside, on and on the climate and the weather—begins with an introduction and then turns to two or more contributions on particular aspects of the section's theme. The editors offer the more specific chapters as examples of current work in urban environmental history and models for scholars to emulate.

Like many edited collections, this one does not aim for comprehensive coverage of its subject. The history of Boston's "remaking" since the seventeenth century is so complicated that such a book would have to be many times the size of this one to address every aspect. Among the topics this collection does not consider systematically are industrial air pollution, the search for an adequate water supply, and the legal aspects of the cleanup of Boston Harbor. As a transoceanic trading center that focused its manufacturing strategically on light industries, such as textiles and shoes, Boston is distinctive among American metropolises for the relative unimportance of industrial air pollution. Boston avoided the toxic clouds which choked cities that produced iron and steel or refined petroleum. A windy city located on the coast, its upper atmospheric currents carried away what air pollution its factories did produce. Neither mountains nor valleys trapped pollutants causing temperature inversions and stagnation.[5] Moreover, such topics as the provision of an adequate supply of potable water and the judicial history of the harbor cleanup are already the subjects of important monographs and do not require further attention here.[6]

Despite its focus on Boston's urban and environmental history, neither is this collection limited to the research of historians. *Remaking Boston* contains contributions from specialists from such fields as earth science, environmental engineering, biology, cartography, geography, and urban planning as well as urban and environmental history. This multidisciplinary approach to the urban and environmental history of a single city adds contextual richness to the topic. Each discipline possesses its own theoretical structure and glossary of terms, and the contributors studying Boston from different disciplinary perspectives enhance the understanding of place. Similar to other urban places, settlers and then residents reshaped Boston's fragile ecology. Each chapter reformulates prior understanding by crossing disciplinary boundaries with new research findings about the city's history.

The Harbor and Land at the Water's Edge

As a thriving coastal city, Boston's urban and environmental history represents an opportunity to examine the transformation of space from a natural habitat into a vibrant metropolis. In this context its physical space includes the deep-channeled harbor that entrepreneurs developed into a waterway to support flourishing commercial enterprises. Harbors, as the most prominent geographical feature of coastal cities, do not inhabit a fixed economic, cultural, or social space. For this reason geographers, geologists, hydrologists, cartographers, and historians study harbors in contextualized spatial terms. Boston was no exception as its transatlantic commercial partners turned the harbor into a booming port engaged in global transactions.

The beginning of the harbor's spatial transformation predated human habitation by many millions of years. Knowing its prehistory, which extends back in time to its geological origins, pushes at the boundaries of urban and environmental history. This earliest history encompasses the geographical space beyond the water's edge to include more than the commercial traffic in clipper and steam-driven ships on the harbor's surface. In this sense the water's edge, the harbor islands, its channels, and its sediments are manifestations of complex geological and human actions. With the end of the last ice age, 11,600 years before the present (BP), the evidence of 96,800 years of accumulated ice approximately 1.25 miles high at its zenith left a topography and physical space familiar to the region's early Native American inhabitants. Glaciers sculpted the harbor's edge and the region's landscape. Hills of undifferentiated soil composition, sand, clay, and gravel on the surface of the land, in the harbor's islands, and in its bottomlands are reminders of the city and the region's earliest history. Runoff from the land carved the harbor's channels, inundated coastal zones, and created the shoreline that early settlers, both Native American and European, encountered. Rebounding of the land from the weight of mile-high glaciers continued in juxtaposition with slowly rising sea levels.

In the book's first part, "The Harbor," earth scientists Peter S. Rosen and Duncan M. FitzGerald point out in their chapter, "The Drowning of Boston Harbor and the Development of the Shoreline," that "glaciers of the most recent ice age were the indispensable agents in the creation of the harbor."[7] At the water's edge the harbor became the scene of dynamic change rather than a fixed container impervious to natural and human interventions. In geological time, changing sea levels sculpted the boundaries defining the water's edge until a point of stability was reached four thousand to five thousand years ago. The natural world at the water's edge remained relatively unchanged until early English communities intervened beginning in the seventeenth century. The reshaping of the landscape by glacial deposits encompassed both the harbor and the land and defined the topography of the harbor's sediments, its islands, and Boston's hills, valleys, and flatlands as well as its rivers, streams, and ponds.

Analyses of urban land-use patterns have focused on human interventions. Landscape origins, gleaned from the work of geologists, deepen our knowledge of spatial relationships and expand our definitions of landscape patterns. Of similar importance is the way in which geological origins serve to integrate the harbor and the land, to see them as a unit rather than as separate entities. By focusing on the physical environment, Rosen and FitzGerald recognize its fundamental importance in shaping the urban space that would become Boston and its surroundings. Its hills, navigable rivers, soil composition, water quality, and much of the natural world around it shaped its subsequent expansion. The city's future capacity to create a vibrant economy and to expand geographically to accommodate waves of immigrants needed a physical space with abundant natural resources devoid of economic, legal, and social bottlenecks.

Historian Michael Rawson's chapter, "What Lies Beneath: Science, Nature, and the Making of Boston Harbor," reminds us that encroaching upon the harbor's physical space through land making engendered opposition from many of the city's mariners and some of its engineers. Drawing on shifting paradigms in science, the competing interests of stakeholders, and federal and state policy, Rawson constructs a convincing argument about the role of scientific thought in shaping Boston Harbor and the larger subject of the interrelationship of the harbor, the city, and nature. Land making, a nineteenth-century initiative of pro-growth real-estate interests, legislators, and some corporations, violated the fragile ecological integrity of the harbor's physical space, its bottomlands and navigable channels. Rawson points out that urban and environmental historians should study coastal zones from the perspective of shoreline activities from the harbor's waters to ships exchanging cargo at port sites: "They have not dived deep enough to make it to the bottom of harbors."[8]

Boston Harbor was part of a larger complex hydraulic system linking the harbor's waters to the inland with sea levels rising as much as eight to ten feet with the incoming tides. From the perspective of mariners, the filling by landowners of the mudflats and marshes that constituted many miles of shoreline disrupted the natural kinetic energy of the tides to scour the harbor's bottomlands. The result was a degraded and increasingly precarious harbor for entering and departing ships. For much of the nineteenth century, proponents of the science of hydrology shaped public policy regarding Boston's harbor. In filling the tidal flats, the Commonwealth of Massachusetts charged Boston by the cubic yard of tidewater lost. The commonwealth's Harbor Commission used such fines to construct seawalls, engage in dredging operations, and make improvements to the harbor's navigational system.

Powerful commercial and economic opponents waged a sustained campaign against an equally powerful preservationist lobby of shipowners and their allies in science, engineering, and government. A scientific paradigm shift in the early years of the twentieth century revealed the scouring theory to be erroneous, however. The evolving science of geology and its theories about glaciers and their meltwater explain the existence of Boston Harbor's deep channels and not the alleged scouring capability of the tides or the rivers that emptied into it. Rawson offers a compelling argument about the ways in which scientific paradigms along with economic and political factors have shaped the production of urban and environmental spaces.

In his contribution "Remaking Boston Harbor: Cleaning Up After Ourselves," environmental engineer Steven M. Rudnick unravels the long and protracted history of the harbor as a sink for metabolized waste. He details the efforts over time to rid the harbor of its most objectionable matter. Definitions of what constituted waste have changed historically from nuisance to pollution, from solid waste to any decaying odorous biological matter. With the intervention of scientific analysis in the twentieth century, microscopic particles on the surface, in the water, or buried in the sediment at the bottom of the harbor became the domain of newly imposed federal regulations in postwar Boston. While recognizing the growing burden that effluent placed on the harbor, centuries of neglect by mayors, aldermen, state legislators, select committees, and commissions resulted either in denial about the severity of the harbor's health or in shortsighted measures that delayed the eventual and inevitable cleanup. A pattern of discharging effluent further away from its point source in Boston and the surrounding towns into the harbor and its outer islands characterized much of the public policy about human waste disposal. With advances in engineering technology available to public officials in the nineteenth and twentieth centuries, the construction of main drains, submerged tunnels, and combined sewer outfalls pushed effluent so far away from its point source that residents of Cape Cod in the 1990s challenged the legality of an outfall 9.5 miles from the shoreline.

As Rudnick has pointed out, the harbor cleanup required the application of biological and chemical solutions to waterborne pollution, the technology of primary and secondary sewer treatment, the imposition of federal water quality standards, and an edict from a federal judge establishing an independent regulatory authority with powers to levy fees on all municipalities that contributed to using the harbor as an ultimate sink. A contributing factor to consider in the harbor's renewal was the expectation among the commonwealth's citizens for the amenities associated with healthy beaches. At the water's edge beaches had become another physical space dedicated to recreation, much like the emerging park system that offered a respite from the toil of daily life for the city's growing twentieth-century population.

The town's original shoreline configuration has remained a matter of inquiry among cartographers attempting to identify the city's original boundaries and its geographical relationship to the harbor. Mapping Boston is the subject of many books and museum exhibitions. In "In Search of the Shawmut Peninsula: Using Modern Cartographic Analysis to Discover the 'Original' Boston Shoreline," Stephen T. Mague uses the latest scientific methods to consider this ageless conundrum. His contribution addresses the effects of urban infrastructure on flora and fauna from a cartographer's perspective by using historical maps and technical analysis to examine alterations of the land at the water's edges. His investigation of spatial transformations of the shoreline cuts across the disciplinary boundaries of law in property rights and real-estate transactions, of civil engineering, and of urban design to solve such problems as groundwater incursions and land subsidence and the potential for earthquake disruptions of Boston's built environment and its supporting infrastructure. The city's coastal configuration and attention to the relationship of the land to its harbor reflected an ongoing deliberation among Boston's stakeholders engaged in economic and commercial activities. Legal title to land created from tidal pools, mudflats, and marshes defined private waterfront property with warehouses, wharves, docks, and offices. At the same time, regulations guaranteed, restricted, or denied public access to tidelands. Land at the water's edge represented Boston's dynamic expansion, and its newly built environment became a contested space. Mague's chapter provides a nuanced definition of what constitutes urban and environmental space by analyzing how it was created, contested, and maintained.

Implied in Boston's land-making history is its transformation of the land the glaciers left and the effects of a growing population on the surrounding countryside. Initial patterns of settlement and later development began the transformation of the woodlands into fields, meadows, and woodlots. Domesticates and wildlife at life's end became the ingredients of a tanner's vat. Offal, flesh, and hair entered the waterways, providing early examples of the water pollution in a commercial city with a growing population drawing on the countryside's resources to sustain itself. William Wood's 1634 descriptions of New England's natural bounty would become a lost memory a century later, as settlers changed much of nature into salable commodities.

The Interrelationship of City and Countryside

The book's second part, "The Town and the Countryside," analyzes another major theme in urban and environmental history: the relationship between the city and its hinterland. This section elaborates on the expanding metropolitan economy and the relationship between Boston and the countryside. Although much early environmental history focused on the American West, on definitions of "wilderness," and on an agro-ecological interpretation of environmental history, William Cronon's *Nature's Metropolis,* a pathbreaking study of Chicago and its relationship to the vast natural resources of the American West, inaugurated a new and vital period of scholarship into the intimate connections between cities and their outlying regions.[9]

Using historian and critic Lewis Mumford's categorization of energy regimes of the eotechnic, paleotechnic, and neotechnic, historian Brian Donahue, in "Remaking Boston, Remaking Massachusetts," provides a panoramic view of the community's relationship to the countryside. He notes that throughout colonial history, town and countryside inhabited separate economic and social spaces. Donahue points out that Massachusetts yeomen practiced sustained self-sufficient farming, while Boston's merchants with little in the way of domestic produce to sell looked outward and engaged in the vibrant world of Atlantic trade. With the coming of the paleotechnic age, the networked city was connected to hinterlands and other metropolitan areas by coal-burning, steam-powered railroads. Commercial agriculture replaced self-sufficient family farming. Rural space underwent sustained social, economic, and cultural transformations. Nature's forested landscape and abundant wildlife were degraded by the power of market capitalism. In Boston's networked city, land making paralleled similar changes in the countryside. Sustained spatial transformations in the city and the countryside expanded productive processes, making them interdependent yet vulnerable to forces beyond their control.

With the onset of a period of agricultural intensification and the elimination of many less productive farming enterprises, forests regenerated and the forest products industry revived to meet consumer demands for furniture and packing materials. Physical rearrangements of space once again suggest the contextual richness of rural places, neither fixed nor immobile in time or space. The combination of abandoned farms, a growing population, Boston's decaying housing stock, the glitter and affordability of automobiles, and favorable federal housing loan programs turned vacant land into sprawling suburbs. Networks of new roads connected the city and the countryside as auto mobility, despite its dependence on oil, a paleotechnic energy source, represented the neotechnic world of transformed segregated space based on class, race, and gender.

In geographer William B. Meyer's chapter, "A City (Only Partly) on a Hill: Terrain and Land Use in Pre-twentieth-century Boston," the glaciated world of Boston's drumlins mostly disappeared, as leveling the land of the Shawmut Peninsula became public policy. Unlike the lengthy historical analysis Donahue provides, Meyer's contribution remains intensely focused on the geographical alterations of Boston before vertical space—tall buildings and skyscrapers—became symbols of wealth, status, and power. Living and working at ground level dominated life within the city's boundaries. Walking up steep hills, climbing flights of stairs to reach upper-level living quarters, reflected lower-class status. Only the hovels of the poor and elite mansions, which the owners reached by horse-drawn carriages, inhabited vertical space. As Meyer points out, interior spatial representations indicative of social class have been an unexplored domain for urban and environmental historians. The focus has been outward, drawing on the pioneering work of J. Heinrich von Thunen's geographical model of concentric circles and Ernest W. Burgess's model of cities. As such, geographers, sociologists, and historians have written about the physical city as class and racialized spaces.

"A City (Only Partly) on a Hill" explores the spatial patterns of urban land use when the human and economic costs of access to people and products dominated city planning. The city's core was its vital center, where business activity and elite housing dominated the physical space. Its poor, its charitable institutions, and its cemeteries inhabited the city's periphery. Because Boston's hilly core violated this rational spatial pattern, removing the hills and using the fill for ongoing land-making activities served dual purposes. The newly landscaped horizontal center rearranged social and economic space while land making expanded the zone of elite housing and commercial activities. As Meyer points out, Boston's spatial patterns, much like those of other cities, reflected modes of transportation. As these modes changed and mobility by streetcar, railroad, and automobile made the periphery accessible to a growing middle class, suburban sprawl became a new spatial pattern.

The next two chapters—by the historian David Soll, "Reforestation in Norfolk County, Massachusetts, 1850–1910," and by the urban planner James C. O'Connell, "How Metropolitan Parks Shaped Greater Boston,

1893–1945"—are focused on Boston as a central place, showing how the city's influence had intended and unintended consequences for the surrounding communities. Soll's contribution shows market forces at work, transforming economic relationships between Boston and the surrounding towns. As grain from the American West became available to feed Norfolk County dairy cows, the need for extensive pastures diminished. With fewer acres required, the land reverted to woodlands at a time when the production of dairy products increased substantially to meet the needs of Boston's growing population. Again, spatial rearrangements responded to changing economic imperatives. Reforestation became one of its unintended consequences. O'Connell describes the consequences of regional planning in establishing the park and parkway system, the first formal effort nationally. Market forces played little or no role in this endeavor, while growing population density in Boston and its closest towns provided the incentive to establish a network of parkways to connect the city and countryside. In the process, access to the towns transformed their spatial relationship to Boston, making them more urban and less rural, and more densely populated than before.

While the parkways became a vital link in the built environment, intended and unintended consequences resulted from this metropolitan plan. The establishment of a metropolitan commission to oversee construction and maintenance of parks and parkways minimized the collision of municipal interests, reduced delays, and enhanced administrative control over the environment and the social lives of citizens for whom these amenities were intended. The metropolitan park system integrated the natural world into a world of roadways and asphalt pavement. Greening urban space softened a built environment of concrete, steel, and asphalt, as parks beautified urban space and became the "green lungs" of the urban environment. By cutting networks of roads through both pristine and regenerated woodlands, the parkways unintentionally created fragmented ecological space, making it difficult or impossible for wildlife and walkers to traverse this newly developed built environment.

In "Reclaiming the Middle Charles River Reservation," the urban planners Daniel Driscoll and Karl Haglund address the relationship of the river's edge as a vital pathway to metropolitan Boston and the conflicts and compromises encountered in establishing access to all citizens. Efforts to reclaim the river and link it more closely to the harbor and its tributaries were part of a history a century and a half long. The river's potential as a regional landmark and natural pathway connecting towns and neighborhoods became fragmented over time by the built environment of railroad bridges and warehouses. As a result, the physical and visual river with mostly unobstructed views extended from the Harvard University campus in Cambridge and Allston to the esplanades and its basin and no further. Beyond the basin, the river disappeared visually as it approached the harbor. Likewise, the middle Charles, the river's longest expanse, suffered a similar fate, fragmented by residential and commercial development. Renewed efforts by the Metropolitan District Commission to reestablish the river as a major unobstructed and accessible pathway in the late twentieth century faced opposition from a number of abutters and stakeholders. Driscoll and Haglund's contribution documents grassroots democracy in action as local and oftentimes parochial interests came into conflict with regional planners intent upon reclaiming the middle Charles River and linking it through a series of connected paths to the river as a whole.

The Climate and the Weather

The three chapters in the final section of the book address a topic of increasing scholarly interest to researchers from many disciplines—the role of climate and weather in human life. Of the many global challenges facing us in the twenty-first century, the need for knowledge about the Earth's changing climate system is among the most serious. Until recently, this topic received little attention from urban and environmental historians for many reasons, notably a widespread belief in climate's marginal influence on historical developments. In "Boston's Weather and Climate Histories," William B. Meyer focuses on Bostonians' changing understanding of the weather from the colonial period to modern day, perceptions created by the rhythms of economic and social life. Snow-covered roads during the colonial eotechnic era reduced surface friction and therefore required less energy from humans and draft animals to move heavy loads. These same winter roads, however, became hazardous for the drivers of motorized vehicles of the modern neotechnic era. Bostonians imposed different meanings on their weather as the world of human inventions blunted its effects on their daily lives, their need for speed and punctuality in an increasingly competitive world of commerce, and their desire for ambient amenities. Although modern technology has overcome many weather and climate vulnerabilities from the past, Meyer notes, other likely more severe phenomena pose threats as rising global temperatures, high winds, coastal storm surges, and land erosion at the water's edge compromise the city's quality of life.

Meyer points out how preconceptions influence our understanding of natural events, a fact common to every era. The historian Lauri Bauer Coleman, in her chapter "'Rain Down Righteousness': Interpretations of Natural Events in Mid-eighteenth-century Boston," offers three examples of this process drawn from the middle of the eighteenth century: a drought in 1749, an earthquake in 1755, and a comet in 1759. Coleman explains how religious beliefs shaped the popular understanding of these surprising natural events. Drawing on both reformed Protestant scriptural exegesis and on widely held beliefs about God's role in the world, she shows how New Englanders made sense of natural phenomena for which they had no compelling secular or scientific explanations. The result is an account of extraordinary natural events seen through the eyes of those who experienced them.

As she points out, "supernatural rationalism"—a term invented by historians to explain the integration of ideas about natural-world phenomena and the omniscience of God—guided eighteenth-century discourse about natural events. The development of the scientific method and its findings in the centuries that followed make supernatural rationalism appear cumbersome and contradictory. In the mid-eighteenth century, however, New Englanders seeking larger meaning in disruptive events looked to sources seldom used by environmental historians—namely, sermons and discourses on natural philosophy that traced logically the causes of natural disasters back to sources in the supernatural. This worldview established a symbiotic relationship between scripture and the natural world. Although the supernatural-rational approach established a framework for interpreting such naturally occurring events as earthquakes, hurricanes, and droughts, determining their meaning created conflict and controversy among Boston's religious and secular elites. A wrathful God, using natural disasters to impose moral order on the wayward, was juxtaposed with a wise God whose universe inspired awe, not fear. Given the complexity of understanding extraordinary events and the range of attitudes and beliefs that merged to create what we now call supernatural rationalism, Coleman's chapter uncovers the underlying reasons for the intense debates surrounding nature's unpredictable and disruptive events.

The use of much nontraditional data in understanding the degree to which Boston's weather has changed in the past century is evident in the final contribution in this part, "Biological Responses to Climate Change in

Boston," by plant biologists Abraham J. Miller-Rushing and Richard B. Primack. Using herbarium and archival collections that include preserved plant specimens, photographs, and personal journals, they document the magnitude of climate change in Boston since the nineteenth century. Archival collections including photographs and personal journals are familiar source materials for historians, while herbarium collections and preserved plant specimens remain the exclusive domain of plant biologists. Employing these seemingly unrelated materials to construct a coherent and convincing narrative about climate change in Boston represents interdisciplinary writing at its best.

The use of photographs that document the arrival of spring does not diminish the written text; in fact, pictures taken decades apart and juxtaposed one to the other speak particularly clearly about a changing Boston climate. Miller-Rushing and Primack's most convincing evidence of a warming climate comes from the timing of spring events, the melting of snow, the migration of birds, and the flowering of plants. The personal journals of the noted naturalist Kathleen S. Anderson recorded the arrival of birds, butterflies, and amphibians as well as the timing of flowering plants on her forty-hectare farm in Middleborough, Massachusetts, from 1970 to 2003. Analysis of these data led to the surprising conclusion that Boston's temperature rise of 2.7 degrees Fahrenheit in the past century equals the projected rise for the planet later in the twenty-first century. In the wake of a warming climate, their research resonates with the findings of other plant biologists and climatologists who track the movement of the Mountain Pine beetles, currently eating their way across forests from Colorado to British Columbia, killing millions of trees.

REMAKING Boston required reshaping and expanding its physical space and reordering its economic and social priorities. The physical space expanded by making, annexing, and reshaping land. As evidence of changing climatic cycles and human intervention, the movement of glaciers, sea-level rise, and a redefined built environment sculpted the land and the water's edge. Bostonians leveled the land to suit their social and economic interests and dredged the harbor to make it a safer and more inviting space for ships. With twentieth-century prosperity the vertical city of steel and glass skyscrapers replaced the nineteenth-century horizontal city of wood and brick. Past impediments disappeared with the new integrated energy flows of electricity and internal combustion. Roads, turnpikes, canals, and railroads connected the city to outlying farms in nineteenth-century greater Boston, but the arrival of the automobile turned recently abandoned farmland into sprawling suburbs. Low-density developments in Boston's surrounding communities expanded more rapidly than the high-density, high-rise buildings at the city's center.

As many of the contributors have pointed out, the social structuring of Boston underwent a transformation as physical spatial arrangements pushed the poor to the decaying residential center while the affluent moved to the newly designed periphery. As the commercial and financial entrepreneurs of twentieth-century Boston established air rights in the imagined space over the land, definitions of real estate acquired new meaning. In the process of these multiple remakings, the outlines of the city's major spatial relationships remained intact, as suggested by the harbor's enduring connection to the city. Yet each remaking possessed the human dimensions of changing social, cultural, and economic values that transformed Boston's environmental history.

THE HARBOR

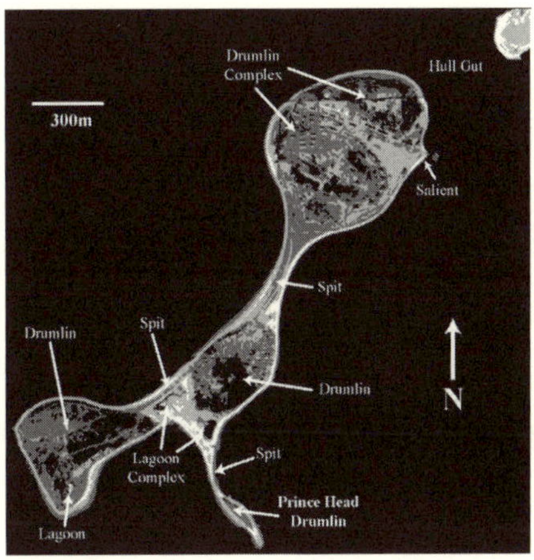
Drumlin Complex
Hull Gut
300m
Salient
Spit
Drumlin
Spit
Drumlin
N
Lagoon Complex
Spit
Prince Head Drumlin
Lagoon

2

The Drowning of Boston Harbor and the Development of the Shoreline

Peter S. Rosen and Duncan M. FitzGerald

IT was the harbor, deep and inviting, that attracted Boston's English founders to the peninsula on which they built their settlement, and the harbor has continued to mold the community's development to the present. Boston's environmental history consequently begins at the water's edge. Before considering how human settlement and the harbor interacted, however, it is important to understand the context. It is imperative to assess the physical characteristics of the harbor and how it assumed its current form. Physical processes produced the land, the shoreline, and the harbor that Boston's Puritan settlers encountered in 1630. The glaciers of the most recent ice age were the indispensable agents in the harbor's creation. As it drove southward, an inexorable wedge of ice plowed the earth before it, depressed the land by virtue of its weight, and sculpted the shoreline. When temperatures rose and the ice receded, the till—the undifferentiated deposits of sand, clay, gravel, and rocks—that the glaciers left behind became hills and drumlin islands as well as underwater mounds on the harbor floor. As the melting continued, the runoff formed riverbeds that became harbor channels and a rising sea

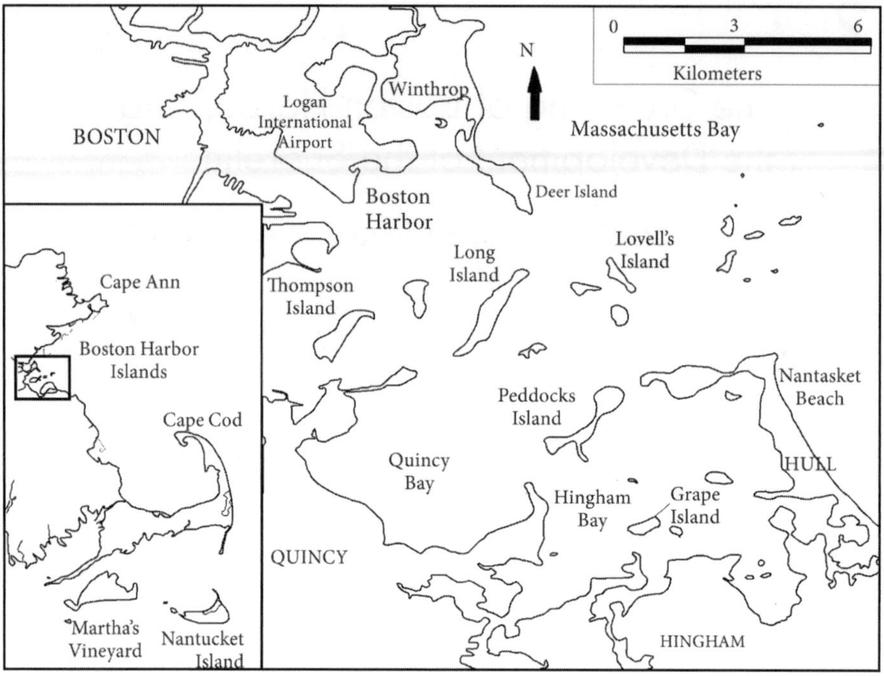

Figure 2.1. Location map for Boston Harbor. Harbor islands are composed of drumlins and tombolos, and major portions of mainland shore are landfill. *Source:* Authors' map.

level drowned low-lying coastal land. At the same time, relieved of the weight of the ice, the ground gradually rebounded. The result was a new shoreline where the land and the sea met (figure 2.1). All these developments established the physical context for the harbor's modern history of commerce, fishing, land-filling, and pollution, which began with the arrival of English settlers.

Ocean Invasion of Boston Harbor

Most modern shorelines in New England have formed as a result of rising sea level and the drowning of the landscape in response to the melting of ice sheets. In addition to changing water levels during this period, deglaciation removed the massive weight of the ice, causing rebound of the depressed land. Thus the position of former sea level (where the ocean meets the land) was the consequence of the concurrent movement of the water level and the land level. Sea level change in Boston Harbor included four major stages: (a) initial glacier retreat, when sea level rose and completely inundated the Boston Harbor region; (b) rapid fall of sea level, moving the shoreline far offshore of Boston; (c) gradual rise of sea level, flooding Boston Harbor; and (d) stabilization of sea level approximately four thousand to five thousand years ago to its current level. During the past two hundred years, sea level has been rising slowly at a rate of 2 to 3 millimeters a year, or about one foot per century.[1]

The record of the early drowning of the Boston region lies in the well-known Boston Blue Clay, which was deposited about fourteen thousand years ago. The clay settled out from a mixture of seawater and glacial meltwater and covers much of the landscape in the Boston region. Up to seventy-five feet thick in the Boston area, it has filled in many low-lying areas. The clay extends inland to an elevation of sixty-five feet, and it can be found as far west as Watertown, which indicates the extent of inundation at its greatest point. In Boston there is clay throughout the city, in the harbor, and at least nine miles offshore.[2]

Between thirteen thousand and eleven thousand years ago, despite ongoing glacial melting, which added water to the world's oceans and raised the sea level globally, glacial rebound of the land overtook the rising water. This caused local sea level to drop about fifty-five feet below modern sea level. During this time the glaciated landscape of Boston was exposed and what is now within the city's limits was well inland from the shore. Between roughly eleven thousand and ten thousand years ago, sea levels rose and land levels rebounded simultaneously, creating a fixed shoreline position. This delicate balance of land and water levels left a distinctive sedimentologic record. As river mouths, along with other shorelines, remained at a fixed position, the sediment they carried was deposited at the shoreline, creating a delta. The Merrimac River Paleodelta, which is currently offshore of the mouth of the modern Merrimac in about a hundred feet of water, is the local reference mark for this period of lowstand sea level.[3] Following the lowstand, sea level rose rapidly, reaching approximately its modern-day position about four thousand to five thousand years ago. Since that time, sea level has risen slightly. During the last period of inundation, Boston Harbor was flooded and the modern shoreline was established (figure 2.2).

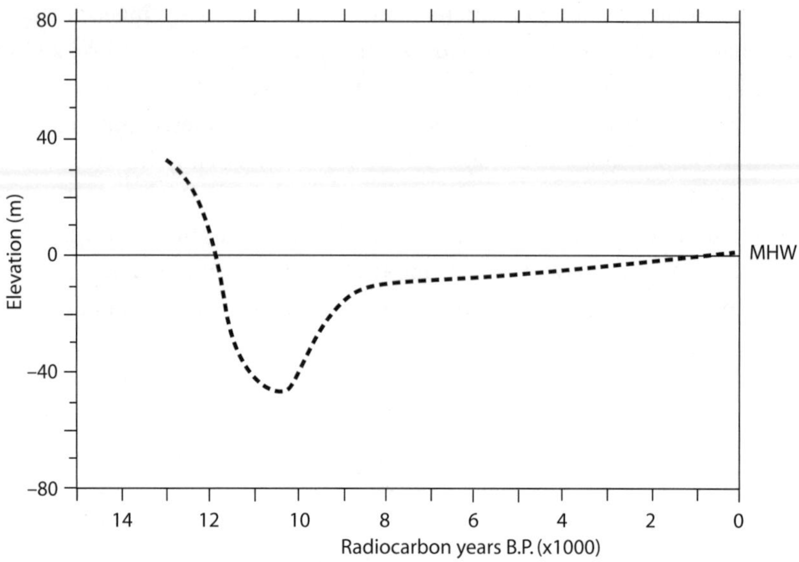

Figure 2.2. Sea-level changes in Boston showing a postglacial highstand and subsequent lowstand of water level, followed by the gradual submergence of the harbor in the past ten thousand years. *Source:* Authors' data and rendering.

Sea Level Regime and Sedimentation in Boston Harbor

Boston is situated in a region known as the Boston Basin, a low-lying area between uplands to the north and south. Elongated, glacial sedimentary mounds called drumlins dominate the basin. These coastal features comprise several hundred hills, islands, and submarine deposits that surround Boston and extend into Massachusetts Bay. The drumlins are generally elongated on a northwest axis, which reflects the direction of glacial ice flow when they were formed (figure 2.3). Drumlins are not isolated features; they occur in groups or fields. The Boston field contains more than two hundred drumlins and many more exist below sea level.[4] Boston is the only drowned drumlin field in the United States.[5] There are others along the Eastern Shore of Nova Scotia and in northwestern Ireland.

Land developers filled in much of Boston's inner harbor during the nineteenth century. The shallowest areas, comprising extensive salt marshes and tidal flats fringing the Back Bay shore, were the most commonly filled.[6]

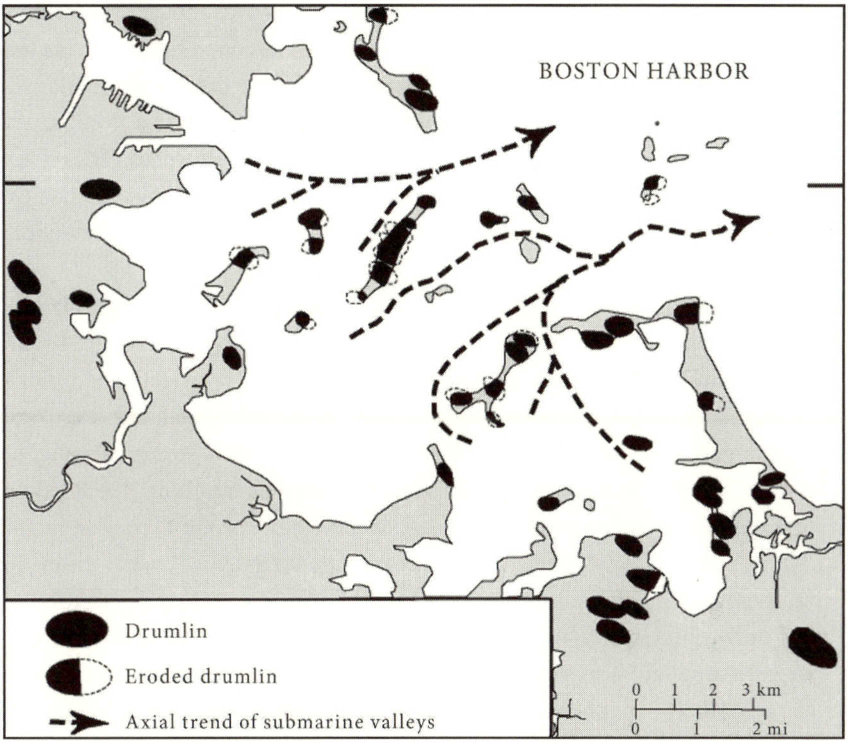

Figure 2.3. Drumlins in Boston Harbor and channels formed from river drainage during lower sea levels. *Source:* Authors' map.

Recent building excavations in this area, which penetrate through the fill to layers below, have provided important direct opportunities to observe the harbor bottom sediments that accumulated throughout its history of submergence. These estuarine and bay sediments cover a weathered surface of the Boston Blue Clay, which overlies the glacial till. The crests of the drumlins are the major glacial features that protrude up through these deposits in the harbor area. Blue Clay forms much of the modern harbor bottom and sub-bottom materials. Excavators in 1987 found a fifteen-foot-thick estuarine deposit at the construction site for the 500 Boylston Street building in the Back Bay. These sediments record changing sea level conditions since ocean waters invaded the inner harbor 5,630 years ago. The earliest tidal exchange in Boston Basin (Boston Harbor) occurred in river channels—the Charles,

Neponset, and prehistoric Weir rivers—where freshwater and saltwater mixed to form estuaries. Flocculation, or clumping of suspended clays in these settings, produced rapid fine-grained sedimentation. A salt marsh peat at the 500 Boylston Street site marks the lowest marine deposit and the first encroachment of the ocean. It overlies the older, weathered Boston Blue Clay. Salt marshes form in the intertidal zone, so it is not surprising that the record of the leading edge of rising sea level would be captured between high and low tides.

Several changes took place over time in the developing Boston Harbor. Initially, sea level rose rapidly, but by four thousand to five thousand years ago the rate had substantially declined. First, water levels deepened quickly in the harbor and encroached into many low-lying regions, but with time submerged areas expanded much more slowly. Early on, the tidal range in the harbor area was small compared with nearby ocean tides (perhaps one to three feet). This vertical rise and fall swelled to the present ten-foot range. The increase in tidal range is attributed to the form and dimensions of the regional marine basin into which Boston Harbor opens (including the Bay of Fundy, the Gulf of Maine, Massachusetts Bay, and Cape Cod Bay). This large basin came into a resonant frequency, or became synchronized with, the periodicity of the astronomic tides as its dimensions enlarged because of rising sea level flooding increasingly larger parts of the basin. The sediments observed at 500 Boylston Street, a record of the harbor's past, reveal four major environments above the lower peat. These are layering muds, tidal channel bottom sediments, sandy mud channel fill, and upper peat:

Layered muds. From 5,630 to 3,570 years ago, sea level was rising more rapidly than later times. Layered muds indicate deepening water, a lack of disturbance, and may reveal thin annual layering. This sequence may reflect the period before the amplified tidal range in Boston, as there is no evidence of significant tidal current deposits.

Tidal channel bottom sediments. Above the lower muds, there is a thin, coarse-grained unit that was deposited in a relatively short time span (3,570 to 3,090 years ago). Geologists interpret this segment as tidal channel sediments, basing their conclusion on a distinct lower layer of broken shells, which are comparable to modern channel bottom deposits. Tidal currents in this region of the harbor probably caused the deposits.

Sandy mud channel fill. Lying on top of the channel lag deposits (or larger pebbles and shell fragments left behind when smaller material is carried away by tidal currents) is a sandy, silty mud that consists of well-mixed mud intercalated with occasional tidalites (or sand-and-mud layers). This unit represents a tidal setting, where sedimentation is a function of tidal energy.

Upper peat. A second layer of salt marsh peat similar to the "lower peat" caps the channel fill sediments, suggesting sedimentation exceeding sea level rise and marsh grasses invading an intertidal mud flat.[7]

Thus, during the past five thousand years after the harbor flooded, channel depth initially deepened as sea level rise was greater than the rate of sediment deposition in the estuary. Sea level rise rates eventually declined and sedimentation caused a shoaling of the estuary. The depositional environment ended, in historic times, in the same type of intertidal setting as the initial environment. The upper salt marsh peat recorded at the 500 Boylston Street site is the marsh equivalent of what existed in the Back Bay before landfill in the late nineteenth century. Historical landfill deposits overlay this sequence. The marine sediments observed at 500 Boylston Street included over sixty-five thousand wooden stakes from the Boylston Street Fishweir, a series of structures built, used, and maintained over a period of several hundred years circa forty-four hundred years ago. The weir is located in the lower unit of the marine sediment sequence. Despite the collection of the stakes, since their discovery during an excavation for the Green Line transit route under Boylston Street in 1913, no other artifacts associated with the structure have been found.[8] The vertical wooden stakes from the weir often exhibit markings indicative of being pounded into the sediment, yet no stones suitable for pounding have been found even though the weirs were hundreds of feet long.

Many fishweirs from historical times in northern New England have a similar construction: vertical wooden stakes fixed in the bottom with branches or brush, known as wattling, interwoven to form a fence to divert fish to one end. They were maintained at low tide by foot, so the lack of artifacts associated with installing or maintaining the weirs over hundreds of years appeared inconsistent. However, during the times when the fish weir was active (in the lower mud unit), Boston had a microtidal setting (less than a three-foot tide range), so it would not have been exposed at low tide like present conditions. The lack of artifacts may be explained by the observation that the fish weirs were probably subtidal devices built and maintained by canoes or rafts, rather than by people on foot. This is similar to records of fish weirs in the microtidal setting of Chesapeake Bay during early historic times.[9]

Development of the Modern Harbor Form

Partially submerged by the rising sea level, drumlin islands dominate Boston Harbor. Waves eroded, transported, and deposited sediment from the drumlins to form a variety of beach types on the modern shoreline, ranging from sandy beach and dune systems like the south shore of Lovell's Island to cobble-boulder beaches like the northern tombolo on Peddocks Island. Rising sea level gradually continued to elevate the position of wave attack on the shore, resulting in a long-term history of erosion and shore retreat of these glacial features. However, in some areas sufficient sediment accumulated to form beaches and other landforms. The earliest stable beaches in Boston Harbor were formed about three thousand to four thousand years ago. Apparently the rate of sea level rise slowed around this time, so that sediment could accumulate, resulting in beach formation. Boston Harbor began looking similar to the form and outline of the modern harbor before landfilling at about this time.

Three major shoreline features exist on the drumlin islands: *drumlin bluffs,* which are typically the primary source of sediment for beaches; *connecting barriers* formed by spit/tombolo processes; and *salients,* which are triangular-shaped accumulation forms on the lee, or sheltered, side of islands.

Drumlin Bluffs

The earlier stages of bluff erosion occur on the flanks of drumlins and are dominated by slumping. Wave erosion at the base of a bluff oversteepens the slope and makes it unstable. Blocks of material slide down the slope and accumulate at the base. The sedimentary material protects the base of the slope from further wave attack. Waves eventually erode this "sacrificial sediment" and transport it either offshore or along the shore. Once the protective sediment is removed, storm-induced wave erosion undercuts the slopes, producing another slump. The bluff is typically a bare slope over which the slump block slid downward.

As a bluff erodes deeper into the drumlin's interior, its height increases, reaching a maximum at the crest of the drumlin. During this phase the volume of material in each slump block also increases. Bluffs greater than about thirty feet high produce large quantities of sediment in a single slump block (several tens of tons), which protect the base of the bluff from wave attack for periods commonly spanning decades. This provides time for other processes, such as rainsplash and runoff (or surface flow), to play a role in eroding and modifying the slope. These processes also continue to feed some sediment to the base. Erosional rills and gullies (incipient channels) are most commonly found along these higher bluffs.[10]

Eroding bluffs are the source of gravelly sediment that dominates beaches in the harbor. Sediment is eroded from the face of the bluff by both runoff processes and slumping. The glacial till composing the drumlins consists of cobbles and boulders in a matrix of sand, silt, and clay. As the bluffs erode, most of the clay and silt are suspended by the waves and carried away and deposited in quiet, deeper water. Most of the sand, pebbles, and small cobbles are transported along shore by breaking waves. Beaches at drumlin headlands typically consist of a thin veneer of the coarse gravel and boulders overlying a glacial till retreat platform. The retreat platform is the denuded surface of the original drumlin. The thin but coarse beach layer protects the remaining portion of the drumlin from further wave attack, so that the platform can endure long after it is submerged below sea level.

Connecting Barriers (or Tombolos)

Most drumlin shores in Boston Harbor have been modified through sediment deposition and the formation of spits or tombolos. Sediment moved along shore by wave energy builds spits that link nearby drumlins to one another. Up to five drumlins have been connected by spits (tombolos) to form a single island (for example, Peddocks Island) (figure 2.4). Within the harbor each drumlin-tombolo island system is located within a single glacial till platform. It appears that the control over which drumlins are connected by tombolos is influenced by the preexisting bedrock and/or glacial topography. As sea level rose, sediment accumulated and spits formed on these drowned retreat platforms from the glacial topographic highs. Although the tombolos are accretionary features that have been built by wave action during the past three thousand years, most tombolo shorelines are currently retreating (that is, eroding), as the longshore supply of sediment may not keep pace with the renewed rising sea level.[11]

Salients

Modification of the eroding drumlin islands within the harbor also typically includes the formation of salients. These are triangular-shaped accre-

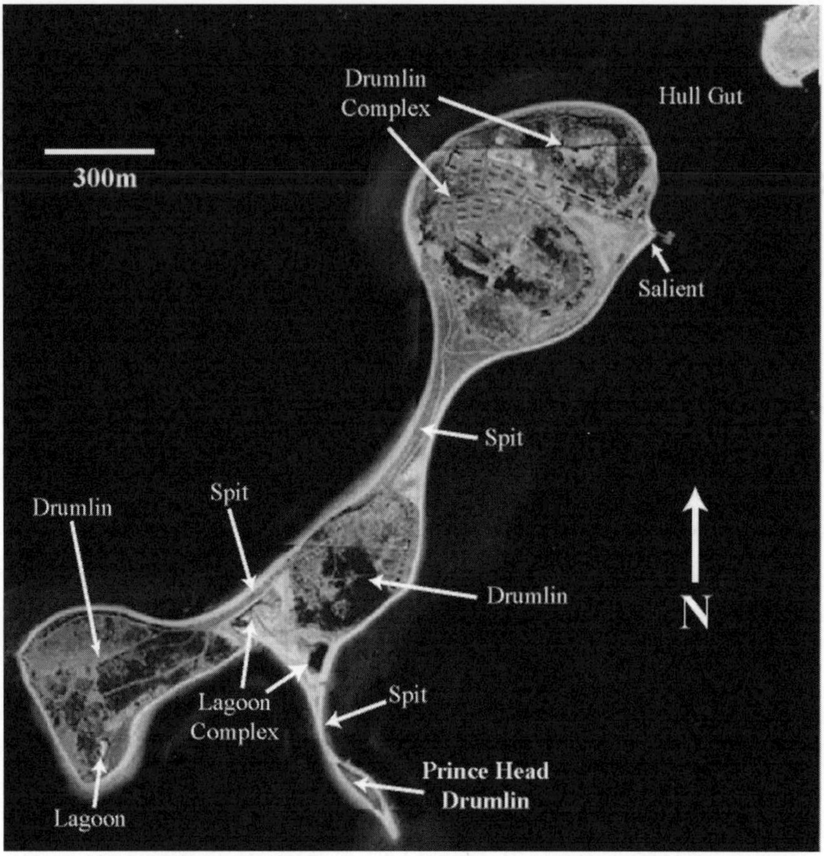

Figure 2.4. Aerial photograph of Peddocks Island showing major features comprising the shoreline form. *Source:* Authors' map.

tionary landforms that develop because of converging longshore sediment transport systems. Their cuspate features consist of a variety of sediment ranging from mostly sand to entirely gravel and are found on the low wave energy, lee-side of islands. The salients can project from either the drumlin or tombolo portions of the island shorelines. Salients may move along shore forming "traveling forelands," where unequal opposing longshore transport rates result in a feature that migrates in the dominant transport direction.[12] Good examples are found on the northern sides of Long and Thompson islands. There is no consistent direction of salient migration among the islands,

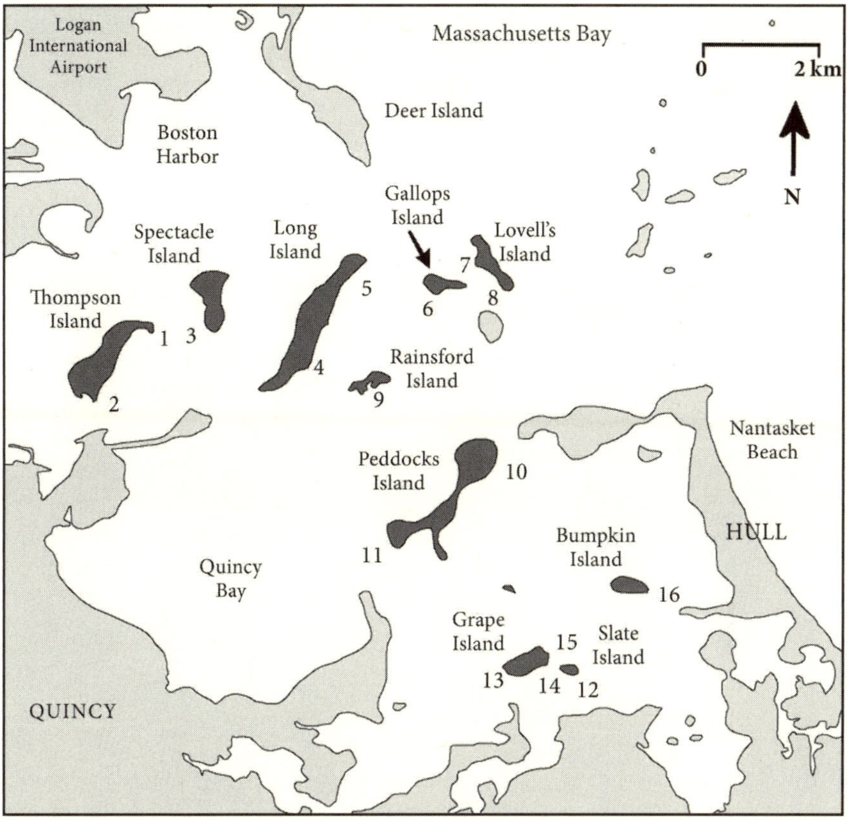

Figure 2.5. Location of salients in Boston Harbor. *Source:* Authors' map.

since both sediment supply and wave climate are localized within the harbor. Rates of migration of salients range from 0 to 1.5 feet per year (figure 2.5 and table 2.1).[13]

Armoring the Shore with Seawalls

Since the late seventeenth century, the harbor shoreline has been changed by ongoing landfill to create real estate. Although most of the new lands were created in the City of Boston, landfilling also took place extensively throughout the mainland portions of the harbor in populated areas.[14] Land-

TABLE 2.1

Characteristics of salients in Boston Harbor.

Location number	*Island*	*Position*	*Migration direction*	*Approximate long-term migration rate (m/yr)*	*Terminal bar*	*Preserved ridges*
1	Thompson	N	SW	0.28	N	Y
2	Thompson	S	stable	–	Y	Y
3	Spectacle	SW	stable (altered)	–	N	N
4	Long	SE	SE	0.38	N	N
5	Long	NE	SW	0.62	N	N
6	Gallops	S	stable (altered)	–	N	N
7	Lovell's	W	S	1.42	N	N
8	Lovell's	SW	NW	0.56	N	N
9	Rainsford	S	W	0.37	N	N
10	Peddocks	NE	NE	0.45	N	N
11	Peddocks	SE	stable	–	N	N
12	Slate	S	stable	–	Y	Y
13	Grape	SW	stable	–	Y	N
14	Grape	S	NE	0.05	N	Y
15	Grape	SE	stable	–	N	Y
16	Bumpkin	E	stable	–	Y	N

Source: Air photo interpretation, field inspection. Migration rates derived from Massachusetts Shoreline Change Project, Office of Coastal Zone Management, available online at http://www.state.ma.us./czm.

filling was often used to extend wharves seaward to reach deeper water in order to stimulate commerce as well as to create new coastal land. As land was created and as existing land became more valuable, major segments of the shoreline became protected with seawalls.

Records of colonial seawall or wharf construction in New England are limited. It may be that the construction of seawalls was considered too commonplace to merit description by historians. The earliest colonial seawalls in the Boston region were constructed from logs with open crib form, known as "cobb cribs." A wharf of cobb construction was built on Boston Neck around 1676.[15] Cribs were constructed of logs joined by trunnells until around 1840, when iron nails and spikes came into use. The cribs had an open framework that could readily be floated into position and sunk with rock from local sources, such as the rounded glacial cobbles and boulders found along the shoreline. This technology was very established, as demonstrated by an eighteenth-century crib excavated at the Charlestown Navy Yard in Boston, which was similar in its style of construction to a second-century crib excavated in London.[16]

As local fill material became scarcer because of ongoing wharfing and landfilling activities, solid wood cribs, filled with a large variety of materials, came into use. Boring logs from landfilled waterfront areas in Boston revealed that fill material typically contained bricks, asphalt, wood, cinders, ash, coal, ceramics, glass, and leather, along with soil. These wooden crib structures required constant maintenance as a result of the rapid decay of wood that marine borers caused before the era when water pollution kept the population of these organisms in check. Stone seawalls date from as early as 1784 in the Boston area, although construction was difficult. Rounded glacial boulders were difficult to lay, so early stone seawalls often incorporated wooden piles as part of the structure. Efficient methods to cut or hew the local granites were not widely used until about 1830. Some early stone seawalls were constructed on wood platforms, which would sink into the mud as the weight of stone was added. By the mid-1800s stone seawalls typically stood on wood pile foundations.[17] Although modern concrete was developed in the mid-1800s, the harsh environmental conditions in Boston may have resulted in the common practice of constructing a seawall of concrete and continuing to face the structure in stone.

The Future

Sea level is likely to continue to rise for the foreseeable future. The rate of rise has been measured directly for nearly a century on the Boston Tide Gauge in the inner harbor. These records show a long-term trend of rising water levels at about one foot per century. Although there is debate whether this rate will increase in the future, it is likely that water will continue to rise at least at this rate. Seawalls now protect most urbanized areas in Boston Harbor. Here, the major effects will be increased frequency of flooding and repairs of the coastal structures. If the natural shorelines are permitted to continue to erode, the mixed gravel beaches along their base will shift in a landward position over time, but their forms will remain. The mud in the eroding glacial sediments will continue to be put into suspension by waves, and they will settle out in deeper or protected waters, as in the past. Eroding drumlins will also provide sediment carried along shore to maintain tombolos and salients.

3

What Lies Beneath

SCIENCE, NATURE, AND THE MAKING OF BOSTON HARBOR

Michael Rawson

RETIRED sea captain John Sleeper made little effort to hide his frustration as he addressed the assembled members of the Boston Marine Society. Asked to deliver a paper at the society's annual meeting in 1872, Sleeper chose a subject that weighed heavily on the minds of the gathered mariners: encroachments on Boston Harbor through land making. The state had recently approved yet another new land-making project, this one to fill almost a thousand acres of the South Boston flats, and the society's members were united in their belief that making new land where there once had been water was irrevocably damaging the harbor. Sleeper agreed, and so did the federal engineers who had spent years conducting scientific studies of the harbor's hydraulics. But development of the city's coastal flats seemed to accelerate with each passing year nonetheless, and Sleeper railed against the "soulless corporations" and irresponsible government officials who ignored science in a headlong pursuit of profit. "In this age of mammoth corporations and bold speculations," said Sleeper in a conclusion that still feels relevant today, "the past seems forgotten, the present only is regarded as of importance, and a veil is drawn over the future." Before the meeting was over, the members of the society had unanimously endorsed Sleeper's remarks and appointed a committee to coordinate their printing and distribution.[1]

But neither dry land nor wave-capped sea lay at the heart of this debate. Bostonians were most concerned about what lay beneath the water: the harbor floor. From the 1830s to the end of the century, many Bostonians believed with good reason that the harbor's ship channels were shoaling, and the merchants, ship captains, and engineers who knew the harbor firsthand embraced a scientific theory that blamed land making for the deterioration. According to the theory, Boston Harbor existed in a fragile equilibrium of natural forces that excessive land making could disrupt. If thrown out of balance, the huge and complex harbor might prove impossible to repair. So with the help of the state and federal governments, and over the opposition of real estate developers, railroad corporations, and pro-growth legislators, these defenders of Boston Harbor used the theory to regulate land making and direct the construction of harbor improvements into the twentieth century. Their efforts to protect the harbor gave Boston's coastline the shape it has today.

This controversy was long, loud, and central to Boston's development. Yet it remains entirely unexplored, in part because harbors have fallen between the cracks of urban and environmental history. Harbors are coastal areas used to shelter ships and are distinct from ports, which are the facilities where ships load and unload. Urban historians have paid some attention to America's ports, but they have largely ignored harbors. In fact, urban historians tend to treat cities as if they stop at the water's edge. A curious scholar might follow ships in and out of the harbor, but the land disappears as a topic of study when it disappears beneath the waves.[2]

Environmental historians have paid more attention to water. They have written extensively about rivers and are beginning to study the history of the oceans. But they have not dived deep enough to make it to the bottom of harbors.[3] Even historians who have made a special study of urban harbors have focused mostly on shipping and pollution. In the case of Boston Harbor, several histories describe its romantic clipper ships and commercial activity, and two recent books explore how it became one of the most polluted harbors in the nation and what the state did to clean it up. But none of these studies look beneath the surface of the water at the submerged landscape that nineteenth-century Bostonians knew so well and cared about so deeply.[4]

Although it might be tempting to dismiss as unimportant the parts of a harbor we cannot see, this invisible environment can have an enormous impact on urban development. On the one hand, a harbor plagued by shifting sandbars, submerged rocks, or rough anchorage, or cursed with a ship channel that is too shallow, narrow, or unstable, is unattractive to shipping. On the other hand, a harbor that lacks these environmental limitations is a natural highway that can lead the wealth of the world right up to a merchant's door. The importance of a good natural harbor to nineteenth-century cities cannot be overestimated: cities as diverse as Boston, New York, and San Francisco would not exist in their current forms—and might not exist at all—without their harbors.

In fact, any historian who wants to understand how immense, dynamic, and complicated ecosystems have shaped cities would be hard pressed to find a more revealing topic of study than urban harbors. What might look like a simple sheet of water when viewed from one of Boston's wharves is really a vast hydraulic system whose mix of saltwater and freshwater responds to directions given by wind and moon, islands and headlands, channels and sandbars, rivers and bays. Even forces as distant as the lakes that feed the rivers emptying into the harbor and as minute as the composition of the sand that covers a particular part of the harbor floor affect how the system functions. Such enormous and complex ecosystems are so difficult to understand and control that they become places of accommodation where humans and the natural world are both changed by the encounter. That give-and-take relationship makes harbors one of the best places to study the relationship between cities and nature.

The complex ecologies that define many urban harbors, and the human desire to control them, also make harbors particularly good places to study the environmental impact of scientific thought. Science has been one of the foremost tools for managing America's harbors since the nineteenth century, when engineers not only studied scientific principles and applied them to their work, but also considered themselves scientists. Urban harbors are therefore understudied laboratories of the human effort to employ scientific thought toward the use and simultaneous conservation of natural systems. On a more site-specific level, harbors can provide evidence of the profound impact that larger shifts in scientific thought have had on particular urban environments. Just as the evolution of scientific ideas has led to "scientific revolutions" (new scientific paradigms), and social and economic change has produced "ecological revolutions" (new regional environmental relationships), so changing scientific understandings of a specific environment have sometimes revolutionized the human relationship to that place. These latter revolutions are both scientific and environmental, and they demonstrate that evolving scientific interpretations of environmental processes can lie at the very heart of major changes to the urban environment.[5]

That brings us back to Boston Harbor, which bears dramatic and surprising imprints of the scientific thought used to manage it. Environmental historians have begun to pay close attention to the application of scientific ideas to specific natural systems, and they continue to uncover stories about well-intentioned but misguided scientists who destroyed the resources they were trying to save.[6] At first glance, the story of Boston Harbor might appear to follow the same path: in the early years of the twentieth century, a new scientific understanding of the harbor's hydraulics displaced the prior one and demonstrated that land making did not have a negative impact on the harbor and never had. That meant that several generations of Bostonians had used flawed science to decide where to make land, what areas to dredge, and where to build sea walls to train the tide. Boston came to owe its very shape to the faithful application of a scientific theory that subsequently proved to be completely wrong. But this tale contains a distinctive twist, because the old theory—as flawed as it was—had probably done more good than harm.

The Nature of Boston Harbor

In the nineteenth century the voyage of a large ship westward into Boston Harbor was a journey through deep-water channels that meandered through a maze of steep headlands and hilly islands. Travelers approaching the main ship channel, the deepest of three, first encountered the scattered and windswept Outer Islands. They then entered the outer harbor through two headlands that lay seven miles apart. Point Allerton appeared close by on the left, while its partner, Point Shirley, remained hidden behind more islands to the right. Ships followed the main channel past Nantasket Roads anchorage and through the Narrows, a tight passage between the pastures of Lovell's Island on one side and the farm and fort of Gallup's and George's islands on the other. They then passed between the tips of Deer and Long islands, the largest islands in the harbor, and sailed into President Roads anchorage, where the harbor opened up and the channel ran west in a straight line toward the city. To the south lay another dozen or so islands, and beyond them the harbors of Dorchester, Quincy, and Hingham.[7]

Ships continuing on to Boston moved from the outer harbor to the inner harbor by passing between the forts on Governor's and Castle islands. By this point in the journey a ship would have traveled some twelve miles since first encountering the Outer Islands, but it still had at least three miles to go, depending on where its captain planned to dock. Travelers would begin to see the scores of busy wharves that bristled around the shorelines of Boston proper, East Boston, South Boston, and Charlestown, and if the tide was low, they could see hundreds of acres of mud flats extending into the harbor. Finally, their eyes might climb the rising mound of roofs and spires on Boston's peninsula until they rested on the dome of the State House. By the time the ship was nestled into its berth, the passengers would have seen only part of the forty-eight square miles of water and land that composed Boston Harbor. Residents of the city liked to brag that their harbor was big enough to hold all the nation's shipping at once.[8]

Boston Harbor was not only a place of natural beauty and human activity, however, but a complex natural system as well. The Outer Islands, together with the headlands of points Allerton and Shirley and the large islands strung between them, sheltered the harbor from the large waves, relentless winds, and fierce storms of the open ocean. The remaining islands divided the outer harbor into tranquil basins, creating almost seven square miles of anchorage deep enough for large ships and calm enough for frailer vessels. The main ship channel—an underwater trench that snaked between the islands and linked the anchorage grounds—provided deep water for large ships. Without its island buffers the harbor would have suffered from the worst of wind and wave and provided a much less friendly environment for commerce. Without deep-water channels through its shallow bed, the harbor would have lacked the highways that enabled goods from all over the globe to reach Boston. This fortuitous arrangement of environmental features gave Boston one of the finest natural harbors in the nation.[9]

The harbor was also part of a larger hydraulic system that extended far inland. Twice each day, flood tides rolled in and raised the water level around the city's wharves between eight and eleven feet, depending on the point in the monthly tidal cycle. Seawater filled the large tidal reservoirs of the Back Bay, South Bay, and Mill Pond and pushed several miles upstream into the freshwater of the Charles and Mystic rivers. The flood tides alternated with ebb tides that, twice each day, dragged water back out of the reservoirs and rivers, through the harbor, and toward the open sea. Freshwater and saltwater, river and ocean, land and sea—all came together in Boston Harbor through the action of a hydraulic cycle that seemed just as eternal as the harbor itself.

The Theory of Tidal Scour

By the 1830s, however, members of the maritime community had begun to question the health of the harbor, in part because the islands and headlands of the outer harbor were visibly deteriorating. Most of these areas were private property, and their owners inadvertently encouraged erosion by harvesting as much as 150,000 tons of stone and gravel from the beaches each year to sell as ship ballast. The weathered material often found its way into the channels of the outer harbor or impeded navigation by creating dangerous shoals.[10] The Narrows had filled by several feet in places as the islands bled into the channel, and eroded material from Great Brewster Island formed a treacherous spit that stretched for over a mile and lay hidden beneath the waves at high tide. Erosion also threatened the harbor's anchorage, since the wearing away of Deer Island risked exposing President Roads to the full force of ocean storms. In short, the erosion of the islands and headlands was causing a massive migration of material from where it would preserve the harbor to where it would cause the most harm.[11]

Many observers insisted that the inner harbor was filling as well. Winslow Lewis, who had been sailing in and out of Boston since 1783, claimed that the main ship channel off Governor's Island had shoaled four feet in the thirty-five years between 1814 and 1849. Samuel Sewell, agent for the Cunard steamship line, complained that the slips between the line's wharves had required dredging twice in just nine years, and wharf owners all over the harbor experienced a similar maintenance burden.[12] Comparisons of old charts with more recent ones supported some of these claims, although the charts were of varying degrees of sophistication and reliability, and a few dated to before the Revolution. Overall, the evidence suggested that the inner harbor might be deteriorating as fast as the outer harbor, and an 1852 pamphlet expressed a common fear when its title prophesied "The Destruction of Boston Harbor."[13]

Observations of physical decay in the harbor were difficult to separate from widespread anxiety about the decline of Boston's commerce. The completion of the Erie Canal in 1825 had catapulted New York City's growth well beyond its rivals and made much of the West its hinterland. Although Boston remained one of America's busiest ports, its relative decline wounded the city's psyche and prompted a search for causes and solutions. Some Bostonians blamed shoaling in the upper harbor for the port's commercial problems, and many more experienced the harbor's deterioration as a physical manifestation of the port's economic erosion. Both the economic and environmental problems required specific solutions. To energize their faltering commerce, Bostonians turned to railroad building with mixed success. To repair their decaying harbor, they turned to science.[14]

Merchants, ship captains, marine insurers, engineers, and other close watchers of the harbor interpreted the shoaling observed in the inner harbor through a scientific paradigm they sometimes called the "theory of tidal scour." Drawing on observation of the harbor and analogy with other estuaries, the theory described Boston Harbor as a series of channels created and maintained by the scouring force of water moving in and out of the harbor and river systems and the tidal reservoirs. A trio of engineers wrote as early as 1837 that "the harbor of Boston is not an open broad bay, surrounded on all sides by the sea shore, where the tides simply flow and ebb with a gentle and almost imperceptible current." Rather, they explained, the harbor "is wholly made and continued as channels, through which the tides ascend into immense basins and rivers . . . and from which the tides descend again to the ocean, and in their progress scour out the channel." More water would deepen the channels, and less water would encourage shoaling.[15]

Although the theory represented the best science of its day, the scientific understanding of hydraulics was still in its infancy. Engineers expressed considerable uncertainty over how water and land behaved when they met. As late as 1874, Thomas Stevenson, a British authority on harbor design and maritime engineering, acknowledged that many seemingly basic questions about how scouring forces worked awaited definitive answers. Some engineers, for example, believed that only freshwater had scouring power, while others insisted that saltwater did most of the work. Such disagreements arose from the incredible complexity that inhered in any large estuary. The volume and velocity of water and the width and depth of channels constantly fluctuated, and the materials that composed the beds of rivers and harbors existed in endless combinations. Stevenson confessed that the number of variables at work presented "an almost hopeless complexity for the mind to grapple with, or for even elaborate observations to unravel." As Heraclitus might have said, no one can step twice into the same harbor.[16]

Boston Harbor not only seemed too complicated to understand fully, but it assumed the added burden of appearing to be in peril. Many longtime observers felt sure that the harbor had existed for uncounted years in a happy equilibrium that some force had recently disrupted. This sense of lost stability lent the impression that the harbor was fragile, and it encouraged the fear that making land in the wrong area, or letting a key island erode too far, might disrupt the balance of natural forces even further. How such a destructive mechanism might operate remained unclear. "No one," noted Captain Sleeper in his address, "however learned and skilful as an engineer, can predict the consequences which will result from any considerable change in the character or direction of the forces which nature has provided for keeping the channels open."[17] Similarly, no one knew whether engineers would be able to repair whatever damage might occur. The harbor's hydraulic secrets, still only partly understood today, remained largely hidden in the nineteenth century. That suggested to many that the city should approach all development with caution.

The theory of tidal scour made intuitive sense given the city's geography, and it gained additional credibility by suggesting a plausible cause for shoaling in the harbor's channels: land making. Thousands of acres of mudflats extended from the city's shoreline into the harbor, and thousands more stretched from the banks of the Charles and Mystic rivers to the edges of their channels. Most flats were hidden at high tide by shallow water but visible when the ocean receded, and they often nourished large growths of sea grass. Building up these areas with dirt, gravel, ashes, and other material was a long-accepted way for owners of waterfront property to expand their real estate. But seen through the theory of tidal scour, filling the flats was not as harmless as previously thought because it diminished the amount of water in the harbor. Less water meant a weaker scour, and a weaker scour could not maintain the ship channels at the needed depth. Some observers believed that extensive land making in the upper harbor was so disruptive that it was contributing to the filling of the Narrows in the lower harbor ten miles away.[18] Such long-distance connections were difficult to prove, but the notion that filling the flats was upsetting the harbor's natural equilibrium and diminishing the scour seemed logical and even likely. In the cloudy context of inscrutable natural forces and an uncertain commercial future, the theory of tidal scour clarified the harbor's origins, identified the source of its problems, and armed efforts to curb land making with the authority of science.

To Fill, or Not to Fill

There was no question that the city's residents had created an enormous amount of land by filling flats. The first settlers began the process by building solid wharves to accommodate their ships. Over time, many wharf owners expanded their properties by extending their wharves outward or filling the docks between them, a process called "wharfing out." With the exception of a few large projects, the filling of flats progressed gradually in the seventeenth and eighteenth centuries. By the beginning of the nineteenth century, however, the acceleration of urban growth had dramatically increased the need for more land and the desire to create it. Pressed particularly by the need for railroad terminals along the already congested waterfront and the desire to cover the industrial and domestic waste flowing onto many of the city's flats, Bostonians made more land in two generations than they had in the previous two centuries.[19]

The harbor's geography had inspired the law that made land making such a widespread activity. From the earliest years of settlement, the preponderance of gently sloping flats had made it difficult for Boston's merchants to build wharves long enough to reach deep water. So in a law sometimes called the Ordinance of 1641, the Massachusetts General Court extended the property rights of riparian owners to the line of *low* tide, up to a maximum distance of one hundred rods from the line of high tide. Most other states, in contrast, drew the line of private property at the line of *high* tide. As a result, many riparian owners in Massachusetts had the right to build and fill far beyond the state's natural coastline, enabling property owners along Boston's waterfront to convert enormous amounts of water into land at will.[20]

Despite this peculiar property law, Boston was just one of the many nineteenth-century cities that extended their shores into the lakes, rivers, and bays they bordered. Chicago filled land along Lake Michigan, New York built into the Hudson and East rivers, and San Francisco reclaimed part of its bay. Charleston, New Orleans, Seattle, Washington, D.C., and many other cities made more or less land depending on local desire and local geography. It was a transnational phenomenon: urbanizing places from Toronto to Tokyo that wanted more space and bordered on shallow water made a great deal of land, and they continued to do so into the twentieth century.[21]

Although the theory of tidal scour encouraged its adherents to view any changes to the harbor with suspicion, the harbor's defenders did not oppose all development. Merchants, ship captains, marine insurers, and the members of the influential Boston Marine Society had tremendous financial incentives to expand port facilities, and the business of civil engineers was to harness the power of nature for human use. These groups insisted, however, that any changes to the harbor safeguard its natural hydraulics. They conceded that the future needs of a growing commerce might require the filling of certain carefully chosen flats to supply additional wharf space and other harbor infrastructure. But they were more likely to advise that, if the city's flats must be changed, they should be dredged rather than filled. Dredging would expand and deepen the harbor's anchorage so it could accommodate more and larger ships, and the greater volume of water entering and leaving the harbor would enhance the scour and keep the channels clear. London and Liverpool provided models, since these cities created docks by excavating their riverbanks rather than building out from shore. The guiding principle of those concerned with the harbor's protection was that the state should not permit any development that decreased the total volume of water in the harbor, setting them firmly against a tradition of land making that stretched back two hundred years.[22]

In contrast, those who privileged land over water, like powerful railroad and real-estate interests, tended to see land making as indispensable in a city with Boston's geography. Economic and population growth required more land, much of it on or accessible to the waterfront, and the only way to get such land on Boston's narrow peninsula was to fill flats. Those who promoted land making recognized the harbor's complexity and supported efforts to save the islands from erosion, but they denied that the harbor was fragile or difficult to repair. Placing their faith in technology and the ability of humans to manipulate nature, this group insisted that any problems caused by land making would be reversible. Accordingly, they criticized preservationists for following a "let-every-thing-be-as-it-is" policy that favored established economic interests and inhibited the city's growth.[23] Their formula made humans masters of nature rather than slaves to its fickle whims, and it encouraged individual and uncoordinated expansion into what they presumed to be a malleable natural world.

Preserving the Harbor in Fits and Starts

Any kind of large-scale preservation effort would require governmental oversight and funding, and the federal government had the motivation and authority to take a leading role. The United States military viewed Boston as a key link in the nation's system of coastal defenses, and it already managed an extensive navy yard in Charlestown. As early as 1825, the Army Corps of Engineers began building seawalls to preserve some of the islands, and in 1833 they began constructing forts to defend the harbor. The federal government also purchased George's and Lovell's islands outright, and the city bought Gallup's Island in 1860, bringing all three of the islands that bordered the Narrows under public control. But partisan wrangling in Congress over the wisdom of internal improvements led to chronic underfunding. By 1854 most federal work in Boston Harbor had come to a stop regardless of its state of completion.[24]

The city of Boston also had an obvious stake in defending the harbor, but unlike the federal government the mayor and council had to balance competing demands on the city's flats. As a result, the city pursued inconsistent policies that worried the harbor's defenders. Although it created a standing committee to monitor the harbor's problems and energetically lobbied the state and federal governments to take action, the city also began a massive land-making project in 1845 that required the filling of sixty-seven acres of city-owned flats in the South Bay. Acting against the advice of several state commissions, the city hoped to create a residential neighborhood for native Bostonians tempted to escape foreign immigrants by fleeing to the suburbs. From the perspective of city leaders, the deterioration of the harbor was no more pressing a problem than the deterioration of the city's tax base and the threat that immigrants might win the reigns of government. But such a policy seemed hypocritical to the harbor's defenders, who questioned the city's commitment to their cause and resisted its efforts to gain control of state-owned flats.[25]

The state held the most control over land making. It exercised the rights of a property owner, because it owned many of the flats surrounding the city, and it had the power to restrict the building of structures over privately owned riparian areas if they would impede navigation. But most important, the state could use its police power to protect the harbor if it determined that the health and welfare of its citizens depended on such action.[26] Armed with this extensive authority, the state appointed several commissions over a period of years to draw "harbor lines" around the city's waterfront areas. First adopted for parts of the peninsula in 1837 and later supplemented and adjusted, the lines set boundaries beyond which no property owner could fill. Their purpose was to protect the harbor from encroachments that might block navigation or reduce the volume of water available to scour the channels.[27] The legislature defended the lines aggressively against unlawful encroachment, instigating legal action in 1846 in twenty-three cases in which wharf owners had extended their wharves to or beyond the lines without permission from the state.[28] Although subsequently altered on various occasions, the harbor lines—and the theory of scour that informed them—determined the shape of much of the city's coastline.

Like the city government, however, the legislature did as much to facilitate development of the flats surrounding the city as it did to preserve the harbor. Petitions for permission to fill various flats streamed into the legislature, which readily authorized the extension of wharves or the creation of new land to the established harbor lines. But the state's generosity did not end there. Not legally bound to observe the harbor lines, the legislature repeatedly granted permission to fill beyond them. In the 1850s it authorized the Mystic River Corporation to build a massive wharf on flats off Charlestown despite strong resistance from the maritime community, and it permitted the Boston Wharf Company to extend its wharf far into the South Boston flats. Although the harbor lines did constrain development, there were ways around them.[29]

Growing interest in filling the immense basin of the Back Bay encouraged further public discussion about the future of the flats surrounding the city. As its name suggested, the Back Bay was an inlet of the Charles River on the west, or back, of Boston's peninsula. About eight hundred acres of tidal water covered it until 1820, when entrepreneurs completed a series of dams that cut off the bay from both river and ocean and divided it into two basins. The "full basin" received water from the flood tide, and the power created by slowly draining it into the larger "empty basin" was supposed to run a series of factories. The project was an early experiment in industrialization, but a failed one. The owners built only a few factories, and over the next twenty-five years the empty basin became a noxious cesspool. By the time the filling and development of the Back Bay began in earnest in the 1850s, the bay had not provided a beneficial scour to the harbor for many years and few supporters of harbor preservation objected to the project. But the discussions highlighted the fact that the state lacked a clear policy for managing the thousands of acres of flats under its control.[30]

With no guiding principles in place, the state continued to act in a piecemeal fashion that sometimes pitted different branches of government against each other. One particularly rancorous confrontation occurred in 1851, when the legislature approved the construction of a solid roadway across the South Boston flats and Governor George Boutwell vetoed the bill. A veto was a rare thing in Massachusetts politics at the time: no governor had vetoed a bill in the previous twenty-five years, and the last one to do so had cited constitutional grounds for his reason. But Boutwell believed that the proposed roadway failed to give due regard to "the rights of individuals, the rights of the Commonwealth, and above all, to the safety of the harbor." The road would connect Boston's peninsula with South Boston and enclose thirty-eight acres of flats. Although drawbridges and sluiceways would give the tide access to the flats, Boutwell feared that the area would become another Back Bay, largely cut off from the sea and eventually so filthy that the state would have to fill it. His decision drew considerable fire from those who believed that the project would not harm the harbor, but he successfully killed the plan. The idea that the harbor was a complex and fragile system governed by the theory of tidal scour was slowly shaping the course of the harbor's development.[31]

Calling in the Troops

By the late 1850s, however, the harbor's defenders could claim little progress in halting the harbor's deterioration, and Mayor Frederick Lincoln began looking for ways to involve the federal government more closely in its management. He suspected that Congress would make a greater commitment to the harbor if advised to do so by its own impartial experts, so in 1859, Lincoln asked three federal engineers to form a commission to study the problem. After receiving their consent, he petitioned their superiors to reassign them. Lincoln sweetened the deal with the offer of city funds to support their work, and the federal government formed the officers into the U.S. Harbor Commissioners on Boston Harbor. It was a distinguished group, composed of General Joseph G. Totten, chief engineer of the United States; Professor Alexander D. Bache, superintendent of the U.S. Coast Survey; and Captain Charles H. Davis, superintendent of the U.S. Naval Academy. They were impartial, however, only with regard to local political and economic interests. Their professional training gave them a decided preference for water over land.[32]

For the next seven years, including the period of the Civil War, the U.S. Commissioners studied the harbor closely. They reviewed old charts and coordinated a series of detailed new studies to clarify the harbor's condition. Within a few years they had completed a meticulous survey of the inner harbor, a hydrographic survey of the main ship channel, and a topographical survey of the shores and islands of the outer harbor. They also took detailed measurements of the currents in the Charles and Mystic rivers and answered queries about projects extending from Mystic Pond in suburban Medford to Cape Cod. Between 1860 and 1866 the commissioners published their findings and recommendations on the harbor in a series of ten reports.[33]

Based on their studies, the U.S. Commissioners endorsed the theory of tidal scour, opposed most land making, and insisted that the harbor's survival required the restoration and strengthening of existing natural systems. They also confirmed that the depth and width of the main channel in the upper harbor depended on the scour produced by the flow of the Charles and Mystic rivers and the tidal waters that entered and exited the rivers and bays. Although the U.S. Commissioners admitted that the river currents were feeble compared with those in other harbors, they believed that any river current, no matter how small, helped to sweep sediment out to sea by producing stronger ebb than flood tides. Further shrinking of the harbor's reservoirs and rivers through land making would therefore ruin the harbor, especially since preliminary data already showed substantial shoaling. With no sense of exaggeration, the U.S. Commissioners concluded that the inner harbor "cannot, in our opinion, afford to lose another cubic yard of tide water."[34]

Moments such as this one, when Boston Harbor's enigmatic hydraulics shaped the scientific ideas used to interpret them, were critically important because many subsequent changes to the harbor would flow from them. Like past observers, the U.S. Commissioners were using information gathered from the harbor to construct their theories of how its hydraulics worked. "Science," writes environmental historian Carolyn Merchant, "is an ongoing negotiation with nonhuman nature for what counts as reality."[35] That aspect of "negotiation" gave the harbor a big role to play in its own future. Every day, the harbor produced a never-ending stream of observable phenomena, such as the rise and fall of tides, the erosion of sandy beaches, the shifting of material on the seafloor, and the mingling of freshwater and saltwater. Scientists reduced these phenomena to data that they used to reshape their understanding of how the harbor worked, and they then used their modified ideas to make recommendations about how to reshape the harbor. There was a constant feedback loop between the harbor's natural hydraulics and the scientific understandings that pointed toward particular courses of action.

After interpreting the data they collected, the U.S. Commissioners designed their specific recommendations around the principles of "preservation" and "improvement." Their first goal was "to preserve the great physical features in their ancient order, and to hold on to the old landmarks."[36] Accordingly, they recommended a renewal of efforts to protect the key harbor islands and headlands by building seawalls. They also advised against the continued filling of South Bay and Fort Point Channel, which some Bostonians believed must inevitably disappear to form a land bridge between South Boston and the main part of the city. The bay was essential to commerce, argued the U.S. Commissioners, and both were too important to the scour to tolerate continued encroachment. In general, they strongly advised against haphazard land-making schemes that did not consider the needs of the harbor.[37]

They also recommended a series of improvements to the harbor that would offset prior damage and increase its commercial potential. To enhance the natural scour, they proposed a seawall around the South Boston flats that would concentrate and accelerate the tidewater passing into the main channel, and they suggested converting the lower Mystic Pond into a tidal reservoir. Gates could hold freshwater back in the pond to allow more tidal water into the river, and then they could open to release the water down the Mystic River toward the harbor. Finally, the U.S. Commissioners recommended a radical restructuring of the network of bridges that crossed the Charles River into Boston, since their piles did not align with the current and interfered with the river's flow. With the "great physical features" of the harbor preserved, and some of their natural functions augmented through engineering, the U.S. Commissioners claimed that the harbor's scour would function even better than it had before.[38]

Central to the U.S. Commissioners' plan was their proposed system of "compensation in kind." They were not blind to the pressure for land making, and they acknowledged that developers could fill some of the state's flats with little damage to the harbor. But they were adamant that the volume of tidal water not be further reduced and proposed that no flats be filled without creating an equal volume of water elsewhere through dredging. If the state selected the site to be dredged carefully, "a new receptacle of tide water may perhaps be chosen possessing all the advantages of position to the natural one, or even more."[39] The state alone should perform the work to ensure its proper execution, although developers would pick up the cost by paying a fixed amount into a "compensation fund" for every cubic yard of tidewater they displaced. Compensation in kind was a form of "water mitigation" that anticipated by more than a century the "land mitigation" so popular today. Where the former would compensate for displaced water by dredging in another area, the latter compensates for the loss of wetlands and farmland by enhancing, restoring, or creating new wetlands and by using conservation easements to protect equivalent areas of farmland elsewhere. All manage the environmental costs of development by exchanging one resource for another and placing the financial burden on developers. The idea of a resource exchange might be older still, since compensation in kind was not a new idea even in the 1860s.

Support for the U.S. Commissioners' plans was strong and extended well beyond the maritime and engineering community. In 1863 the noted astronomer Benjamin Apthorp Gould threw his considerable reputation as a scientist behind the U.S. Commissioners' efforts by publishing a series of six articles in the *Boston Daily Advertiser* that translated their technical findings for a popular audience. Five years later, Charles Francis Adams Jr. promoted the commissioners' plans in a now classic essay on Boston's transportation problems. Bringing his exceptional rhetorical powers to the harbor's story, Adams penned a brief and damning history of the damage caused to its rivers, bays, and channels through development. But he reassured his readers that hope existed. "Scientific men," Adams wrote, had "sought out at once the cause and the remedy of the evil" and showed how the harbor could be saved. Adams's simple but powerful story of environmental decline and scientific redemption cast the U.S. Commissioners in the role of saviors and suggested that only science could rescue the harbor.[40]

So encouraged was the legislature by the work of the U.S. Commissioners that in 1866 it finally acted on the past recommendations of several ad hoc harbor commissions and established a permanent Board of Harbor Commissioners. The board had the power to draw new harbor lines and to reject or force alterations in any plans for development that involved filling flats or building structures in or over tidewater. The work of the U.S. Commissioners had shaped the debate over the harbor's future so deeply that the legislature wrote the concept of compensation into the statute that created the board. From that point on, when developers displaced tidewater in Boston Harbor, the Harbor Commissioners were to make them restore the volume through dredging in a carefully selected site, or compel them to pay into a compensation fund so that the board could do the work itself.[41]

That same year the U.S. Commissioners ended their six-year study of the harbor, leaving only one task unfinished. Yet its completion by other federal engineers two years later threatened to unravel all of their previous recommendations. The U.S. Commissioners had begun to compare soundings they took of the inner harbor in 1861 with a similar survey performed in 1835.

They expected the results to give a picture of the inner harbor's deterioration over that period. But when finalized after their departure, the data revealed that the volume of water below the line of low tide had not changed significantly during that thirty-six-year period. The massive shoaling described by close observers and the U.S. Commissioners themselves had never taken place.[42]

Instead, the data showed that the changes many had sensed over the years represented a much less threatening shift of material from one place to another rather than a generalized process of accretion. The constricting of the river mouths through filling and their obstruction by bridge piles had increased the velocity of water traveling through them. As a result, the water's enhanced force had excavated half a million cubic yards of material and distributed it unevenly around the harbor floor. The new data explained years of anecdotal evidence for shoaling and suggested that massive land making over the past third of a century had done little if any harm. Boston Harbor's impenetrable hydraulics had dealt its friends a powerful blow.[43]

Reshaping the Harbor

Those who favored land over water used the new findings to assail the harbor's defenders. In 1868 and 1870 two legislative committees reported on the practicability of filling various flats around the city, including the massive tract off South Boston. Francis W. Bird, a paper manufacturer and state legislator with a passion for polemic, chaired the committees. He also wrote both reports, which strongly criticized "the received doctrine of those who claimed to be the special protectors of Boston Harbor."[44] Bird complained that all of the opposition to filling in South Boston that he had encountered rested on an attachment to the idea of compensation in kind, which he resolved to discredit. He reviewed the U.S. Commissioners' writings on compensation and concluded that true compensation could only take place in the tidal reservoirs and rivers that drained into the harbor. The U.S. Commissioners themselves, however, had eliminated South Bay as a site for dredging, and the new data eliminated the Charles and Mystic rivers since their rapid flow was already moving material around the harbor floor. If anything, Bird claimed, the tidal reservoirs were already too big. So by his reasoning, compensation, which "never had any scientific basis or fair and thorough practical test," was unnecessary and maybe even counterproductive.

"Compensation in kind," he concluded, "may be consigned to the receptacle of things lost on earth."[45]

But Bird did not stop there. He also used the new data to argue that land making itself did not hurt the harbor. The doomsayers, he lamented, had done their work so well that their prophecies had gained the status of common knowledge. To believe a fraction of their predictions, however, would be to conclude that in barely a generation or two, the city's residents would see "a vast mud-hole, or at best, a larger-sized frog pond" where their harbor used to be.[46] Bird claimed that the evidence showed something different. The city had made more than thirteen hundred acres of new land since 1835, an amount that looked alarming "when read by the light of a theory." In the light of hard data, however, "it is readily seen that there has been much more fear than danger."[47] Even cutting off the huge Back Bay from the tides and later filling it seemed to have had no effect on the harbor, leading Bird to view the project as a model for future action.

As a result, Bird's committee proposed a radical amount of land making, advising the state to fill South Bay, Fort Point Channel, and parts of the Charles and Mystic rivers. It also suggested that the state cut off the Charles River to navigation, since the drawbridges inconvenienced the railroad corporations, and foretold a day when the Charles would become a narrow, two-way canal covered with a continuous bridge from East Boston to Brookline, a distance of about three miles. Bird and his committee concluded that Boston's greatest need was land, and that as long as the state preserved the ship channel and sufficient anchorage, it could fill flats with no worries about injuring the harbor. If damage did occur, the state could simply repair it.[48]

But Bird's reports failed to shake the legislature's confidence in the theory of tidal scour. To the harbor's defenders, the new data actually confirmed their suspicions that development was in fact affecting the harbor floor in negative ways. The harbor may not have suffered as seriously as they had feared, but the damage was worrisome and continuing to take place. Most important, the new data failed to suggest an alternative theory for how the harbor's hydraulics functioned. The theory of tidal scour still seemed to provide the only explanation, and until a better one came along, land making would continue to seem like a threat.

Undeterred by their critics, the Harbor Commissioners began reshaping the harbor with the theory of tidal scour as a guide. One of their first challenges concerned the South Boston flats, some nine hundred acres of land that stretched from shore to ship channel at low tide but lay beneath ten feet of water at high tide. A state commission had advised as early as 1847 that building a seawall along the flats would improve the scour by training the water from South Bay directly into the channel. The U.S. Commissioners had recommended a similar plan, and the state decided to build the wall and pay for it by filling the flats behind it, since engineers believed that the shallow water flowing over the flats contributed little to the scour. The project took decades to complete and is easy to mistake for just another landmaking scheme. But the Harbor Commissioners and legislature consistently viewed the project as a harbor improvement. The creation of new land was of secondary importance.[49]

The Harbor Commissioners also regulated other development around the harbor just as closely. When approached, for example, with a proposal to fill several hundred feet into the Charles River from Cambridge to build housing and parks, the Harbor Commissioners refused to approve the plan. The commissioners explained that they did not sanction lightly the replacing of natural scour by dredging, since natural forces were cheaper and required no human supervision. Only developments that required access to water, like commercial facilities, warranted such a change. Because the developers could build houses anywhere, they should build them elsewhere.[50]

When the Harbor Commissioners did allow developers to make land, they zealously collected compensation money from them. In 1870, after Boston filled a sizable area to create Atlantic Avenue, the Harbor Commissioners sent the city a whopping bill for $61,663.46 payable to the compensation fund. The city appealed the bill to the legislature and even tried to escape payment by adopting Francis Bird's argument that the theory of compensation had been discredited. But the Harbor Commissioners refused to reduce the amount.[51] They faithfully collected thirty-seven and a half cents per cubic yard of tidewater displaced from everyone who made land, and by 1878 the compensation fund contained more than $130,000. The commissioners used the interest to make improvements around the harbor.[52]

Neither did they hesitate to use their power to force changes from the railroad companies. In 1869, when the legislature required the railroads to widen drawbridges crossing the Charles and Miller's rivers, the Harbor Commissioners used their regulatory powers to force additional improvements. They offered to accept the proposed designs only if the railroads improved some of the bridges as well by realigning the piers and draws with the current.[53] They also used legal proceedings to prevent the Boston & Lowell Railroad from filling flats without authorization.[54] Moving with scientific certainty and legal authority, the Harbor Commissioners halted abuse, compelled improvements, and became the most powerful guardians the harbor had ever had.

Paradigm Shift

But it was not long before the Harbor Commissioners had to face a threat from an entirely new direction. By the 1880s the maritime and engineering communities, along with their supporters in state and federal government, believed that they had made substantial progress toward restabilizing the harbor. Federal engineers were working particularly hard to shield key islands and headlands with granite seawalls. Then, in 1894, a joint board composed of the State Board of Health and the new Metropolitan Parks Commission proposed that the state dam the Charles River to raise the water level above the river's polluted flats and create a lake for recreational boating. Rather than battling real-estate developers and railroad companies as in the past, the harbor's defenders found themselves opposing clear public interests in community health and recreation.[55]

The joint board used the findings of Frederick P. Stearns, a well-respected engineer, to challenge the Harbor Commissioners' understanding of the harbor's hydraulics. After examining the available data, Stearns concluded that damming the Charles River was unlikely to cause any damage to the harbor because the river carried little silt and was too weak to produce a significant scour. He even claimed that the past thirty years of land making had not harmed the harbor. Such statements were unprecedented from an engineer and reflected how difficult it had become to fit new data from the harbor into the established theory of its origins and function. This was an important moment, because it was the first time that someone trained in the engineering sciences had contradicted the dominant understanding of the harbor in such a high-stakes fight.[56]

The Harbor Commissioners, heavily invested in both their policies and their underlying science, objected to the plan and defended the river's role in maintaining the harbor. They noted that, in the thirty years since the U.S. Commissioners had written their reports, every federal engineer who studied the harbor had agreed that scour maintained it and that the rivers made important contributions to that process. The Harbor Commissioners recognized that large amounts of dredging in the harbor made new studies necessary, but they saw no reason "to adopt the theory of novel impression." Their opposition played a major role in the proposal's defeat.[57]

Advocates of a dam tried again a few years later, but this time the legislature hired John R. Freeman to study the problem. The most influential hydraulic engineer of his day and a tireless worker, Freeman supervised a biologist, geologist, chemist, medical doctor, and two engineers who prepared a wide variety of reports. Freeman himself contributed a number of additional studies on topics ranging from the velocity of harbor currents to air temperatures in the Charles River basin. In 1903 he summarized the collective findings in a report still admired for the rigor of its analysis. Freeman noted in the report that his team had "found itself face to face with a long-accepted theory of the maintenance of Boston harbor." That theory, he believed, was entirely erroneous.[58]

Freeman concluded that the theory of tidal scour that had dominated thinking about the harbor since the 1830s did not apply to Boston Harbor. With the aid of far more sensitive and reliable equipment than previously available, Freeman determined that the inner harbor had experienced no substantial shoaling in the past seventy years, despite an enormous amount of land making. He also found that the rivers carried very little silt, the currents were too feeble to perform any meaningful scour, and the water velocities needed for scour that the U.S. Commissioners had relied on in their work were far below modern figures. In short, no large source of silt confronted the channels, and they required no scour to maintain them. Boston Harbor was not a naturally scouring harbor and never had been.[59]

Freeman proposed a radical new explanation for the harbor's origins that amounted to a paradigm shift in how scientists understood its formation and hydraulics. Drawing on advances made in the science of geology, particularly glacial theory, he explained that the Charles and Mystic rivers had not dug the channels of Boston Harbor because they had not existed at the channels' creation. When the glaciers retreated from the Boston area thousands of years before, they melted into giant rivers that carved the channels from the blue clay now forming the bed of the harbor. The present rivers could not possibly have moved such large quantities of material. Since that time, the entire Boston area had subsided below its former level and the old river channels had become harbor channels. "In other words," concluded Freeman, "the harbor channels are strictly what may be called a series of drowned valleys." In one stroke the new paradigm brushed aside the theory of tidal scour that had dominated the development of Boston Harbor for three quarters of a century.[60]

The new understanding of the harbor proposed by Freeman allowed construction of the dam to move forward and justified additional land making into the twentieth century. The Charles River Dam, which began operation in 1910, turned the river into a lake and made possible the development of parks along its banks. Two years earlier, the state had completed a smaller dam across the upper Mystic River. Where damming the rivers had seemed like madness a few years before, it now appeared logical in the context of a new scientific paradigm. Freeman's explanation of the harbor's origins and hydraulics also made later land-making projects, such as the construction of a massive airport on the East Boston flats, seem much less threatening to the harbor. A different reading of data from the harbor had led to a different scientific theory that engineers would use to produce a different harbor.

The Legacy of a Scientific Theory

After seventy years or more as a dominant scientific paradigm, and half of that period as the foundation for official state and federal policy, the theory of tidal scour as wielded by the harbor's defenders had left Boston and its harbor transformed. The harbor lines drawn early in the century to protect the scour defined the city's shape. Land-making projects had lived and died at the pleasure of the Harbor Commissioners, who accepted, denied, or modified proposals based on how they might affect the scour. South Boston, the city's largest land-making project before the twentieth-century construction of its airport, began as a harbor improvement intended to train the tide and increase its force. The construction of the Charles River Dam was delayed by a decade because the Harbor Commissioners opposed it. And Fort Point Channel exists today because nineteenth-century engineers thought its current necessary to preserve the main ship channel. Only when these effects on Boston's development are added together does it become clear just how deeply the city was shaped by the relationship between a complex and inscrutable natural system and a scientific theory designed to explain its mysteries.

But how should historians assess a well-intentioned but misguided group like those who tried to defend Boston Harbor by applying a scientific theory that was later discredited? Were they the unwitting victims of their own scientific hubris? Were they projecting their anxiety over the port's economic decline onto the harbor's environment? To date, most historians have simply ignored them. One of the few historians even to mention the efforts of this group discounts all of the scientific research and theory behind their thought and maligns them with the label "traditionalist maritime alarmists."[61] But they were more than reactionaries panicked by the tides of change. Working toward what they sincerely believed to be the greatest public good, the harbor's defenders used the best science of their day to encourage urban growth while safeguarding one of the key natural systems that supported the city's economy.

In fact, the members of this group actually accomplished some good. They prevented the indiscriminate expansion of the city's shoreline, brought the erosion of the harbor islands and headlands under control, and forced the legislature to create a permanent board to manage the harbor. They also avoided some of the more radical suggestions for reshaping Boston's estuary, such as Francis Bird's hope that development would fill all of the Charles River except for a narrow ship channel. Their unheralded legacy, then, lies in the profound and often positive changes they brought to the harbor, their prevention of more development than Bostonians might be happy with today, and their example of commitment and stewardship—all long obscured by flawed science.

Long-forgotten stories like this one shed important light on the relationship between science, nature, and cities, and more such stories await scholarly attention. What lies beneath the waves that lap at so many of our cities is a world of channels, currents, rocks, sands, seawalls, jetties, and wharf pilings that underpin much of the human activity that goes on above them. What lies beneath our knowledge of such large natural systems are scientific ideas that might be far less reflective of reality than we like to think. Understanding how science has shaped harbors, and how harbors have shaped urban areas, represents an important new direction for understanding the relationship between nature and cities. Historians of the urban environment are perfectly poised to plumb its depths.

4

Remaking Boston Harbor

CLEANING UP AFTER OURSELVES

Steven M. Rudnick

ON September 2, 1988, as television cameras rolled, Republican presidential candidate George H. W. Bush clambered aboard a boat in Boston Harbor and charged his Democratic opponent, Michael S. Dukakis, with responsibility for the port's deplorable condition.[1] Although Dukakis ultimately played a major, and perhaps negative, role in the cleanup of Boston Harbor, his 12 years as governor of the commonwealth do not explain 350 years of pollution that has plagued the harbor. From Boston's earliest days of settlement, its proximity to the Atlantic Ocean has suggested an easy way to dispose of its inhabitants' waste. The resulting problems have been apparent more or less continuously throughout the following centuries, but only rarely has the issue reached high enough on anyone's policy agenda to result in significant action. Almost immediately after their arrival in 1630, the Puritans began to fill the closest flats with their waste. The refuse hardly went unnoticed, but the first successful steps toward dealing with it did not occur until 250 years later.

When the ecologist Ellen Swallow Richards and others began the study of sanitary chemistry in the late nineteenth century, they established many of the scientific underpinnings that would allow the effects of waste disposal in the harbor to be measured and addressed.[2] Miasmic theory, which held that disease occurred as a result of the foul vapors from decaying matter, drove attempts as late as the start of the twentieth century to clean up the pollution along the shores of Boston Harbor, but only by dumping the refuse into the water. Objective standards for water pollution and water quality that would move the focus to the waters of Boston Harbor would have to wait. The harbor was remade constantly at the shoreline over the course of more than three centuries to solve waste disposal problems on the land. The realization that harbor pollution is a water pollution as well as a land pollution problem did not emerge until the mid-twentieth century, however, and only informed the most recent major cleanup. Before this, almost all efforts to take care of the harbor were to solve its "aesthetic" problems.

Early Concerns

Garbage and excrement on the streets of Boston became a public health problem from the first years of settlement as the population increased. The need for the proper disposal of feces and urine led to town rules regarding outhouses, privies, and specific exclusions from sewers, but refuse was less regulated except for some prohibitions in common areas. Before the mid-nineteenth century, when abundant new water supplies changed the nature of sewerage in the city by allowing for flush toilets, public concerns about waste disposal mostly involved food waste, or what we generally call "garbage." Later, fecal waste also became an issue.

During the seventeenth century the choice of disposal seems to have been an individual one. Before 1700, Bostonians built their first sewers themselves or with neighbors as they identified the need. Some of these conduits were above the ground; others were underground pipes for carrying rainwater, kitchen waste, and water pumped from flooded cellars. None of these systems was for carrying human waste.[3] The thinking behind the exclusion of human excrement does not seem to be fully documented, but concerns over the disposal of this material were well known. Outhouses and privy vaults, the regulation of which occupied much of the time of Boston's selectmen during the eighteenth and nineteenth centuries, were the designated repositories for fecal discharges. It is highly unlikely that human wastes regularly found their way to the drains that were carrying off kitchen refuse, although sometimes rainwater running out of yards probably contained overflow discharges from the outhouses and privy vaults. It was the kitchen and commercial animal wastes that caused the greatest stench and consternation. Because of Boston's geography in the seventeenth century, drains and sewers did not have far to go to reach the harbor, or at least to a point where the effluent would easily flow downhill to the shore. Indeed, the word "sewer" derives from the Old English for "seaward."[4]

Fear of epidemic disease—widely seen as an indication of the wrath of God—was a major impetus for early colonists to remove waste from their households and yards. Cholera and yellow fever epidemics seemed to appear most frequently around dock areas where there were outfalls.[5] As early as 1634, only four years after the initial settlement of Boston, officials prohibited residents from throwing fish or garbage near the common landing (most likely the area that later became the town dock and the Faneuil Hall marketplace). The miasmic theory informed this action as well as many subsequent ones.[6] The growth of Boston, both as a residential community and as a commercial center, continued to create waste issues for the selectmen throughout much of the eighteenth century. Every available selectmen's annual report in the early part of the century records several rulings allowing additional sewer connections to the common drains that discharged at various places along the harbor. The selectmen's records from 1716 to 1736 suggest that the only issue that took up more of their time was the granting of permits to sell "beer, ale, cider and strong drink."[7]

Throughout the early eighteenth century, Boston continued to expand seaward by filling lands on the eastern and northern shores.[8] Because all of the early conduits discharged on the existing flats, each time the town's area grew the sewerage lines needed to be extended to the new shoreline. Residents' complaints in the selectmen's records offer a clear indication that much of the sewage was remaining on the flats and not being washed out by the tides; the vast majority of these complaints were about the stench that arose from these areas. The selectmen's rulings on development included requirements such as that of April 30, 1724, which stipulated that a strip of land be left clear of development to allow wastewater to drain to Ship Street, the most seaward street at the time. A ruling in 1726 required an open watercourse from Hanover Street to Union Street on the north side. Several selectmen's orders forbid the filling of docks such as Winthrop's and Leverett's because that would stop the natural watercourse of sewerage.[9] Neither the officials nor the residents worried very much about the quality of the water around the docks. The selectmen appear to have granted every request to ex-

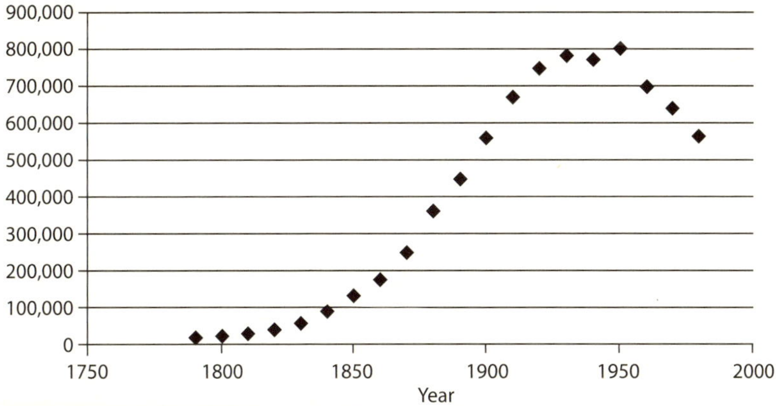

Figure 4.1. Population of Boston, 1750–2000. *Source:* Sidney Redner, "Boston Population," Department of Physics, Boston University, May 28, 2007.

tend sewer outfalls to the docks, as well as extensions to the new flats that were created in other areas. Boston encouraged the extension of the flats because filling the docks and extending the wharves was important for a town that depended on its maritime activity for its growth and wealth.

In many ways sanitary practices in the eighteenth century were a continuation of those of the seventeenth century. In 1744, after the town had paid attention to sewers and the problems that necessitated them for a century, Dr. Alexander Hamilton described the Boston town dock as "a veritable stinking puddle."[10] The overall condition of this sink during the period undoubtedly became progressively worse, particularly with actions such as those of the Boston Board of Health, established in 1799 as the first in the nation, requiring covered scavenger carts for privy vault cleaners and disposal in deeper waters from various wharves and bridges.[11] This probably alleviated the most noxious disposal along the flats and shallow areas, but only slightly. Sewage near the shore continued to increase with population (figure 4.1). Boston had approximately seven thousand residents in 1700. The number grew to about fifteen thousand by 1765 and then decreased markedly during the Revolutionary War. Soon after the war, however, the population quickly recovered to prewar levels and then began doubling every twenty years until about 1880, when the increase slowed.[12]

One can assume that as the population doubled, household wastes and human waste did too. In addition, there were the horses. Using the estimates of historians Joel Tarr and Clay McShane, one can calculate that in Boston in the mid-nineteenth century, horses produced an average of three quarters of a pound of manure per day for each human inhabitant.[13] The sale of manure for fertilizer, a good source of income for some colonial Bostonians, became less profitable during the nineteenth century, as farms were moved farther away from the city and transportation costs rose. Thus, one can safely assume that more and more manure ended up in the harbor as the century progressed.

The Nineteenth Century and Continuing Fears of Disease

When Boston was incorporated as a city in 1822, more than forty-three thousand residents were contributing to its waste disposal problem. Josiah Quincy, who became mayor in 1823, made improved sanitation a focal point of his administration. Years later, in his municipal history of Boston, he wrote, "no subject . . . [has] more importance than the adoption of systematic plans for effectively removing the various accumulations and nuisances in them, which are incident to a populous city."[14] An "out of sight, out of mind" mentality grew unabated, however, as the administration and aldermen passed numerous ordinances designed to remove waste from the city into the common sewers, and thence into the harbor, still mostly in the shallow areas.

In 1824 an ordinance required all wastewater to be conveyed through drains underground to the common sewer. The regulation was expanded nine years later by requiring all tenements to be provided with sufficient drains and privies, and mandating the connection of vaults, via drains, to the common sewers. This ordinance prohibited solid wastes from being introduced to the sewers, but it allowed the liquids from the privies. New rules in 1834 permitted the introduction of rainwater from the roofs of buildings into the common sewers, probably to aid in transporting sewage to the harbor.[15] At about the same time, fears of cholera increased with its appearance in 1831 in Russian goods, which led to quarantines of these imports. Boston was spared major epidemic outbreaks, presumably because of its quick response and increased vigilance and attendance to cleanliness.[16] Elsewhere—New York, Philadelphia, and Baltimore, among other cities—major pandemics occurred, and even in Boston the commitment to cleaning up and the Board of Health that had been established was short-lived.

When Josiah Quincy Jr. became the mayor of Boston in 1847, he noted in his inaugural address that the increasing number of residences in the Back

Bay was creating a problem on the west side of town at the confluence of the Charles River and Boston Harbor. Almost stagnant water here would "soon require a remedy." Quincy particularly liked the possibility of collecting the sewage and selling it as fertilizer, making the operation economically self-sustaining.[17] Despite a report in 1849 stating that Back Bay was nothing less than a great cesspool—and in line with the continuing prevalence of the miasmic belief that "daily accumulations of decaying [materials] would . . . shortly reach a point of malignity that would propagate disease of a general and fatal character"—little was done other than to extend the sewers further into the harbor and fill the flats.[18] Indeed, the records indicate that miasmic theory dominated continuing discussion of the problem in 1861, after a new Mill Pond seawall further restricted the tidal flushing in this basin and again in 1903 when the legislation that authorized the Charles River Dam at least kept the flats from being uncovered.

At the same time that this city report on the Back Bay was being prepared, the Massachusetts legislature appointed a commission to perform a sanitary survey of the state. The panel, headed by Lemuel Shattuck, a member of the legislature, produced a five-hundred-page report with fifty recommendations, but the topics of sewers and sewerage occupied fewer than seven total pages, with six of them devoted to discussing the agricultural use of human waste. The report contained no specific recommendations for sewers or treatment.[19] Many of the Shattuck Commission's proposals echoed those of a report by Sir Edwin Chadwick, who had performed an extensive survey of the sanitary conditions in England on the relationship of sewers and disease. One of Chadwick's main recommendations had been a "hydraulic system that would bring potable water into homes equipped with water closets, and then would carry effluent out to public sewer lines, ultimately to be deposited as 'liquid manures' onto neighboring agricultural fields."[20] While the disposal aspects of this proposal were never realized, the supply side became a significant reality in 1848, when Lake Cochituate water began to flow to Boston households. With the water supply came water closets. It is estimated that by 1863 eighty-seven thousand Boston households had piped-in water, and fourteen thousand of them had toilets.[21] It is not clear when an official authorization of solid human waste delivery to the sewer system took place, but from these statistics it is apparent that Boston's sewer system experienced a significant increase in load during this period.

Filling the flats around Boston Harbor created a twofold problem. Because with added length the slope to the harbor decreased, the sewer lines became more level and could not empty as easily. Since the tides backed up the sewers toward the residences more easily, some sewers received "tide gates" to try to prevent this problem. As a consequence, the flow of water and waste often stopped long enough to allow the pipes to settle and clog. The resulting decay created gases that flowed back into Boston's houses. Similarly, the decreased slope allowed the tides more easily to deposit sewage on the flats. Several observers began to postulate the role of tidal scour—cleansing by the swift-moving waters of rising and falling tides—in removing sewage from the flats. As early as 1851, Governor George S. Boutwell vetoed a bill to establish a roadway across the South Boston flats, claiming that it would reduce scour and cause the creation of another filthy Back Bay that would have to be filled.[22]

During this period, tidal scour emerged as a point of contention for a very different reason. Shipping interests worried that filling the flats and other land-making projects were changing the tidal flow patterns of the harbor, decreasing the depth of the shipping channels. Throughout the early nineteenth century, competent observers and engineers had suggested that the port was deteriorating, that shipping channels were silting up, and that a major cause of this was land making, which was decreasing the volume of water in the harbor, thereby limiting the ability of the tides to scour the bottom and the shores. The city government was able to persuade the federal government to establish a U.S. Harbor Commission on Boston Harbor mostly on the basis of national security. The commission endorsed the concept of tidal scour in 1866, despite the fact that its own measurements and hearings showed that silting had never taken place. In 1903 a comprehensive and rigorous study conducted by John R. Freeman and funded by the Massachusetts legislature, completely disproved the commission's findings.[23] In terms of sewage disposal, landfilling probably exacerbated the problem because it reduced the slope of drainpipes, not because it reduced tidal volume.

With the conclusion of the Civil War, a new public awareness of the problems of infectious disease and its relationship to sanitary issues also began to emerge, partly as a result of the evolution of medical practices and understanding that developed during the conflict. Before the 1860s the dominant and almost singular theory of disease was that of miasma, but this approach only worked if something smelled bad. In fact, it was easy to forget sanitation during the winter. Germ theory began to gain adherents in medicine in the 1860s. It held that microorganisms—organisms too small to be seen except through a microscope—caused certain diseases. The French chemist and microbiologist Louis Pasteur, the English surgeon Joseph Lister, and the German physician Robert Koch have received much of the credit for development and acceptance of the theory. In the mid-nineteenth century Pasteur showed that airborne organisms caused fermentation and putrefaction; Lister revolutionized surgical practice in the 1860s by using carbolic acid (phenol) to exclude atmospheric germs and thus prevent putrefaction in compound fractures of bones; and in the 1880s Koch identified the organisms that cause tuberculosis and cholera.[24]

Although the emergence of bacteriology and the concept of microbial infection were integral to the acceptance of germ theory, the immediate consequences for sanitation were at most subtle. The Boston Board of Health, reestablished in 1873, still observed two years later that the "filthiness of the air, water and soil . . . may be a more or less necessary condition for the evolution of the specific virus to which the disease owes it origin."[25] By then, however, over significant local resistance, the board was already beginning to effect a significant change in the way the city thought about its waste and sewer problems.

The Main Drain

In its first annual report the newly reestablished Board of Health recognized the need to pay "the promptest attention . . . to stagnant and waste water, effective drainage and sewerage." The "troublesome question of sewerage" had grown so serious that it was apparent that "the time must come, sooner or later, when sewage . . . must be disposed of in some manner independent of and disconnected from either the sewers, or something must be added to our present sewers." Noting the backup problem on the flats, the board added that the "Danube or the Hudson flows quickly, but the Charles does not—nor do the sewers."[26]

If the need to adopt new measures for waste disposal was apparent to the Board of Health, it did not seem at all clear to Boston's aldermen. In what was to become a discernable pattern of accusation and denial, the Board of Aldermen's Committee on Sewers replied sanguinely that "we have the perfect means of collection and the problem is merely one of appropriate discharge into the surrounding waters." Were the Charles River and the harbor becoming cesspools? There was nothing to worry about, concluded the aldermen. To the extent that they admitted the existence of any problems at all, the solution lay in pushing the drains further out on the flats.[27] The aldermen's conclusions were indefensible as far as the members of the Board of Health were concerned, and they specifically expressed this in their second annual report in 1874: "the whole system of sewerage is clearly wrong."[28]

Over the next two years city officials developed an increasingly sophisticated understanding of Boston's waste problems, one that resulted by 1876 in a plan to resolve the issue. The Board of Health's second annual report began to focus the discussion. Noting that Miller's River (which seems to have been in what is now East Cambridge) was a source of stench and cholera, the board posed the question: "should not all sewers be turned southeasterly into deep water or would it be feasible to construct a large marginal sewer into which all others discharge, and this one be carried away into deep water from a point nearing the bay?"[29] Here was evidence that water quality as a modern concept had risen to the level of public consciousness. Later in 1874, the Board of Health also recognized the necessity for improved sewerage in the Roxbury Canal, Stony Brook, and Muddy Brook—all areas that did not benefit from any significant tidal flushing. Miasmic theories of the origins of disease continued to govern the board's analysis of the waste problem—"our prevalent summer diseases are largely influenced by this poisoned atmosphere there can be no sort of doubt," they wrote—but their proposed solution was markedly novel. The members advocated a "main drain" that would carry beyond the flats to the harbor channel, "where in deep water and strong currents, the material will be so dissipated and acted upon by the salt water as to become harmless and unobjectionable to the senses before it can be lodged about the city by the flood tide."[30]

By 1875 the board was beginning to express frustration over the city's inaction. Its annual report detailed the importance of sewerage, the specific problems of discharge on the flats at numerous locations, and emphasized its concern over the relationship between disease and the improper disposal of waste.[31] Perhaps this attention to disease finally moved the Board of Aldermen to action. That year the city fathers appointed a commission to propose a solution. It included Ellis Chesbrough, who had gained a national reputation for his work in draining Chicago sewers in the deep part of Lake Michigan. The commission's report in early 1876 proposed what later came to be known as the Main Drain. The drain would intercept the sewers that extended along the waterfront from the Back Bay around the peninsula through the South End and down to Roxbury and collect all of the sewage flowing in the individual pipes before it emptied into the harbor. The effluent would flow to a point in Dorchester, where large pumps would push it through pipes to holding tanks on Moon Island, a mile across Dorchester Bay. At Moon Island it would be discharged, but only at the beginning of the ebb tide, in theory taking it out to deep water.[32]

The commission's solution had not addressed every issue, however. An editorial in the *Boston Daily Globe* in September 1875 had asked for the definition of "deep water." It cited a study of disposal in the tidal reaches of the Thames in London and concluded that "organic sewage . . . delivered anywhere within our inner harbor will be sure to find its resting place within the limits of tide water."[33] Nevertheless, Chesbrough's success in Chicago convinced him of the common wisdom that disposal in deep water, however defined, was sufficient to dilute the sewage and eliminate the problem of disease as well as the noxious odors. Objections to the city's plans came from two perspectives. Some municipal officials still felt that there was economic value in the sewage and that sea disposal was a waste of resources. And one official, Councilor George L. Thorndike of East Boston, opposed the ocean disposal because he believed that "wastes would accumulate among the islands . . . and sooner or later . . . cause trouble." This prescient view, as well as the profit perspective, was rejected, and the city passed up possible profits from the sale of sludge in favor of the proposed drain.[34]

In addition to the Main Drain, Chesbrough's commission also proposed a Northern Drain, which would serve Charlestown, Chelsea, East Boston, and Winthrop and would be discharged at Shirley Gut, approximately at Deer Island. This location would play an important role in the later development of Boston Harbor's sewer system. Chesbrough's panel also recommended the continued use of combined sewers, carrying both stormwater and wastes, even though some visionaries had foreseen the problems that this would ultimately create. No large city had attempted to separate the two forms of sewage, and the commission was not willing to advocate for it. The recommendations carried a projected price tag of $3.7 million for the Main Drain and $2.8 million for the North Drain. In April 1876 the Massachusetts legislature chose to fund only the Main Drain.[35]

Out of Sight, out of Mind?

The Main Drain was completed and put into operation in 1884. *Main Drainage Works of the City of Boston*, by Eliot C. Clarke, the project's principal design engineer, contains an elegant description of the entire development and construction process, including detailed technical drawings of the sewers and the pump works.[36] Although increasing population and concomitant sewage problems led to the completion in 1895 of the North Metropolitan Drainage (which terminated at Deer Island) and in 1904 of the South Metropolitan system (which extended to Nut Island), these discharge points were apparently far enough out in the harbor to create little concern. At least there seems to be little in the record to suggest any problems.

The Nut Island pipe came about as a result of a 1900 resolve of the General Court, which directed the State Board of Health to "consider the general subject of discharge of sewerage into Boston Harbor and report a plan for an outlet for a high-level gravity of other sewers for the relief of the Charles and Neponset River Valleys."[37] The legislature's action was in response to stress on system capacity. By 1900 the pipe ending at Moon Island was discharging fifty million gallons of wastewater a day, resulting in a slick a mile and a quarter long. The board curiously concluded that there was "no reason to anticipate any problems for years to come"—the only objections had come from those sailing through the effluent—but it still predicted that discharging on the incoming tide would bring sewage ashore and judged that "the increasing volume creates a need to establish another route to the harbor."[38]

In 1914 controversy over landfill, scour, and flats erupted again when the Metropolitan District Commission (MDC) began planning for the construction of the Old Colony Parkway (now Morrissey Boulevard). The City of Boston filed numerous objections to these plans based on its worry that the construction would destroy the Savin Hill bathing area on which the city had recently expended nearly a half million dollars to build a beach and dredge the channel. In 1928 the commissioner of Public Works for the City of Boston complained to Governor Alvin Fuller that errors in construction had caused shoaling of the channels, decreased tidal wash with no scour, and unsanitary conditions on the beach. The MDC responded that studies done in 1924 had shown that there were no tidal currents sufficient for scour in the area and that the City of Boston sewer discharge in the area was responsible for the foul conditions on the beach. Recent elevation studies and flow studies showed that there were no differences on opposite sides of the roadway. In December 1928 a legislative commission sided with the MDC after reviewing four years of documents and charges claiming that the unsanitary conditions existed in Savin Hill Bay before the construction and suggested removal of the sewers and dumps in the area.[39]

As a result of ongoing complaints from a number of different constituencies, the legislature ordered studies of the condition of sewage in Boston Harbor in 1900, 1917, 1929, 1930, 1935, and 1936. None of these investigations resulted in any significant changes in the sewer system from what had been put in place in 1884 and upgraded in 1895 and 1904. In 1937 a study by the state House of Representatives reported that "objectionable conditions [were] confined to the immediate vicinity of Boston Main Drainage outlets but that under certain conditions of wind and tide, problems could develop on the beaches." It suggested preparing plans be for treating sewage and for extending disposal to deep water, specifically perhaps near the Graves (the outermost islands in Boston Harbor).[40]

Whether the cause was parsimony or lack of will, the legislature was very slow to authorize the advanced methods of treating sewage that became available in the early 1900s. The first of these were sedimentation chambers that removed a significant amount of sludge before disposal. Activated sludge systems, which resulted in the removal of more than 90 percent of the solids and bacteria from the sewerage effluent, soon complemented the sedimentation chambers. Boston did not adopt these advanced technologies, however, until the mid-twentieth century. Despite ongoing demands by residents of Boston, Quincy, Winthrop, Weymouth, and Hingham based on concerns of health menace, aesthetics, and loss of opportunities for swimming, the commonwealth did little to alleviate the conditions.

The Beginning of the Modern Era in Boston Harbor

In 1938, near the end of the Great Depression, the legislature authorized the creation of a major commission to conduct a comprehensive study of the sewage issues in Boston Harbor. In addition to holding local investigations and hearings, the commissioners visited Baltimore, Buffalo, Chicago, Cleveland, Elizabeth (New Jersey), Milwaukee, Newark, New Britain, New York City, Niagara, Perth Amboy, and Providence. Their report, submitted in June 1939, detailed problems of grease balls on the beaches, floating matter, and foul odors in Boston Harbor. The committee's realization that "recreation is big business" drove the inquiry, not residential concerns about health or aesthetics. During the hearings several people challenged the commission to recommend advanced treatment of the sewage before discharge, but these requests did not get far.[41]

The commission did recommend the appropriation of $14 million for primary treatment plants for solids and grease removal, sedimentation, and chlorination. Barges would dispose of the resulting sludge in the deep ocean. The commission considered secondary treatment with an activated sludge process but deemed it to be unneeded: "We are unanimous in our opinion that for the present, only partial treatment is necessary for the sewage discharged into Boston Harbor. We are certain that by taking this first step for elimination of pollution, the menace to public health under present conditions will be sufficiently reduced to allow us to safely postpone the additional treatment of raw sewage for a number of years." Engineers designed the system to serve local requirements based on population projections for 1955. In fact, the population of the service area grew from only 2 million people in 1935 to 2.2 million people in 1955, and if the criteria and underlying assumptions used in 1935 were the only considerations, the system as proposed would have been more than adequate.[42]

The underlying assumptions were not correct, however, and it is not clear that the commission even understood what would constitute safe bathing beaches, a growing consideration as Bostonians increasingly turned to the shoreline for recreation. The American Public Health Association Committee on Bathing Places issued its first standards in 1924, but they only covered swimming pools. The committee did not consider natural bathing beaches until 1933, but even then it only issued a classification system, not a standard. The committee did not feel that the epidemiological evidence suggested a major public health problem. There were no marine standards until 1972.[43] Studies conducted throughout the 1930s had indicated that the bacterial counts on Boston Harbor beaches were among the highest in the nation, however, exceeding even the most lax of the American Public Health Association's classification levels. Surveys in 1935 and 1936 consistently showed counts on Boston beaches, mostly at ebb tide, at one thousand times the standards that other states had proposed. By eschewing secondary treatment, the recommendations of the commission would probably have had little impact on this. In any case World War II apparently preempted any action on the commission's recommendations. Beginning in the early 1950s, the Metropolitan District Commission began the construction of swimming pools as a recreation replacement for polluted beaches.[44]

In 1952 primary treatment was finally applied to the South Metropolitan Sewerage District flows at Nut Island. The effluent still flowed into the relatively shallow waters of Quincy Bay. Sixteen years later, similar treatment was applied to the North Metropolitan flows at Deer Island. At the same time normal flows to the original Main Drain outlet at Moon Island were diverted to Deer Island. In fact, the flows to Moon Island were generally above the capacity of Deer Island, particularly when storm runoff added to the volume, so raw sewage continued to be diverted to Moon Island and out into the harbor. During storms, additional raw sewage poured into the harbor from approximately eighty-four combined sewer overflow (CSO) outfalls scattered around the waterfront.[45]

An estimate in the 1970s suggested that 15 percent to 20 percent of the sludge discharged on the outgoing tides ended up back on the shores and the floor of the harbor, resulting in severe pollution, but there was no remedial action. To a large extent, inaction on cleaning up Boston Harbor was a consequence of the fact that Massachusetts tax revenues funded the MDC, the agency responsible for sewerage. Legislators from cities and towns outside of Boston were reluctant to commit their constituents' tax dollars to a cleanup and operations that would primarily benefit the citizens of Boston. As a result, not only were capital funds insufficient for improvements, there was often not enough money even to replace broken equipment to keep the system running on a daily basis. Over the years the MDC had also become a haven for political patronage with the consequence that many of its personnel were not fully qualified to keep the system running.[46]

Scientific Solutions

Secondary treatment processes, which use enhanced bacterial populations and air or oxygen to separate the biological content of waste into a settling solid fraction and an easily handled and less noxious liquid fraction, were well developed in the early twentieth century. They did not see large-scale application, however, until after midcentury. In Boston, moreover, the high cost of the processes caused those responsible for the harbor to reject them long after other communities began to adopt secondary treatment. The condition of the harbor as a result of waste disposal became a public issue only gradually. Perhaps initiated by the publication of Rachel Carson's *Silent Spring* in 1962, nurtured by the many protest movements of the 1960s, and culminating with the celebration of the first Earth Day in 1970, Americans' awareness of pollution grew to a point that set in motion the Green Decade of the 1970s.

A new era of activity in Boston Harbor began with the passage in 1972 of amendments to the Federal Water Pollution Control Act of 1948 (FWPCA). The amendments required all publicly owned sewage treatment works, such as those at Nut Island and Deer Island, to install both primary and secondary treatment equipment. While Boston was not the only city in violation of the law, it was slower to comply than others such as Seattle and Miami. A 1976 study by the MDC and the Army Corps of Engineers recognized that secondary treatment should be implemented at both Nut Island and Deer Island, but politics and related financial concerns produced inaction.[47]

Five years after the 1972 legislation, additional amendments to the FWPCA in 1977 were a disaster for Boston Harbor. They contained a waiver provision, a result of negotiations and machinations in Congress that are worthy of a book by themselves, which let cities with deep ocean discharge potential apply for an exemption to the requirement for secondary treatment. Although the Environmental Protection Agency (EPA) was unlikely to give Boston a waiver, the city and the state chose nevertheless to ask for one. The EPA did not deny this request until 1983, by which time an offer from the federal government to assume 75 percent of the cost of a cleanup had disappeared even though the sewage had not. In the end the communities the treatment plant served paid 92 percent of the cost. Curiously, the MDC application for a waiver indicated that $144 million in construction costs and $14 million in annual operating costs would be saved by not implementing secondary treatment. The ultimate cost of the construction was $3.8 billion. By 1988 the continuing pollution had earned Boston recognition as "America's Dirtiest Harbor."[48]

In the meantime there were other developments. The human waste problem, which Boston had often ignored over the course of 350 years, met its match in the person of William Golden, the city solicitor for Quincy, Massachusetts. Golden was jogging on Wollaston Beach one day in 1982 when he realized that what he was stepping in was disgusting. On behalf of the city of Quincy, he filed suit in Massachusetts Superior Court, claiming that the MDC had violated its federal and state discharge permit as well as the state law prohibiting the discharge into coastal waters of material that would be injurious to public health or would contaminate shellfish. What followed was a wonderful study in sticks and carrots.[49]

The concept that natural resources have an economic value other than as raw production materials is a relatively new one and still very much in its infancy. Had anyone said to taxpayers before 1983 that spending $3.6 billion on cleaning up Boston Harbor would result in between $200 million and $500 million in annual benefits to recreation, fishing, and commerce, it is unlikely that they would have enthusiastically put up their tax dollars. It was the unwillingness of the taxpayers—or the unwillingness of politicians to risk proposing substantial cleanup costs—that led to most of the inaction over the harbor. The entry of the courts removed the excuse that the cleanup was unaffordable. Both Massachusetts and federal law clearly stated that the pollution of the harbor was illegal and had to be stopped. The Federal Clean Water Act had to be enforced and the resources had to be found.

At first the Massachusetts Superior Court tried to forge agreement among the parties. It appointed a special master to discover the facts and resolve disputed issues. A report in August 1983 by Professor Charles Haar of Harvard Law School led to an agreement that temporarily suspended the court proceedings. One of the report's recommendations was to remove the responsibilities from the MDC and place them in an independent authority. On the one hand, this expedient had considerable appeal for the politicians, since they would no longer be responsible for funding the cleanup through tax appropriations. On the other hand, the move cost potential patronage.[50]

A lack of progress brought the parties back to court in October 1984. Through a series of measures, including a threat to place the MDC in receivership, Judge Paul Garrity ultimately forced the Massachusetts legislature to create an agency independently financed through fees and bonds to oversee the cleanup of Boston Harbor and subsequent sewage treatment operations. Thus the Massachusetts Water Resource Authority (MWRA) came into existence. There was little optimism, however, that the new agency would achieve an effective cleanup, so the EPA and the Conservation Law Foundation, a regional environmental advocacy organization, filed suit in Federal District Court to force a court-monitored cleanup schedule for the harbor under the Federal Clean Water Act.[51]

The federal case was assigned to Judge A. David Mazzone. In December 1985, Mazzone established a cleanup schedule for the MWRA. It included a plan to complete a new primary treatment plant by 1995 and a secondary plant by 1999. The sites of these plants were still undetermined, and they would cause much legal wrangling and public dissent. At several points taxpayers dumped sewer bills in the harbor to protest the rising costs of the cleanup. These demonstrations produced some small changes in the program's funding, but they did not have major consequences for its progress. A group from Cape Cod brought one of the major challenges to the cleanup plan. It claimed that the new effluent outfall, 9.5 miles out to sea from the Deer Island treatment plant, would increase pollution in Cape Cod Bay and threaten the breeding grounds of the right whale, an endangered species.[52] The challenge was ultimately turned aside, primarily on the basis of extensive modeling by the U.S. Geological Survey.

In 1997 the first phase of the secondary treatment plant was completed. It provided for removing 80 percent of solids from dry weather flows at Deer Island and allowed the plant to meet the requirements of the Federal Clean Water Act for the first time. Significant milestones quickly followed. In July 1998 an interisland tunnel that transferred all of the flow that previously went to Nut Island was completed, allowing the closure of that facility. In 2000 the outfall tunnel was completed, and on September 6, 2000, the new system began to send the effluent off to Massachusetts Bay.[53] Judge Mazzone's management of the cleanup provides a study in mediation and negotiation. Nearly every decision affecting the project was reached through consensus or by granting priority to the entity that was responsible for the design element. The result was the largest public works project in New England up to that time and one that came in essentially on time and under budget.

The Final Chapter or Just Further out of Sight?

So, after twenty-two years and a $3.8-billion expenditure, is Boston Harbor cleaner? Even the most casual observer would have agreed that the harbor became noticeably cleaner in 1989, when both the Deer Island and Nut Island primary treatment facilities stopped discharging floatable pollution, such as paper, oil, grease, and plastics. A second major improvement came two years later, when the sludge-to-fertilizer plant came on line and the system stopped dumping settled materials in the harbor.[54] The MWRA sells the pelletized fertilizer to feed citrus crops in Florida and golf courses and other landscaping enterprises in New England and elsewhere, except when testing shows that the metals' content is excessive. In the latter case, which does not appear to occur very often, the sludge goes to landfills in Utah. How ironic that more than one hundred years after the construction of the Main Drain ended the sale of Boston's waste for use on crops, the MWRA has gone back into the fertilizer business.

Historians often maintain that it is dangerous to use the past to project into the future. Although it helps to understand what has happened, the lessons are only guidelines. Such is the case with Boston Harbor. Chemical and biological measurements taken between 1991 and 1998 should show the results of stopping the most noxious past practices. The results of these studies are not definitive, however. Heavy metals in the sediments of Boston Harbor show a slight decrease but certainly not as dramatic as one would hope for.[55] Early biological studies after the cessation of sludge deposition suggested that the mix of benthic biological faunal organisms is moving toward less pollutant-tolerant animals and therefore are indicative of a cleaner harbor.[56] More recent analyses of the infaunal communities strongly support the conclusion that the environment on the harbor floor is indeed recovering from decades of pollutant input.[57]

Has the harbor cleanup had a noticeable effect on Boston's beaches? Between 2002 and 2004 the number of days during which the beaches were closed due to high bacteria counts more than tripled.[58] The causes of variations in water quality are not obvious. For example, Carson Beach in South Boston is the second cleanest beach in the harbor, but it has seven CSO outfalls and storm drains. Wollaston Beach in Quincy—one of the most contaminated beaches in the harbor—has no CSO or other sewage discharges but has eight storm drains that discharge along its shoreline. So it is probable that factors unrelated to the cleanup may be responsible for the ongoing concerns in the harbor. The courts and the MWRA have made closing the combined sewer overflows a major design goal, but it is certainly not clear that this will completely solve the problem.[59]

One consideration that is often overlooked in talking about Boston Harbor is the continuing influence of the rivers that flow into it. After all, concerns about the Charles and Neponset rivers led to an expansion of the sewer system in the early 1900s and thereafter. Before the treatment plant outfall pipe was moved offshore, the rivers contributed as much as half of the freshwater flow into the harbor. These sources, and storm drains, still represent a major source of pollutants to the harbor. Polycyclic aromatic hydrocarbons, an organic pollutant with a considerable level of carcinogenicity, come almost entirely from the rivers by way of runoff of atmospheric deposition in the watershed.[60] The rivers are also probably the primary source of elevated metal content in the harbor sediments.

A large-scale monitoring program in place in Boston's inner and outer harbor will ultimately assess the consequences of moving the outfall offshore. The court cleanup order specifies this monitoring and requires the MWRA to have a plan should the pollution at the outfall prove to be more damaging than anticipated. If it is, then one of the possible remedies would be to restore the outfall to the inner harbor, probably with additional required advanced treatment processes. The citizens of Boston spent more than 350 years remaking their harbor from a pristine near-wilderness teaming with fish into the nation's dirtiest. Although the provisions of federal and state clean water legislation have generally had positive effects on harbors across the country, the special attention paid to Boston Harbor since 1983 has remade it again. Today, perhaps we have come closer to contemporary handbook writer M. F. Sweetser's 1882 description of "the perfection of physical comfort" that comes from passing Boston Harbor on a warm day in the summer.[61]

5

In Search of the Shawmut Peninsula

USING MODERN CARTOGRAPHIC ANALYSIS TO DISCOVER THE "ORIGINAL" BOSTON SHORELINE

Stephen T. Mague

It is not down in any map; true places never are.

Herman Melville, *Moby Dick, or The Whale*

Surveying a place, according to my Idea, is takeing a Geometrical Plan of it, In which every place is to have its true situation.

J. C. Beaglehole, ed., *Journals of Captain James Cook*

THE shape of the Shawmut Peninsula before the bays and marshes surrounding this fist of land were filled in to create present-day Boston is of both academic and practical interest. For scholars, the subject illuminates early attempts to modify the terrain, with implications for the landscape, property rights, and the built environment. For those with more immediate concerns—including civil engineers interested in the potential effects on development of groundwater seepage and earthquakes and lawyers litigating real-estate cases dealing with ownership interests in filled tidelands—the issue has direct consequences. This chapter offers one answer to the question of the peninsula's shape, describes modern techniques for determining it, and relates the result to one of the most pressing reasons for investigating the subject: the public's rights to the shoreline.

Mapping the Elusive Shoreline

The accurate identification, location, and representation of natural, ephemeral features such as shorelines on maps are subjects of much debate in current scientific journals. Shorelines move constantly in response to waves, tides, storms, relative sea-level rise, and human alteration. To complicate matters, this movement occurs along several time scales (for example, daily, yearly, and geologic, extending thousands of years), making graphical representations valid for only discrete periods and presenting unique challenges to surveyors and cartographers charged with mapping the coast. According to the historical geographer Conrad Heidenrich, "All maps are graphic interpretations of spatial information [and] . . . abstractions of spatial reality at a point in time when reality was perceived by a particular author."[1] The mapping of natural features such as shorelines, and the spatial accuracies associated with them, also depends on additional factors ranging from the definition of the shoreline to the reputation of and methods used by the surveyor performing the work.[2] The environmental characteristics or context in which the survey was conducted can also significantly affect the reliability of shoreline depictions.

The cumulative effect of these factors on the cartographic accuracy of shoreline position is often difficult to quantify, particularly for the period before the standardization of the coastal mapping methodologies, procedures, and symbologies that the U.S. Coast Survey introduced in the early 1840s.[3] Indeed, for areas of shoreline filled or otherwise altered before the work of the Coast Survey (such as large areas of Boston Harbor), assessing the reliability of early shoreline positions can be both an art and a science. Where tidelands alteration began early in the development history of a waterfront, as was the case in Boston, the cartographic limitations of seventeenth-, eighteenth-, and early-nineteenth-century mapping methods and technology typically frustrate assessments of the positional reliability of pre-altered shoreline conditions. For such areas contemporary, plan-based efforts to identify accurate or reliable pre-alteration shoreline positions frequently depend on charts and maps depicting multisource reconstructions. These historical shoreline reconstructions are rarely the result of a single actual survey but are compilations from a variety of available sources, both scientific and anecdotal, and traditional cartographic error assessment techniques that quantify the accuracy of these early shoreline positions statistically can be misleading.

Variations in the position of the Shawmut Peninsula (the "original" shoreline) are depicted on several eighteenth-, nineteenth-, and twentieth-century maps and plans.[4] Figure 5.1 depicts one version of the former penin-

Figure 5.1. A version of the Shawmut Peninsula superimposed over the city of Boston in 2001. *Source:* Image from Massachusetts Geographic Information System (MassGIS).

sula, superimposed on a 2001 aerial photograph of Boston. State-of-the-art shoreline change analysis software have generated a spatial database or catalogue of former shorelines and quantified positional comparisons of reconstructed "original" shorelines depicted on several plans of the Shawmut Peninsula. Areas of spatial agreement and disagreement have subsequently been identified to support a qualitative assessment of the most reliable reconstructions using basic statistical measures such as average and standard deviation. These disagreements are further explored in the context of additional factors that might help account for areas of significant disagreement. Recommendations are provided for a plan-based approach to the challenge of identifying the most reliable source or sources of shoreline information where cartographic limitations and early human alteration preclude quantitative assessments of map accuracy. Reliable determinations of historical shoreline position play a fundamental role in modern public policy as it relates to protecting and promoting the public's rights to and along the waterfront.

Multidisciplinary Applications of Historical Shoreline Maps

The ability to assess maps depicting the same shoreline location at different points in time provides fundamental information to scholars from a variety of disciplines (from the physical to the social sciences) interested in the causes and consequences of changing features and landforms. Whether shoreline changes were the intentional product of active land filling or the natural by-product of coastal processes (the movement of landforms associated with the interaction of wind, waves, and water), reliable mapping provides valuable insight into the nature of these transformations and their reasons. Generally, cartographers define the shoreline as the intersection of a water surface (typically at high water) with the land. In the absence of human occupation or real-estate interests, precise knowledge concerning this line's location is rarely critical. In the presence of development (in particular, valuable coastal development), however, documentation of present and former shoreline locations is often fundamental to contemporary legal, regulatory, scientific, and engineering considerations. From a legal and regulatory perspective, comparisons of reliable historical shoreline positions frequently form the basis for establishing private title to, and the property lines of, valuable waterfront parcels and for the demarcation and delimitation of municipal, state, federal, and international boundaries.[5] These same lines define the jurisdiction of many coastal state regulatory agencies charged with protecting and promoting the public's rights in tidelands.[6]

Time-specific comparisons of shoreline positions are also important to engineers and scientists seeking to identify areas of filled tideland, to characterize coastal processes affecting past and present nearshore movement, and to predict future movement to ensure the safe design and construction of waterfront development, infrastructure, navigation, and shore protection projects. Since the early 1990s, partially in response to increased public demand for more and better information concerning coastal flooding and erosion, comparisons of accurate coastal maps over long periods of time have been used to develop a baseline for understanding and assessing the effects of shoreline change and sea-level rise.[7] In the area of environmental science, coastal ecologists are using accurate historical maps in their work to improve the water quality and habitat value of coastal resource areas through the identification and restoration of extensive areas of filled salt marsh. Similarly, hydrogeologists are employing historical maps to help with the detection and monitoring of underground hazardous waste plumes and the design of remediation solutions.

Perhaps one of the more interesting uses of geographically reliable historical coastal maps is the ongoing project of the city of Boston to address the subsidence of structures located on the filled tidelands portion of the historical Back Bay area. Filling of this former tidally influenced area of Boston began in earnest in the mid-nineteenth century and constitutes one of the largest areas of filled tidelands in the commonwealth. Today it remains an area known for its beautiful historical structures and residences. Over time, groundwater infiltration into the deteriorated pipes of the old stormwater and sewer systems serving this area contributed to a lowering of the water table that exposed many formerly submerged beams and piles to the air. More susceptible now to rot, these beams and piles no longer provide adequate support, threatening the structural integrity of increasing numbers of original buildings. Historical shoreline plans are being used in the development of the city's long-term strategy to deal with this problem.

Reliable historical mapping is also emerging as an important tool within the humanities and the social sciences. For example, many historians use accurately georeferenced maps to contribute an important visual component to studies recreating period-specific landscapes of battlefields, maritime ports, and coastal settlements.[8] Where scholars traditionally relied on narrative descriptions and sketches, these maps provide an added dimension to interpretations of significant events.[9] Historical maps offer an accurate spatial record of wharves and filling that scholars have used to explore the history of waterfront development and maritime commerce of many important ports, as a baseline for initiating underwater archeological investigations, and more generally to document and record the cultural, land use, and development history of coastal communities.[10] From a Massachusetts policy perspective, the reliable re-creation of the Shawmut Peninsula is important because the location of its former shoreline serves as the basis for determining the nature and extent of the public's rights to the waterfront pursuant to the commonwealth's Public Trust Doctrine as implemented through the Department of Environmental Protection's (DEP) Waterways regulations.[11] All jurisdictional inquiries pursuant to these regulations necessarily begin with the identification of the shoreline that existed before human alteration.[12]

Challenges to a Cartographic Approach

The shoreline frequently defines the geographic extent of state regulatory jurisdiction associated with the protection and promotion of the public's rights within, along, and to the coast of the commonwealth. Knowledge of the methods, equipment, and standards that mapmakers used can serve as the basis for qualitatively evaluating plans of the locations of such features as historical shorelines. When plans' qualities are suitable for the particular inquiry at hand, one can georeference them to contemporary coordinate systems or datums for the purpose of transferring historical spatial information. Although cartographers are able to "rubbersheet" any map, modern computer georeferencing or plan registration techniques can help to establish the best geometric fit with known geographic points and remove distortion related to the expansion and contraction effects of humidity on the original plan medium.[13] From the perspective of the individual plan, the georeferencing process also provides a measure of the geometric quality of the original survey in terms of a "registration solution" reported in the form of root mean square (RMS) or residual values.[14] One can further evaluate the positional reliability of the survey by comparing the computer-generated positions of well-defined, discrete points on the historical plan with higher-accuracy information for the same points on the contemporary datum.[15] Based on the limitations in equipment and technology, it is not surprising that seventeenth-, eighteenth-, and early-nineteenth-century maps georeferenced to a contemporary datum frequently result in registration solutions of questionable utility.[16]

There does not appear to be a seventeenth-century map based on an actual survey that, by itself, is of sufficient scale and accuracy to depict the Shawmut Peninsula shoreline before alteration. Cartographers must, therefore, reconstruct the location of this shoreline from multiple sources.[17] This chapter compares several such reconstructions prepared from various sources and with varying methods during the nineteenth and twentieth centuries. Although the accuracy of the base map for each reconstruction can be assessed by georeferencing, there is no practical means of verifying the actual shoreline reconstruction. These reconstructions are typically compiled from multiple sources and documents with varying levels of reliability. In addition to knowledge about the quality of the original contributing surveys, contemporary users who seek to form an opinion about the accuracy of the reconstructed shoreline must depend heavily on the professional reputation of, judgment of, and instructions to the surveyor. Simple statistical comparisons of the shoreline depicted on extant reconstructions of the Shawmut Peninsula can, however, provide interesting and useful information to support plan-based efforts to assess the reliability of reconstructed shorelines along waterfront areas that experienced early and intense alteration.

Comparison of early historical reconstructions, particularly where the feature of interest is ambulatory or constantly moving, is difficult because of the lack of formalized mapping standards before the 1840s. Today, cartographers generally define the shoreline as the local mean high water level (MHW), representing the average of the daily high waters over a nineteen-year period known as the Metonic cycle.[18] The use of this long-term average of high water heights compensates for short-term astronomical and meteorological conditions that may affect day-to-day tidal heights, and it was recognized legally as the accepted datum for shoreline mapping in 1935.[19]

The work of the U.S. Coast Survey, begun in earnest in the mid-1830s and 1840s, represents the first scientifically based effort to record daily tidal observations continuously at various locations along both the East and West coasts to determine an MHW shoreline as closely as possible.[20] With MHW as the standard, therefore, it is possible to identify and compare time-specific shoreline positions for the same areas depicted on various U.S. Coast Survey T-sheets and H-sheets.[21] By contrast, many maps and plans produced before this standardization, while depicting shorelines, provide little supporting information defining this term. For example, until the 1830s the standard of accuracy for measuring United States mapping efforts was the eighteenth-century work on the North American Atlantic coast by the renowned cartographer J.F.W. DesBarres.[22] DesBarres, a British Army engineer and surveyor, tells us that he "went along shore, and reexamined the accuracy of every intersected object, delineated the true shape of every head land, island, point, bay, rock above water, etc and every winding and irregularity of the coast."[23] He does not, unfortunately, provide us with any indication of exactly which shoreline he mapped.[24]

The lack of a common shoreline definition for early coastal mapping efforts contributes to the uncertainty associated with positional comparisons of early mapping efforts, particularly where the use of a different local tidal datum (whether knowingly or unknowingly on the part of the cartographer) resulted in the mapping of different shorelines, the location of which varied significantly depending on the relief of the area being mapped.[25] For example in Boston Harbor, where the average difference in height between the mean high water and the mean high waters of spring tides is approximately 0.7 feet, the mapped shoreline position varies significantly, particularly where nearshore relief is relatively flat.[26] In 1852, Boston city engineer E. S. Chesbrough, charged by the Boston Harbor Committee with reconstructing the shoreline of the Shawmut Peninsula before human alteration, highlighted this difficulty when he remarked in his final report: "The greatest cause of doubt [in reconstructing the shoreline of the original Shawmut Peninsula] is owing to the flatness of the original margin of a great part of the City, in consequence of which the differences between the line to which ordinary high water and that to which very high water flowed must have been considerable, in some places probably three to four hundred feet . . . particularly . . . at the foot of the Common."[27] Today the problem of identifying and mapping a standardized shoreline continues to be a challenge and is subject to significant debate in contemporary shoreline change analysis, particularly in areas of extensive tidal flats with moderate to low relief.[28]

The problem of mapping a poorly defined and ambulatory feature such as the shoreline becomes even more complicated when the high water line meanders through flat areas of dense vegetation, such as in the extensive salt marsh systems that typified seventeenth- and eighteenth-century Boston Harbor and fringed much of the Shawmut Peninsula. The identification of this line within the extremely flat New England salt marshes, where marsh grasses often obscure the actual intersection of sea surface with land, proved troublesome even for the standardized methods and procedures of the Coast

Survey, which typically depicted the seaward edge of vegetation as an "apparent" shoreline.[29] Indeed, the identification of the historical high water line in salt marshes continues to present a problem for contemporary surveys and shoreline change studies, frequently contributing to erroneous or erratic rate calculations.[30] The presence of salt marsh along the Shawmut Peninsula could, therefore, be expected to contribute similarly to some variation in the mapped shoreline position of most early efforts.

Reconstructions of the Shawmut Peninsula

This chapter considers seven reconstructions of the "original" Shawmut Peninsula shoreline. The maps depicting these shorelines were registered or georeferenced to the contemporary North American Datum of 1983 (NAD83). T-229, the 1846 U.S. Coast Survey T-sheet covering the area of the peninsula, was also registered to provide an accurate mid-nineteenth-century base map. This base map facilitated the registration of the reconstruction plans, since it depicted numerous geographic features that appeared on earlier plans for which there is little or no accurate extant positional data.

The seven plans registered for initial assessment and positional comparison consisted of (1) Samuel Clough's "Map of the Town of Boston 1676" (Clough); (2) Abel Bowen's "Map of Boston in 1824" (Bowen); (3) George Baldwin's 1846 "Plan of Part of the City and Harbor of Boston Showing Lines of High and Low Water, Determined from Recent Soundings, Abstracts of Deeds, Plans, and Other Sources" (Baldwin); (4) Boston City Engineer E. S. Chesbrough's 1852 "Map of Boston Harbor Showing Commissioners Lines, Wharves, etc." (Chesbrough); (5) George Walker's 1824 "Map of Boston Proper" (Walker); (6) Charles Perkins's 1895 "Plan of Boston Proper Showing Changes in Street & Wharf Lines 1795–1895" (Perkins); and (7) the Department of Environmental Quality Engineering's "Earliest Known Shoreline of Boston Harbor as Compiled from Plans of Henry Pelham 1775, A. S. Wadsworth 1817, John G. Hales 1813, and U.S. Coast Survey 1847" (DEQE). Justin Winsor's map entitled "Boston Old and New," included as the frontispiece in volume 1 of his *Memorial History of Boston,* published by James R. Osgood & Co. 1881–1883, was also registered. Although not included in the comparative analysis, the Winsor map was used to help estimate the location of early salt marsh along the peninsula. A detailed discussion of the methodology used to assess and compare these seven reconstructions of the Shawmut Peninsula shoreline, along with additional information about the plans, appears in the appendix at the end of this chapter.

Identifying the Most Appropriate Cartographic Source

Figure 5.2 depicts the approximate areas and extent of tidal flats (the intertidal zone or the area of low relief situated between the high and low water lines) and salt marsh along the Shawmut Peninsula shoreline in the seventeenth century. The bases of this depiction are the eighteenth-century maps and plans of J.F.W. Desbarres, John Hills, and Samuel Holland; the plans of Clough, Chesbrough, Baldwin, and Winsor; and the historical narratives of Nancy S. Seasholes, Walter Muir Whitehill, and Esther Forbes.[31]

The dominant landforms of the peninsula that its first English settler, the Reverend William Braxtton, viewed in 1625 differ considerably from those of the densely developed city so familiar to contemporary visitors. In the early seventeenth century Boston was a peninsula connected with the mainland to the south by a narrow, mile-long isthmus or neck (Boston Neck) that storm or spring tides often covered at high water.[32] The area to the west of the Boston Neck (including present-day Back Bay) was a vast expanse of tidal flats that extended upward of three thousand feet seaward of the shoreline and salt marsh that the tide periodically covered at many stages.[33] Moving north, along the westerly shoreline of the Neck, the area of salt marsh narrowed, closing with the shoreline approximately 350 feet to the south of today's Commonwealth Avenue and Common. Continuing northerly, the shoreline meandered along and through an expansive salt marsh (formerly known as Round Marsh) into which two streams flowed from the Common and the area of the Frog Pond. Tidal Flats, while less expansive, extended two hundred to three hundred feet from the shoreline. North of the Common, areas of salt marsh occupied shallow coves situated between stretches of relatively marshless shoreline. Expansive areas of tidal flats continued to parallel the northwesterly facing shores, extending approximately three hundred to five hundred feet seaward.

Running somewhat northeasterly, the peninsula shoreline then meandered through a marshy cove in the vicinity of today's Cambridge Street and Massachusetts General Hospital, before turning northwesterly around the former Barton's Point, an area characterized by fringing marsh and an

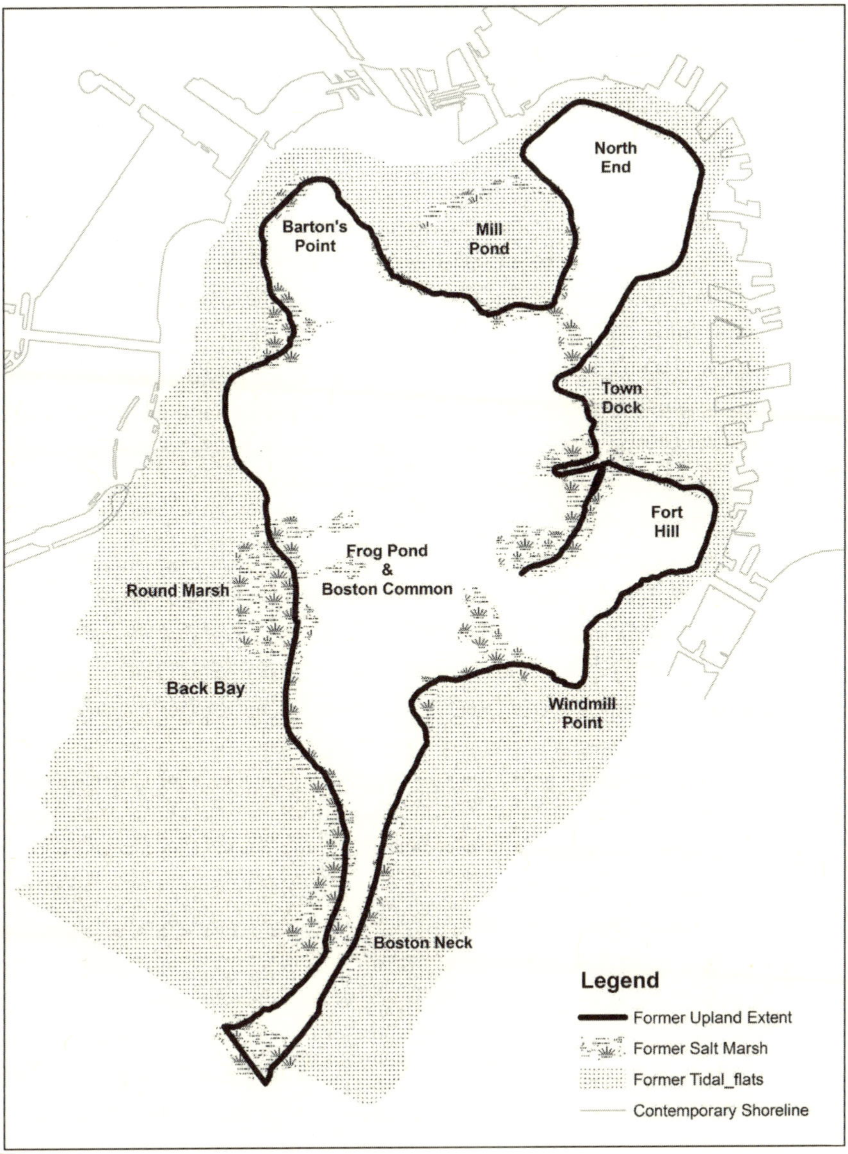

Figure 5.2. Dominant landforms and geographic names of the Shawmut Peninsula, ca. 1630. *Source:* Author's map.

intertidal zone on the order of two hundred to three hundred feet wide. From Barton's Point the shoreline turned sharply to the southeast, continuing almost to Hanover Street before swinging northwest to the North End to form a shallow and marshy cove, later enclosed by a causeway and referred to as the Mill Pond. Leaving the cove and continuing easterly along the peninsula, low water closed to within approximately a hundred to two hundred feet of the shore. Fringing salt marsh in this narrow intertidal zone was less prominent. Proceeding southerly, the low water line once again moved seaward to create an expansive intertidal zone that Baldwin's 1846 work illustrated.

A relatively extensive area of salt marsh that was tributary to at least two streams draining to the east complemented the flats, which extended approximately a thousand to eighteen hundred feet from the original shoreline. Turning easterly, away from the area once known as the Town Dock, the shoreline left behind fringing salt marsh, circumscribing Fort Hill, an area to the west of the present-day Old Northern Avenue Bridge, characterized by a relatively narrow intertidal zone and little or no salt marsh. Finally, the shoreline turned southwesterly and southerly around Windmill Point (to the west of South Station and well landward of today's Fort Point Channel) to form the easterly shore of the Boston Neck, paralleled by an extensive intertidal zone (in excess of fifteen hundred feet wide) and salt marsh.

Figure 5.3 combines the information from figure 5.2 and the appendix (see figure 5.7, below) to depict a line representing an average position of the Clough, Baldwin, Chesbrough, and DEQE reconstructed shorelines. The line segments illustrate the degree of agreement or disagreement (represented by the standard deviations) between the four reconstructed shoreline positions superimposed on the approximate areas of seventeenth- and eighteenth-century tidal flats and salt marsh. Areas of greatest agreement in reconstructed shoreline position (that is, those areas represented by smaller values of standard deviation) are along stretches of shoreline with little or no salt marsh and relatively narrow intertidal zones. The line segments along the shorelines of the northerly and easterly projections of the peninsula identify these areas. Conversely, the areas of largest disagreement (that is, relatively higher values of standard deviation) between the reconstructed shorelines appear to correlate with extensive areas of salt marsh, such as in marshy coves and other protected environments, or tidal flats. This is particularly evident along the westerly shores of the Boston Neck and in the area to the west of the former Town Cove/Dock.

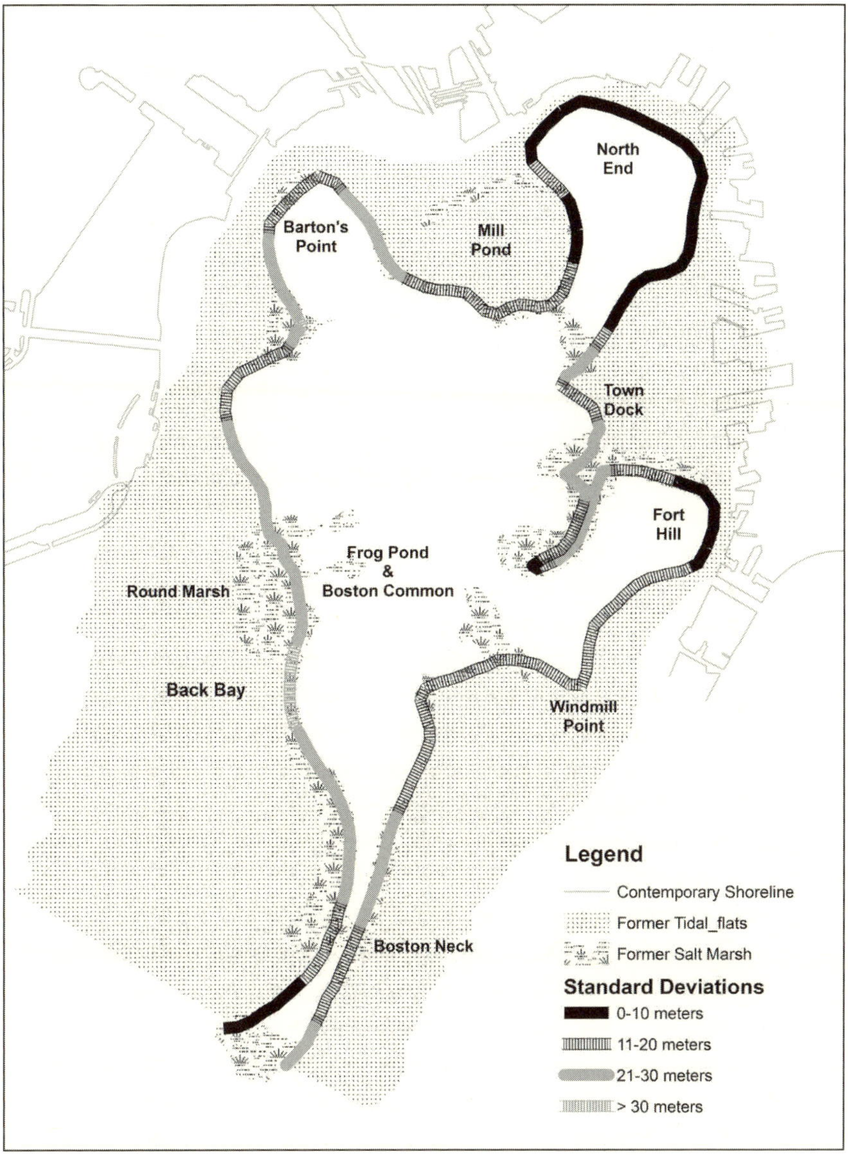

Figure 5.3. Averaged Shawmut Peninsula shoreline compiled from four independent cartographic sources. Line segments represent standard deviations illustrating the degree of agreement or disagreement between the cartographic sources. *Source:* Author's map.

The emergence of well-documented procedures and standards in the mid-nineteenth century supports contemporary research efforts to develop period-specific estimates of mapping accuracy or reliability. The lack of such universal standards for seventeenth- and eighteenth-century shoreline mapping efforts frequently makes the development of even coarse estimates of positional accuracy for the shorelines these maps depict extremely difficult. In the absence of quantitative evidence, the contemporary researcher must rely therefore on a more qualitative assessment strategy and professional judgment to identify reliable sources of cartographic information for the Shawmut Peninsula shoreline.

One can quantify the reliability or accuracy of individual shoreline positions from estimates of the georeferencing error, the survey error, the digitizing error, and the shoreline variability or interpretation error.[34] With contemporary, computer-assisted georeferencing techniques, one can minimize the often significant effects of plan distortion and the error associated with survey and registration control points. The survey error reflects the ability of individual surveyors and mappers and is typically compiled from estimates of equipment quality and of the accuracy achievable for survey work performed in accordance with documented procedures, standards, and methodologies. Digitizing error is the quantifiable error associated with digitally tracing the shoreline of interest. Finally, the shoreline variability error is an estimate of the potential positional variation encountered when different shoreline definitions guided the mapping process (for example, the location of the ordinary high water line versus the location of the spring high water line). The total uncertainty of a shoreline position can be quantified as the square root of the sum of the squares of the error for each individual error.[35] In this way error assessments (often referred to as uncertainty budgets) can help to identify the most reliable plans or the best available cartographic evidence of former shoreline position.

Typically, the assortment of methodologies and equipment used and a lack of uniform standards preclude the determination of dependable estimates of the survey and shoreline variability components of shoreline mapping before the nineteenth century. This is particularly true in the case of maps and plans depicting reconstructed shoreline positions that rely on multiple cartographic sources, anecdotal evidence, and local knowledge of natural topographic conditions. Although it is not possible to quantify the spatial veracity of much of the coastal information depicted on these maps, it is possible to compare various waterfront reconstructions and analyze the degree to which they agree and/or disagree using an average shoreline and related standard deviations. When combined with a qualitative assessment of the possible reasons for agreement or disagreement—including the purpose of the plan, extant instructions to the preparer, the sources and methods relied upon to produce the plan, the reputation of surveyor, and, as in the case of the Shawmut Peninsula, the environmental setting in which the work was performed—a qualitative approach can be used to develop subjectively the level of reliability or degree of confidence associated with each plan.

Applying this qualitative analysis to the reconstructed shorelines assessed in the appendix suggests that Chesbrough's plan is one of the more reliable reconstructed shorelines for the Shawmut Peninsula. Specifically, use of this plan is endorsed by Chesbrough's intimate knowledge of Boston Harbor as city engineer, his recognized abilities and prominence as a civil engineer in the mid-nineteenth century, his stated purposes for preparing his plan, the degree of documentation provided in his accompanying report, the quality of the registration solution afforded by his choice of base map, *and* the reasonable level of agreement with similar shoreline reconstructions using different sources and methods. For similar reasons the use of the Baldwin survey, for localized inquiries in the area of Long Wharf and the former Town Dock, deserves equal consideration, particularly given its relatively larger compilation scale.

One can further refine the location of the shoreline of the former Shawmut Peninsula as these plans depict and supplement them from noncartographic sources depending on the researcher's interests and objectives. For example, real-estate attorneys routinely supplement extant maps and plans with careful research of deeds and legislation to establish and verify the ownership extents of valuable waterfront property.[36] Coastal geologists interested in studying land formation frequently supplement their inquiries with smaller-scale, geologic maps to identify ancient depositional patterns and the emergence and disappearance of such landforms as rivers, hills, and tidal flats to predict future shoreline movement. Coastal scientists and engineers typically supplement cartographic data with information from soil borings to document the presence of fill on former tidelands or the location of former salt marshes in order to determine rates of relative sea-level rise, ensure that structural components of future waterfront development are designed safely, and refine strategies that address the sinking of existing structures such as those that are being encountered in Boston's Back Bay. Although these additional information sources help to refine many shoreline representations, they cannot substitute for reliable cartographic sources that place this line in its real-world context. Indeed, identifying the best available cartographic evidence in the form of reliable historical maps and plans is a critical first step, where the location of former shorelines are the determining factors in furthering public policies designed to ensure the public's rights to access and enjoy the waterfront.

Public Rights, Private Ownership, and the Shawmut Peninsula

As in many coastal states, in Massachusetts the clash between private ownership and public rights emerged as a major policy issue in the latter part of the twentieth century. With escalating waterfront property values and a corresponding increase in private ownership along the four-thousand-mile shoreline, the public's ability to enjoy the coast has become increasingly limited. Indeed, the desire of waterfront property owners to control and limit access to the seashore, until recently an area most assumed to be a common natural resource, has since the early 1990s created a pressing need for accurate historical shoreline mapping to substantiate the extent to which public rights exist in areas of filled tidelands.

Unlike inland real estate, where property ownership is typically static and determined by precise mathematical lines and points represented by such fixed objects as concrete bounds, stone walls, and even trees, the seaward lines of ownership along waterfront parcels (the lines that separate public and private interests) are often dynamic and represented by ambulatory and meandering shorelines. To the infrequent visitor, the coastline appears relatively static year to year. In reality, however, it is constantly moving in response to wind and waves, carrying suspended sand eroded from coastal banks and dunes for deposition elsewhere along the shore. These natural processes of erosion and accretion are ongoing, in some areas resulting in the loss of upland and in others the creation of new upland. Absent development (and ownership interests), the natural movement of shorelines in response to storms and relative sea-level rise is of little import and often goes unnoticed despite its significance to the long-term sustainability of such coastal landforms as beaches, dunes, and habitat for nesting shorebirds and shellfish beds. In the presence of development, however, the shoreline and its position often serve as the battle line for the clash between public and private rights in the coastal zone.

The seaward limits of shorefront properties are generally recognized to be ambulatory boundaries and subject to significant long- and short-term spatial variability. While not universal, private ownership (or title) typically extends seaward to the high or low water shoreline. Upland owners stand to gain when the natural and gradual process of accretion creates new land. Similarly, they lose land when it is submerged or eroded.[37] Shoreline changes, however, that are not gradual, that occur suddenly, or that are the result of human-made alterations associated with filling, dredging, or the placement of coastal protection structures do not affect the title if the owner or a predecessor caused the change.[38] Although the property laws of most states recognize the ambulatory nature of shorelines, the use of dynamic and naturally shifting topographic or hydrographic features to define the extent of private ownership or the location of regulatory boundaries that affect the profitability, and in some cases the viability, of waterfront development often results in contentious debate and confusion, particularly in contrast to the long-term stability of property and regulatory lines encountered in the relatively static framework of traditional upland development.[39]

In Massachusetts it is a well-settled legal precept that property lines along the shore migrate in response to the natural processes of accretion and erosion.[40] Unlike most states, however, according to the colonial ordinances of 1641 through 1647, the seaward property line of land bounded by tidal waters in Massachusetts extends to the low water mark or 100 rods (1,650 feet) from the high water mark, whichever is farther landward.[41] Under the colonial ordinances the area of land in private ownership remained subject to the public rights of fishing, fowling, and navigation. The waters and the land under them, beyond the line of private ownership, continue to be held by the state in trust for the public. As a property- or ownership-based doctrine, the geographic scope of the public trust doctrine migrates in the same manner with the high water or low water lines that define the limits of private littoral ownership.[42]

Although the enactment of the colonial ordinances marked a departure from the English common law public trust doctrine, it still serves as the point of reference for all Massachusetts intertidal law as codified in Chapter 91 (c.91) of the Massachusetts General Laws and implemented through the Department of Environmental Protection's c.91 Licensing Program and the Waterways Regulations. Significantly, the Massachusetts Supreme Judicial Court's 1979 Boston Waterfront Development Corporation (BWDC) decision concluded that the geographic area of both *flowed tidelands* and *filled tidelands,* defined as former submerged lands and tidal flats that are no longer subject to tidal action because they were filled, are still impressed with public rights.[43] Today, in response to this decision, the Waterways Regulations define tidelands as "present and former submerged lands and tidal flats lying between the present or *historic high water mark,* whichever is farther landward, and the seaward limit of state jurisdiction."[44] Therefore, the ability to determine a reliable position of the historic high water mark as the landward boundary of regulated tideland is fundamental to the c.91 licensing process and its goal of protecting and promoting the public's rights to the waterfront.

Early and extensive filling of tidal areas accompanied the development of many of the commonwealth's urban waterfronts and ports. Of approximately 13,600 acres of formerly flowed tidelands along the coast, almost 60 percent (8,000 acres) of this total is within present or former commercial port areas developed early in the commonwealth's settlement history, with more than 5,500 acres (40 percent) alone found in the city of Boston.[45] Thus, in the context of the BWDC decision, the identification of reliable and reproducible historical high water lines in filled tideland areas, in addition to contemporary high water lines in flowed tideland areas, is critical for determining the geographic scope of the c.91 licensing process. Defined by regulation as the mark that "existed prior to human alteration of the shoreline by filling, dredging, excavation, impounding, or other means," the historical high water line is presumed to be "the farthest landward former shoreline, which can be ascertained with reference to topographic or hydrographic surveys, previous license plans, and other historical maps or charts."[46]

For those waterfronts, such as that surrounding the former Shawmut Peninsula, that have an early history of alteration through the construction of wharves, the filling of adjacent tidelands, or the dredging of dock space and channels, plans depicting reconstructed historical shorelines may be the best available, and frequently the only, source of the "original" or prealtered shoreline position. Clearly, the ability to assess the reliability of such plans is particularly important for contentious deliberations concerning the stewardship of public rights.

Figure 5.4 depicts the approximate extent of state Waterways jurisdiction based on Chesbrough's reconstructed shoreline of the Shawmut Peninsula. The area of filled tidelands in this historical waterfront area, consisting of approximately eight hundred acres of some of the most valuable real estate in the commonwealth, was partially responsible for the creation of regu-

Figure 5.4. Chesbrough's Shawmut Peninsula, ca. 1630, the present high-water line, the landward jurisdiction of Chapter 91, and contemporary Boston.

latory rules that generally limit the landward extent of state Waterways jurisdiction to the public right-of-way that is closest to the contemporary shoreline. Consequently, the "original" Shawmut Peninsula shoreline appears to be the defining line for jurisdiction only for a limited area to the north and for a small area just to the south. It is interesting to note that comparisons of the Chesbrough, Clough, and DEQE reconstructions for these areas exhibit a relatively high degree of agreement in the shoreline position of the former peninsula.

HISTORICAL maps and plans are primary sources of data for locating former shorelines. The reliability of this information, however, depends both on an ability to reference the maps to contemporary, real-world positions and on an understanding of how the maps were produced. From a legal and regulatory perspective, time-specific comparisons of reliable shoreline positions frequently form the basis for establishing title to valuable waterfront property and defining the geographic scope of jurisdiction of coastal regulations. As important planning and design considerations for waterfront projects, such comparisons are equally significant to coastal engineers and scientists seeking to identify and restore areas of filled salt marsh; to develop mitigation strategies for densely urbanized areas of filled tideland that are now experiencing subsidence; to characterize coastal processes affecting past, present, and future shoreline movement; and to determine rates of shoreline change and potential impacts of relative sea-level rise. Coastal managers also employ accurate historical maps to document cultural and land-use patterns along the shoreline. Historians and archeologists rely on them for the identification and preservation of maritime-related cultural resources. For the Commonwealth of Massachusetts historical maps depicting the location of former shorelines frequently form the basis for inquiries seeking to strike a balance between competing public and private rights in the coastal zone.

When plans were prepared in accordance with post-1840s standards and methods, estimates of the accuracy of the location of former shorelines could be developed to support selection of the best available cartographic evidence. For many earlier mapping efforts, however, where information on methodologies and standards is rare, spatial accuracy is difficult to verify. For instances such as these, and in the case of maps depicting historical shorelines constructed from a variety of sources (that is, reconstructed shorelines), it is necessary to assess reliability based on more qualitative methods. Specifically, one can compare various shoreline depictions statistically using a computed average and the related standard deviations to depict areas of agreement or disagreement. When combined with a professional assessment of available information about plan preparation and the environmental setting in which the work was performed, one can evaluate the reliability of individual maps qualitatively to help identify the most suitable for the inquiry at hand. Historical shoreline maps and plans continue to play an important role in many contemporary issues and across many disciplines. Indeed, as the example of the Shawmut Peninsula shows, the best available historical maps and plans of the former shoreline are critical to determinations of the extent of jurisdiction for state public trust authority responsible for protecting and promoting the public's rights in, to, and along the waterfront.

APPENDIX
More on the Science

This chapter considers seven reconstructions of the "original" Shawmut Peninsula shoreline. The maps on which these shorelines were constructed were georeferenced, or registered, to the North American Datum of 1983 (NAD83) using the ESRI, ArcGIS 9.0 georeferencing extension, set for a First Order Polynomial (Affine) transformation. A minimum of twenty-four registration points initiated each registration exercise. Where appropriate, those points with excessive residuals were removed to improve the mathematical veracity of the registration solution. A minimum of six points, however, was retained for each registration solution with the goal of minimizing the root mean square (RMS) associated with each registration solution while, to the extent possible, ensuring equal point distribution across the plan.

To assist with the registration process, T-229, the 1846 U.S. Coast Survey T-sheet for the Shawmut Peninsula, was registered using a mathematical transformation of geodetic coordinate values derived from the Coast Survey's Bessel's Spheroid datum to the contemporary NAD83 datum.[47] In this way T-229 served as a reliable historical base map to facilitate the registration of those reconstructions whose base map features, while still evident in 1846, are no longer visible on the contemporary Boston landscape. In addition to the U.S. Coast Survey's T-229, the following seven plans were registered for positional comparison of the depicted reconstructed shorelines:[48]

1. *"Map of the Town of Boston 1676, Scale 1" = 100', Drawn By Samuel C. Clough" (Clough).* Samuel C. Clough, a retired civil engineer, surveyor, and cartographer, who worked as a draftsman for the Boston Edison Company, pre-

pared several period-specific plans of the seventeenth-century Boston peninsula with reconstructions he compiled from "the records of the Colony, Town, Registry of Deeds, Suffolk Probate, and Supreme Court; Book of Possessions, Winthrop Journal, Lechford Note Book, Aspinwall's Notes and City Surveys." This chapter uses Clough's 1676 depiction, as opposed to his 1640s reconstructions, because this later map provided the best registration solution. It is interesting to note that despite the early date, a significant amount of alteration had already occurred along the easterly shoreline of Boston, especially in the area of the former Town Cove/Dock. Areas of significant alterations were removed from this analysis since they did not reflect presettlement shoreline conditions.

2. *"Map of Boston in 1824," prepared by Abel Bowen (Bowen).* This map, published in Caleb Snow's *History of Boston* (1825), was one of those Chesbrough used in his reconstruction of the Shawmut Peninsula.[49]

3. *"Plan of Part of the City and Harbor of Boston Showing Lines of High and Low Water, Determined from Recent Soundings, Abstracts of Deeds, Plans, and Other Sources," Prepared by: George R. Baldwin, 1846 (Baldwin).* In 1846, pursuant to an order of the Massachusetts Supreme Judicial Court (SJC) for the resolution of a property dispute, surveyor George R. Baldwin prepared his plan showing the lines of original high and low water in the area of the former Town Cove (Long Wharf).[50] This plan formed the jurisdictional basis 130 years later for the landmark 1979 BWDC SJC decision. Significantly, Baldwin's base map depicted 1846 waterfront conditions closely resembling those on T-229, greatly facilitating the registration process.

4. *"Map of Boston Harbor Showing Commissioners Lines, Wharves, etc. Prepared by order of the Harbor Committee of the City Council of 1852. Scale, Under the Direction of Ellis S. Chesbrough, Civil Engineer (Chesbrough)."* Chesbrough conducted his efforts in response to the instructions of E. H. Eldredge of the Harbor Committee to reconstruct the original high water line on the map "as accurately as the sources of information obtainable at this office would allow."[51] Chesbrough's accompanying report to the committee provides valuable insight into the development of his reconstructed shoreline, stating that he based his work on "the plans found in Snow's History [Abel Bowen's *Map of Boston in 1824*]; the maps in Frothingham's Siege of Boston [*A Plan of the Town of Boston with Intrenchments etc. of His Majesty's Forces in 1775* by Lieutenant Page and *A Plan of the Action at Bunkers Hill* by Page and Captain Montresor], besides those of Osgood Carleton and S. P. Fuller."[52] Chesbrough also recognized the title work of E. S. Rand, Esq., for information relative to the location of the original high water line in the western part of the city.[53] Chesbrough notes that he based his maps largely on the 1846 cartographic efforts of the

U.S. Coast Survey (represented by T-229), updated for more recent improvements and changes.[54]

5. *"Map of Boston Proper," prepared by George H. Walker, Geo. H. Walker & Co., 1894. (Walker).* One of two plans produced in the latter part of the nineteenth century known to be copies of Chesbrough's reconstructed shoreline, this plan (and that by Charles Perkins, listed below) were included in the assessment for initial comparative purposes and to assess potential differences in the context of a contemporary base map.[55]

6. *"Plan of Boston Proper Showing Changes in Street & Wharf Lines 1795–1895," by Charles C. Perkins, City Surveyor. 1895. (Perkins).* This plan also relied on Chesbrough's reconstructed shoreline. In addition to an "original" shoreline, Perkins also depicts 1795 and 1850 waterfront conditions on his 1895 base map.

7. *"Earliest Known Shoreline of Boston Harbor as Compiled from Plans of Henry Pelham 1775, A. S. Wadsworth 1817, John G. Hales 1813, and U.S. Coast Survey 1847." Geodetic Survey Division of the Massachusetts Department of Public Works (DEQE).* The final plan registered for this study is on file with the Massachusetts Department of Environmental Protection, Waterways Program. It appears that the former Geodetic Survey Division of the Massachusetts Department of Public Works (now the Massachusetts Highway Department) developed the plan. Although a preparation date is not provided, it uses the 1935, 1:10,000 scale, U.S. Coast & Geodetic Survey (successor agency to the U.S. Coast Survey), Chart No. 248 for Boston Inner Harbor as its base map.

TABLE 5.1

Plans depicting reconstructed shorelines and registration solutions.

Map	*Registration points*	*Residual error (meters)*
Clough	48	4.2
Bowen	6	6.2
Baldwin	43	3.5
Chesbrough	17	3.2
Walker	8	2.4
Perkins	42	3.8
DEQE	11	2.2
T-229	7	1.1

Note: T-229 compiled by the U.S. Coast Survey at a scale of 1:5,000 in 1846 and registered using contemporary coordinate values for U.S. Coast Survey triangulation stations.

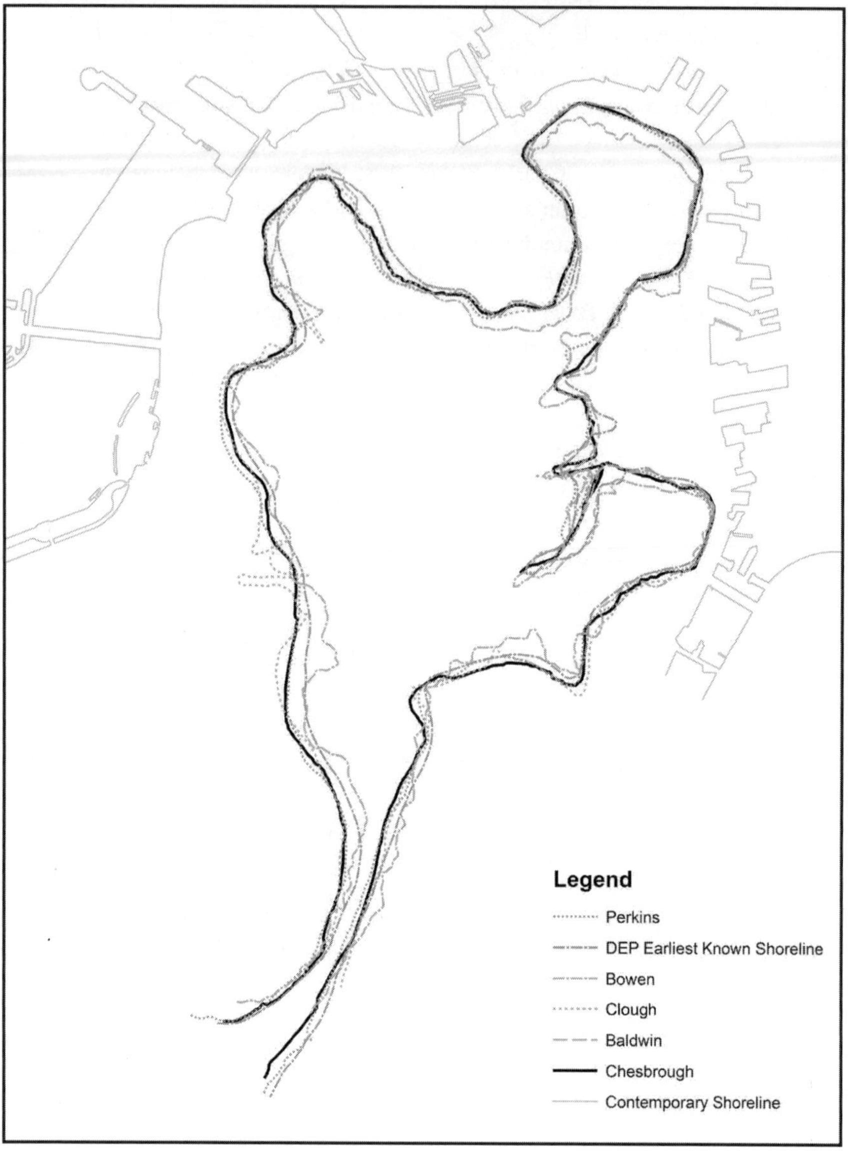

Figure 5.5. Six cartographic representations of the Shawmut Peninsula superimposed on the contemporary shoreline. *Source:* Author's rendering.

Table 5.1 lists the plans considered in this chapter, the number of registration points used to georeference the plan, and the resulting RMS value or residual error (registration solution.) Reconstructed shorelines on the registered plans were subsequently digitized and stored as individual shapefiles. Figure 5.5 depicts the positions of the seven digitized shorelines and that of 2001 for the Shawmut Peninsula area. The Digital Shoreline Analysis System (DSAS), v. 3.0 ARCGIS extension, was used to construct a baseline inland of all digitized shorelines, from which were cast seaward at fifty-meter intervals to intersect each reconstructed shoreline using the built-in 2-Leg Casting Algorithm to maximize transect orthogonality.[56] The method generated a total of 229 transects with associated coordinate values.[57]

These coordinate values, representing the intersections of transect casts with individual reconstructed shorelines, were exported to a Microsoft Excel spreadsheet, where they were sorted and average reconstructed shoreline coordinate values and standard deviations calculated for each transect.[58] Based on these coordinate values, an average reconstructed shoreline position and standard deviation for each transect was imported back into the ArcGIS project as a point file. Subsequently, these points were connected sequentially to create a line representing the average reconstructed shoreline position. Values for standard deviations were smoothed using a 5-transect running average grouped into ranges of 0–10, 11–20, 21–30, and >30 meters. The standard deviations were then merged graphically with the average reconstructed shoreline position line to illustrate the variation in reconstructed shoreline position at each transect. Figure 5.6, depicting the average shoreline position line for the seven shorelines, illustrates areas of relative spatial agreement and disagreement between the plans.

The shoreline reconstructions along several areas of the original peninsula disagree significantly. To assess whether or not closer agreement between any of the shorelines could be established, the process was repeated after removing the Bowen, Perkins, and Walker reconstructions. Visual inspection of figure 5.5 indicates that the Bowen reconstruction (see table 5.1) exhibited the poorest registration solution and appears to be an outlier when viewed in the context of the other shorelines. The Perkins and Walker shorelines were removed to eliminate potential bias in the data since both are known to be based on Chesbrough's reconstruction. With these shorelines removed, a new average shoreline reconstruction and associated standard deviations for the remaining four lines were computed using the same methodology.

Figure 5.7 depicts the new average reconstructed shoreline position line based on the registrations of the four remaining source documents with the computed 5-transect standard deviation running averages. The removal of the Bowen, Perkins, and Walker shorelines results in a significant increase in the level of agreement between reconstructed shoreline positions, particularly along the northerly shoreline of the former peninsula and supports the use of the Chesbrough and Baldwin reconstructions of the Shawmut Peninsula shoreline.

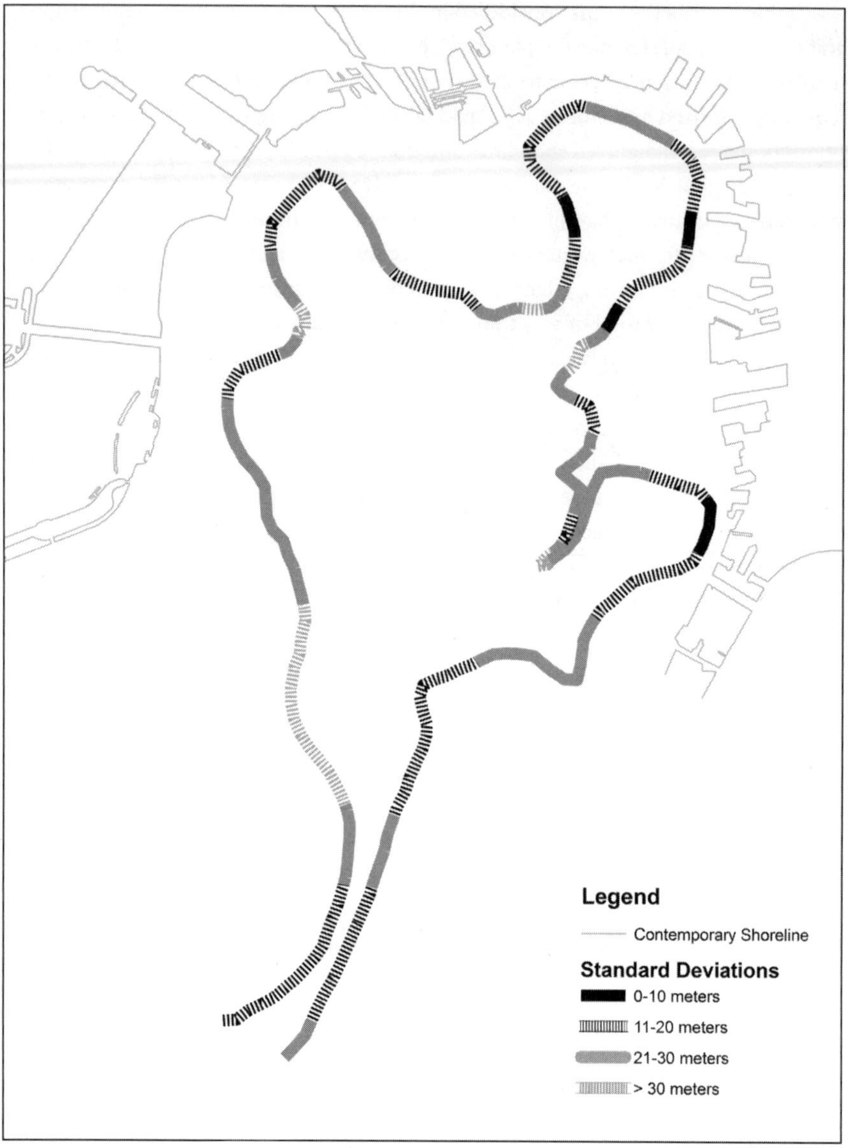

Figure 5.6. The Shawmut Peninsula as an average of shoreline positions in figure 5.5. Standard deviations represent the degree of agreement or disagreement between the cartographic sources. *Source:* Author's rendering.

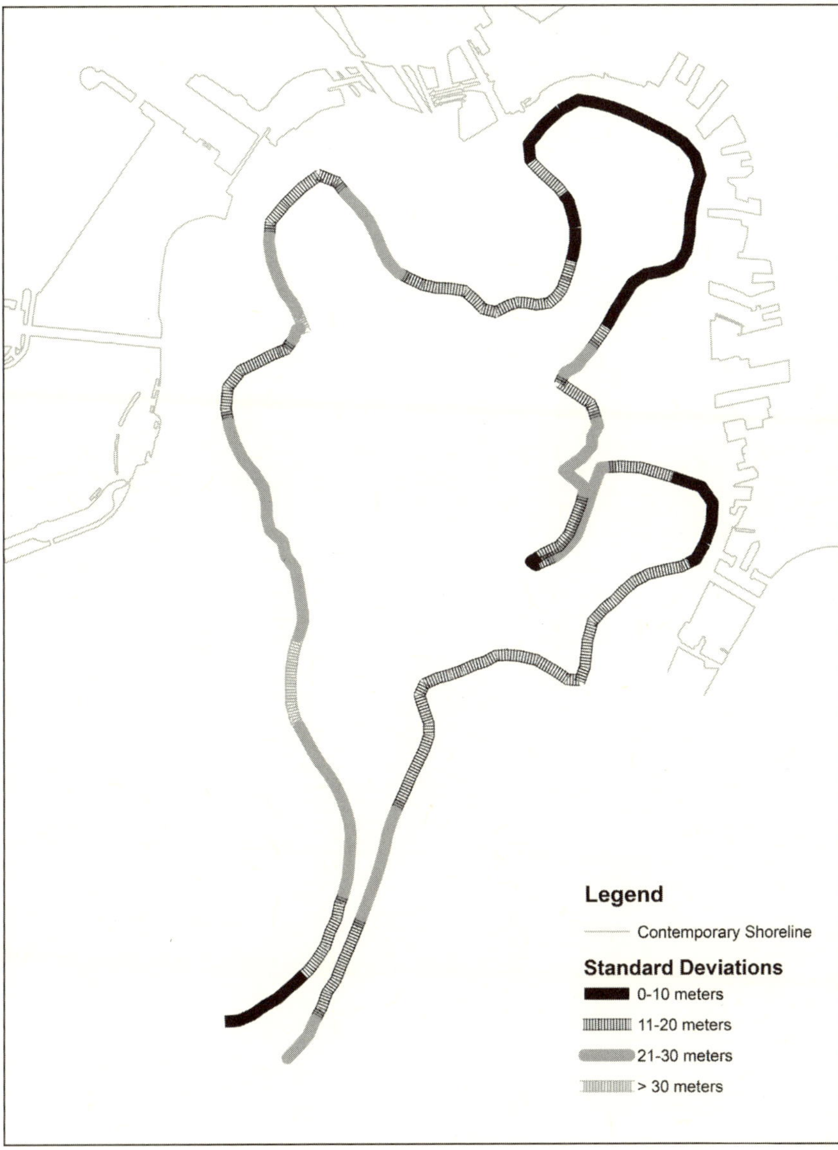

Figure 5.7. The Shawmut Peninsula as an average of the shoreline positions in four cartographic sources. *Source:* Author's rendering.

THE TOWN AND

THE COUNTRYSIDE

6

Remaking Boston, Remaking Massachusetts

Brian Donahue

LOOKING out from the State House dome a century ago, Charles Eliot could have just seen the wooded hills of Weston, perched on the western rim of the rapidly expanding metropolis. Looking back from the top of Doublet Hill today, a resident of Weston can still make out the dome, nestled among the office towers of the modern city. Eliot wanted to protect Doublet Hill as part of a forest reserve for his metropolitan park system, but he did not succeed. Weston acquired the rocky outcrop some eighty years later, and today it maintains a public lookout, but one needs to be either a resident of that affluent suburb or an unusually persistent visitor to find the trail. Blasted into the shoulder of Doublet Hill is a cavernous underground storage tank that distributes water to downtown Boston. Directly below the outlook lies the Lake District of the Charles River, where a century ago one would have seen thousands of canoes on a fine spring evening. One might see a handful today, wending their way beneath the soaring concrete bridge piers of Route 128 and the Massachusetts Turnpike. The view has changed, but so has the vantage. Every remaking of Boston has seen a parallel remaking of suburban and rural Massachusetts.

"Remaking Boston" seems an apt title for an environmental history of that old American city. It is not what one would choose for an environmental history of Chicago, by contrast—a new city rising from the mud on the edge of the prairie, "Nature's Metropolis." True, part of the Chicago story is a famous rebuilding, in steel. But that seems like part of a single making—the city destroyed by the Great Fire had barely been standing long enough to dry out sufficiently to burn. Or Pittsburgh: "Devastation and Renewal." *That* does not sound like Boston, although the city has certainly had its ups and downs, and share of industrial blight. Or Los Angeles: "Land of Sunshine." That story has been one big illusion from the start.[1]

Boston had something to be remade—a distinctive colonial city that was transformed, but whose remains can still be located and whose legacy is still cherished. The city on a hill is old enough to have been remade several times, in fact. In the first remaking, the hill beneath the Puritan city was pulled down and smoothed out to fill the nearest coves of the harbor. In the next remaking, the fill came in from Needham by rail. In the most recent remaking, the buildings rose higher than the hill had ever been. Meanwhile, the Massachusetts countryside was being dramatically remade in much the same way: the forest was pulled down to make farmland, the forest grew back, and now the automotive suburbs advance to take the forest down again. These tales of city and countryside must be connected.

But how? The growth of Boston certainly made and remade Massachusetts, but this city was never really shaped in turn by its hinterland—at least, not in any obvious way. Not like Chicago, whose growth was clearly tied to processing and shipping the bonanza of natural resources being extracted from the middle of the continent in the nineteenth century: wheat and corn, pork and beef, pine lumber. That was a symmetrical, symbiotic relationship between city and countryside. The growth of New York City was also driven by its connection to that same vast American interior. The relationship between Boston and its hinterland is more complicated and obscure. Certainly, Boston is profoundly of New England, and until the twentieth century the New England countryside did provision the city as it grew. But there was never that much *stuff* coming out of the New England soil to drive Boston's growth—something else was going on. One could make a better case for cod as the vital commodity, and the Grand Banks as Boston's first hinterland, than for anything found on dry land. In the nineteenth century the city's most important hinterland was probably Alabama, a supplier of cotton for New England's mills. The Massachusetts countryside never provided anything crucial to make and remake Boston. This relationship between city and country seems oddly asymmetrical and convoluted, like the twisting roads that led into Boston itself. Let's face it: before the city was remade, it was nearly impossible to even *get* to Boston from the country.

The multiple remakings of Boston and Massachusetts were marked by the changing energy base of the economy: biological, coal, oil; or the eotechnic, paleotechnic, neotechnic eras, to employ the great Lewis Mumford's classic terms. These eras have also been called the walking city, the city of streetcar suburbs, and the automobile metropolis. As Sam Bass Warner and other historians have reminded us, running through these revolutions is another, tidal rhythm: the always unequal contest between the push of private economic growth, and the corresponding pull to restrain or guide that growth to serve larger interests, such as public health or the beauty of civic space. The political struggle to harness government to serve these ends has pitted the claim that private economic growth is the primary source of public benefits against the counterclaim that uncontrolled growth can cause public harm, and even undercut the natural and cultural foundations of the economy so severely as to be unsustainable. Who has the power to determine what is in the public interest? These are recurring questions in this story.[2]

The Making of Boston and Massachusetts

The first Boston was the city on the hill—or rather, the small, vigorous seaport on the harbor side of the hill, facing London, connected to the rest of Massachusetts Bay Colony only by a narrow neck and a small fleet of ferries, scows, and sloops. Long before that, of course, the geological maker of Boston was the glacier whose advance had depressed the earth's surface and smeared into place the drumlins that are the hills and the harbor islands, and whose melting away deposited beds of gravel, sand, and clay around those hills—the shifting stuff of the city to come. A few thousand years after the glacier departed, the rising sea level caught up with the rebounding earth and returned to the place where it met three smallish rivers: the Neponset, the Mystic, and the Charles. These rivers fed the great tidal estuary that surrounds the peninsulas that would define the city. For millennia, people thrived on the riches of that estuary, until the advent of Europeans with their relentless economic and religious imperatives. The Native people were able to hold their own against explorers, fishermen, traders, and raiders for a century or so, until disease all but swept them away about a decade before the Puritans arrived in 1630. The resources of the salt marshes remained important to many of the English farmers and fishermen who settled the Massachusetts coast in the following decades, but they had little to do with the making of Boston.

The harbor did. Boston became the leading port on the east coast of America, a position it held through most of the colonial period. This was not by virtue of having a staple crop to export, which New England lacked. It was by virtue of being the first American port to be established, and of being the closest to London. It was also because Boston merchants were part of a keenly driven culture that was not averse to trade, but that remained wary enough of the riches commerce might bring never to rest easy. Boston was close to the banks, and it made the most of salt cod—and later of whale oil—in establishing its key position in the burgeoning Atlantic economy. But as far as any landward backcountry was concerned, there was not much that Massachusetts could offer—nothing like rice, or tobacco, let alone sugar, the central driver of Atlantic trade. Rum, the leading commodity that Boston produced, was made from material that came all the way from the West Indies.[3]

There were a few useful things growing upcountry. There was plenty of white oak in Massachusetts, which was good for making watertight pipes for Madeira or hogsheads for molasses and rum. Red oak went for sugar casks, so there was plenty of work for coopers supplying the West Indies trade, and for farmers supplying them with oak staves and hickory hoops. White oak was also ideal for building ships—another keystone industry of the colonial era. New England farmers provided the Islands with beef, and a few strings of onions; and they provisioned the ships, along with Boston and other seaport towns. But all this seems a bit cobbled together, and one has the feeling that any old hinterland would have done, no matter how rocky and cold. Whatever they could not procure locally, these enterprising merchants fetched from someplace else. And for all its commercial vigor colonial Boston was a small place; perhaps sixteen thousand people in 1750, only a small percentage of the population of the province. We should not exaggerate Boston's importance to the inland economy of Massachusetts. These were, in many ways, two very different and only slightly overlapping worlds.[4]

Cash crops for Boston or export were a small part of most household economies in Massachusetts towns. Inland farmers did produce *something* for market and counted on some imported goods, but few organized their farms and households around maximizing commercial production. But they were not self-sufficient, either. They were deeply embedded in complex local exchange networks with other farmers and craftsmen, and with their neighbors. Marketing surplus produce for cash was a simple extension of production that went mostly to support an increasingly complicated household and artisan economy. By the end of the colonial era, New England towns had formed a dense, successful agrarian society made up of diverse, well-run family farms that were capable of garnering themselves a "comfortable subsistence." These farms produced the bulk of what they needed, plus enough extra to round out their lives and to support a larger economy. The surplus of each farm was not very large, but there were a lot of them, so in aggregate it was considerable. Thus these towns managed slow, steady economic growth over and above demographic expansion, generation after generation. Given the thin ground on which they were standing, it was the first and possibly the greatest Massachusetts miracle.[5]

This was not false magic that depended on depleting the land, either. Instead, it rested on a remarkably balanced, stable form of mixed husbandry, as far as can be told from a detailed study of one such town, Concord. Resources were drawn from every part of a diverse landscape by careful, deliberate management. Intricately drained systems of low-lying hay meadows were replenished by winter flooding, and the manure that resulted from feeding meadow hay to cattle was recycled onto cornfields, maintaining their productivity for generations. Woodlands, vital for timber, fuel, fences, and a host of artisan trades, continued to cover 30 to 40 percent of the landscape, even as the town grew crowded with farms. This was a system capable of going on at a low, steady level of productivity indefinitely. The classic New England pastoral landscape pattern had been set in place by 1750.[6]

By the end of the colonial era, however, it was a system straining at the seams. Older towns such as Concord had been steadily filling with farms for four or five generations; now they were full. A powerful demographic engine was driving their growth, doubling the population of the province every twenty-five years or so. But Concord's population only doubled every fifty years, because some of its children were always leaving for the frontier. This pattern of New England town-making was stable at the core—that is, old farms and old towns did not decline, but held firm—while deliberately expansive at its periphery, steadily spinning off new farms and new towns. With the frontier free from Indian warfare, what happened in the decades surrounding the Revolution was in a way just another doubling: from half of New England, to all of New England. In the next generation these Yankees would sweep across New York and beyond, as well as into the cities and mill towns of the new century. But that surge would take place in a rapidly transforming economic and cultural climate, with the first stages of industrialization and the market revolution.[7] The new generations would treat the land differently and remake the countryside.

The Eotechnic Remaking of Boston and Massachusetts, 1790–1850

The first remaking of the Massachusetts economy and environment took place largely *before* the age of steam. It was driven by water and wind, and by the muscles of horses, oxen, and people. This was "America's Wooden Age"—wood remained the primary construction material, and timber framing imposed a limit on the size of buildings. This remaking was brought about partly by new technologies (that still relied on old power sources), but above all by new ways people went about engaging with the expanding cash economy.[8] Boston grew remarkably during this period, from something less than twenty thousand to well over a hundred thousand people. In 1822, through legal incorporation, the old seaport town became a city. This growth took place even though Boston had lost its leading place in Atlantic commerce to other eastern seaports, first to Philadelphia and then hands down to New York. Those cities had larger, more productive hinterlands to draw on, but completion of the Erie Canal in 1825 really clinched it for New York, opening up the Ohio country and beyond. In essence, the supremacy that Boston had once enjoyed by having the best connection to London passed to New York, which made the first, best connection to the American heartland.

Again, Boston just could not find much to export from its own backyard. Boston merchants built the Middlesex Canal to connect with the Merrimack River in 1803, the first extended canal in the nation—but what, really, can you get out of New Hampshire? Some pine, yes, and naturally some granite. Not quite the stuff of empire. So how did Boston prosper in the new republic? First by the China trade, and then, quintessentially, by reinvesting risky trade profits in manufacturing, particularly of textiles. That, at least for the larger investors who could ride out occasional economic downturns, proved to be stable and profitable.[9] Now what flowed into Boston to be made into finished goods was raw material from a global hinterland that dwarfed New England: hides from Massachusetts but also from Argentina;

wool from Vermont but also from Ohio, and then Australia; and above all, cotton from Mobile and New Orleans. The manufacturing was not done in Boston itself, but out in the mill towns that sprang up across the New England countryside. There was no decent water power in Boston, despite a wretched tidal experiment in the Back Bay that mainly served to trap offal and sewage. The power was all upcountry, in Waltham (a little), and Lowell (a lot), and along every New England river and brook. Boston supplied the capital and the management, sent the raw materials out to the mill sites, and exported the finished goods.

Since Boston merchants did not have anything to export in the way of bulk commodities to balance the lighter merchandise, they had a problem. Ships that had to sail in ballast often stopped at the harbor islands to pick up rocks, because drumlins generate a crop of cobbles and boulders at the waterline. With the rocks removed, the exposed sand and gravel began to wash away at a growing rate, clogging the channel that led to the inner harbor. Now *that* was a trade imbalance with real consequences: because there were no bulk goods to ship, Boston took to dismantling the very islands that guarded its precious harbor! By the middle of the century the islands were being protected.[10] A better choice for ballast was water itself. Of course, in those days it had to be frozen first, but cold was a staple Boston's hinterland did not lack. Seeing an opportunity, Frederick Tudor began shipping ice to the Caribbean—and then increasingly, as the cotton trade boomed in the 1820s and 1830s, to New Orleans and other southern ports so they could make mint juleps. The keys to the growth of the ice trade were first of all dropping the price of production through better harvesting tools, which was accomplished at Fresh Pond in Cambridge by an inventive Yankee mechanic named Nathaniel Wyeth. Second, many ships were going from Boston back to the Gulf of Mexico (or to Calcutta, or wherever) in ballast. Blocks of ice provided a substitute that was at least worth *something,* so Tudor got a cheap ride. All these Yankees could dig up to ship in bulk were rocks and frozen water. Perhaps it is no surprise that after a few years Wyeth tired of ice and headed for the Oregon territory in search of fur.[11]

Boston grew physically by taking down Fort Hill, Beacon Hill, and Mount Whoredom (renamed Mount Vernon by its developers) to heights that were more negotiable, and filling in coves all around the peninsula. Government helped to facilitate the expansion of infrastructure that enabled economic growth by chartering bridges, turnpikes, the canal, and (toward the end of the period) railroads on agreeable terms.[12] For the most part, government simply allowed private development to go forward. Boston at the end of this period was still a "walking city" of one or two miles, but not the most attractive place. It was densely packed with houses, businesses, and shops all jumbled together, with a great deal of poverty and filth and a contaminated water supply. The first era of urban transformation and growth was reaching its environmental limits. This gave rise to two kinds of responses: first, attempts to reassert the public good, and second, new technologies and energy sources that would allow renewed growth on a brand-new scale.

Largely unregulated private growth did meet with a few encouraging efforts to protect public welfare. One was the vote of citizens and the City Council in the 1820s and 1830s to keep Boston Common free of buildings, and to acquire the ropewalks that lay on the flats just beyond the Common to keep the city open to the salubrious bay breezes—land that eventually became the Public Garden.[13] The other was securing a good public water supply, which was accomplished by an effective coalition of working-class demands for a common resource and upper-class desires to combat filth. Water was removed from the pure market economy and made widely available by a municipal authority for all, not just those with access to a private source. The eruption in 1848 of the fountain on the Common thus brought the eotechnic era of urban growth to a close on a high moral note. Reformers saw the fountain as a way for city dwellers to make Romantic contact with nature, through pure water from the countryside, under gravitational pressure. The source of this water was "Long Pond" in Natick, but to give it the sparkle of Native purity, it was rechristened "Lake Cochituate," a completely spurious Indian name.[14]

Oddly enough, this singular victory for public access to water took place at the same moment that the Boston Associates were asserting control over the flow of water throughout the Merrimack River basin to power their mills.[15] Long Pond fed the Sudbury River, and its impoundment to provide water for Boston was thus a factor in a second, rural conflict between private and public interest in water, which had an entirely different outcome. The Sudbury flows north to the Concord, which provided an important part of the water power for the Boston Associates' mills in Lowell, where it plunges into the Merrimack. Along the way the Concord also supplied water to the Middlesex Canal, which had its summit at the Billerica mill pond above Lowell. When the Boston & Lowell Railroad put the Canal out of business in 1851, a textile manufacturer named Talbot acquired the Billerica milldam and its water privilege. This immediately brought him into conflict with hundreds of long-suffering farmers. For half a century farmers upstream on the Concord and the Sudbury had been complaining about increased flooding of their valuable hay meadows along the river, but they had failed in every attempt to dislodge the Billerica dam because of "instrumental" legal protections afforded it by the Canal charter. Although the Canal was dead, the courts ruled that its protection from lawsuits had miraculously survived and passed to Talbot.

Boston's acquisition of Lake Cochituate, which had provided as much as one-third of the summer flow of the Sudbury, added an interesting new dimension to this ancient struggle. Now the mill owners in Billerica and Lowell became alarmed about loss of critical summer water power from the river and sued the Boston Water Board. In response, in 1851 the city built compensating reservoirs at the headwaters of the Sudbury and Assabet rivers to replace the lost flow—and then conveyed those reservoirs to the mill owners. Every summer, when the flow of the river dropped, affording the farmers a chance to cut and cart their hay, the manufacturers would fill the river back up and flood the meadows. The river was "dammed at both ends, and cursed in the middle." This time the outraged farmers took their case to the state legislature, where again, despite an initially sympathetic hearing, they inevitably lost.[16]

This story illustrates powerful new connections between the development of the city and the development of the countryside, and makes it clear that private interests were still firmly in command. The first remaking of Boston was accompanied by a parallel remaking of rural New England. Industrial development in the second quarter of the century dispersed rapidly across the countryside, taking advantage of every available source of water power. Besides the large mills of the Boston Associates, an astonishing proliferation of smaller enterprises began manufacturing just about everything imaginable—some of it made from imported raw materials, but much of it from every local resource that could be found, particularly wood. Out in little Petersham, Massachusetts, miles away from any railroad, an entrepreneur began making ladders, taking advantage of the light, stiff wood of the small spruce trees that grew around local bogs. In Bennington, Vermont, enterprising men hauled cotton from the Hudson over the Taconics in wagons to take advantage of their small but energetic mountain streams. In tiny New Hampton, New Hampshire (which the railroad would never reach), a water-powered mill began turning out boots for markets in Boston and beyond. Everywhere, the old artisanal economy was vigorously expanded to reach regional markets, making use of anything and everything.[17]

Farmers also took part in this transformed economy, and in the process they remade the countryside. Farm families turned themselves into middle-class consumers and entrepreneurs. Many of the basic goods they had formerly procured for themselves or through neighborly exchange, they now began to purchase on the market: they adopted wheat flour in place of cornmeal, factory cloth for homespun, whale oil for tallow candles, coal stoves for fireplaces, coffee and tea for cider and rum. To pay to play in this consumer economy, farmers refocused their agricultural enterprises on specialized production for market. Merino sheep swept the forest from the hills of western Massachusetts, while many other upland farmers took part in the stall-fed beef trade, selling a few head every year to the Boston drovers. Others began growing broom corn on their best land, cutting broom handles in their woodlots, and making brooms. Long-settled towns such as Concord expanded their upland hayfields and pastures to market beef, butter and cheese, oats and hay, and firewood to the city. In neighboring mill towns, there were many more mouths of men, cows, and horses to be fed, but across New England there were also many more farmers vying to feed them. A few did well, not a few went bankrupt, and most struggled to get by, becoming Henry Thoreau's men of quiet desperation, speculating in herds of cattle to buy their shoelaces.[18]

As a result of this vigorous expansion of commercial agriculture, Massachusetts reached its most deforested point, plunging from perhaps 75 percent or more forest cover at the beginning of the century to approximately 25 percent by the third quarter of the century. The western part of the state saw a great surge in both farms and farmland, but even an old town such as Concord, well into its seventh or eighth generation of settlement, slashed its woodland from 30 percent to 10 percent by 1850—without any increase at all in the number of farms. The same thing was going on all across New England: the new, commercial form of agriculture was in many ways *more* extensive, and more exhaustive, than the colonial model. As Henry David Thoreau in Concord and George Perkins Marsh in Vermont both noted, the farm boom put tremendous pressure on the environment, causing not only loss of trees and wildlife but increased erosion and flooding. Continuous mowing and grazing quickly drained the limited native fertility from upland hay fields and pastures, which began to be taken over by brush: the land was being actively unimproved.

Most of the era's agricultural reformers viewed such "skinning the land" as a holdover from the colonial days of crude subsistence farming, rather than seeing it for what it was: the full release of the market's power to turn nature into commodities. A few voices, such as Marsh and Thoreau, protested this wholesale degradation of the land by the combined impact of many private farmers, and spoke in defense of the broader public benefit of land stewardship. No one was yet calling for government action to protect the environment, but some were at least recognizing its value. In the colonial era the land had been husbanded by locally based agrarian economies much more effectively than has been recognized. That sense of the duty of stewardship may not have entirely disappeared from agrarian culture in the nineteenth century, but it certainly was submerged by much more powerful commercial desires and demands.[19]

Perhaps the most important commodity that Boston's hinterland provided for this first remaking of both city and countryside was *Yankees*—lots of Yankees. A dense rural population provided the avid consumers, the provisions, the millworkers, the mechanical know-how, and the innovative drive to transform New England, even before there were railroads and steam to work with. One has to stand amazed at what they accomplished. But in the process of transforming it, they had pushed their natural infrastructure to the brink.

The Paleotechnic Remaking of Boston and Massachusetts, 1850–1920

The paleotechnic era was heralded by the appearance of railroads in the 1830s, and fully arrived with the linking of New England to the national rail network by the bridge at Albany in 1866. Its greatest expression was reached between 1890 and 1920, but it lingered through the Great Depression and World War II. Powerful new forces broke through the old environmental limits to growth in city and countryside, driving a sweeping remaking of both. With this transformation came a host of new social and environmental challenges, followed again by innovative responses to protect the public interest. The paleotechnic was the era of coal, steam, steel, railroads, and, in the later part of the period, of electricity. Boston grew explosively, by another order of magnitude—from just more than a hundred thousand to well over a million inside its metropolitan ring by the turn of the century. The Back Bay, the South Bay, the South Boston Flats, the Charlestown and East Cambridge flats, and the Cambridge Marshes were all filled in. Beyond the small elite downtown neighborhoods and the crowded working-class tenements and factories grew the larger segregated metropolis, with its spreading rings of streetcar suburbs for the middle class, out to about ten miles.

Across the country the growth of such urban behemoths was driven by the railroad, with its year-round, all-weather ability to integrate a national economy and to concentrate resources, especially coal, in the cities. Feeding on that steady diet of coal, steam power allowed industries to cluster near one another as well, rather than needing to be dispersed throughout the countryside near waterpower. The rise of the great industrial cities intersected with the incoming tide of immigrant labor, which could be fed and housed in dense urban quarters, thanks to the influx of food, lumber, and other resources being extracted on an unprecedented scale throughout the nation. The leading industries in Boston remained textiles and shoes, joined by ready-made clothing, machinery, and metalworking—and in the early twentieth century, by the rise of electrical goods as well.

Even as Boston reached metropolitan size, smaller mill towns and cities across New England added steam to their old water power base and still flourished. The transformations of the era are nicely illustrated by the little city of Waltham, which continued to grow in its own right even as it began to take on new metropolitan roles. The Boston Manufacturing Company's celebrated 1813 textile mill kept operating, but now a steam engine powered the spindles and looms, and the workforce was made up almost entirely of Irish immigrants. These had become low-wage, dead-end jobs. Upwardly mobile Yankees found work at a newer enterprise across the river, above the old mill dam. The Waltham Watch company was among the progeny of the Springfield Armory, whose expertise in cutting precise metal parts spawned many new industries. The company made its fortune during the Civil War supplying cheap, reliable pocket watches to Union soldiers, and it ended its operations making bomb timers during World War II. Those two wars defined the arc of its existence and of the paleotechnic era. In its heyday Watch City was home to several watch and clock factories and to many ancillary precision machine shops, which spun off still more enterprises, such as the Metz Company, a pioneer manufacturer of safety bicycles, motorcycles, and automobiles. A retail center developed along Moody Street, serving a dozen surrounding towns. It was anchored by a new department store, Grover Cronin's. Yet most of this commercial and residential development was concentrated in a small area by the river and the railroad. Until the 1950s and the coming of the highways, market gardens and dairy farms still covered most of Waltham north of the Charles.[20]

Waltham was like many other smaller mill cities across New England, thriving into the twentieth century until its older industries began slowly to decline. But Waltham was also close enough to begin to feel the pull of the metropolis. Eight to ten miles from Boston, Waltham was too long a ride to become a residential streetcar suburb. But it was well within the range of the electric trolley lines for recreational outings. The "Lake District" of the Charles River, winding for several miles above the Boston Associates' Moody Street dam, became a popular weekend destination. Amusement parks and other recreational facilities sprang up at the end of the lines in Waltham and Newton. The river was thronged with boathouses and canoes. Across the river, Weston—which never allowed the trolleys to infiltrate its borders—was transformed into a town of grand country estates, the summer places of Brahmins who spent the winter on Beacon Hill. Beyond the streetcar suburbs emerged outer rings of market gardens, dairy farms, piggeries, country estates, athletic clubs, and amusement parks.[21]

The bulk of the inner city's residents, of course, endured difficult lives. Lewis Mumford described the paleotechnic era as an "upthrust into barbarism." Sam Bass Warner and other urban historians confirmed Mumford's assessment forty years ago, and the work of Joel Tarr and environmental historians since then has done little to alter that grim verdict. For that half of the population who were not able to disperse to the suburbs, urban working and living conditions were grimy and dangerous. Boston is the windiest city in the nation, so perhaps the pall of coal smoke was not quite as pervasive as that which hung over other cities of heavy industry (such as Pittsburgh) located in river valleys. But that westerly breeze was not quite as welcome when it blew at low tide over the mudflats that received the city's sewage and the offal of packinghouses. The rivers ran full of dyes and industrial wastes. Air and water pollution, besides making life unpleasant or downright miserable, posed a real threat to public health.[22]

In response came slow and difficult, never adequate, but determined efforts to reassert the public interest in the face of environmental degradation. One was the creation of parkland within and around the city—first with Frederic Law Olmstead and the Emerald Necklace, and then with the work of Charles Eliot and Samuel Baxter to extend that vision to make the Emerald Metropolis. In 1893 the Metropolitan Parks Commission anticipated half a century of suburban growth when it moved to acquire a ring of hills and beaches at the rim of the Boston basin, along with major parts of the three rivers that ran down from that bedrock periphery. The commission also constructed graceful parkways that helped give shape to ensuing suburban growth, until they were overwhelmed by automobiles at the end of the paleotechnic era. Finally, the wasteland of warehouses and stinking flats around the shores of the Back Bay was transformed by the Charles River dam in 1902, which flooded the estuary. Beautiful esplanades were created along both sides of the river.[23]

The struggle to protect those waterways from industrial and municipal pollution and to secure again a safe and adequate water supply was even more fundamental. By the last decades of the nineteenth century, the sheer scale of Boston's metropolitan growth, along with the multiplication of industrial cities along New England's rivers, once again overwhelmed the water and waste infrastructure of the late eotechnic: too many water bodies were serving as both sewers and sources for too many people. Some river water became so foul that even other industries could not use it. A new, regional approach was required—metropolitan water and sewer boards were created alongside the parks commission in the 1890s, and in 1919 all three were combined to form the Metropolitan District Commission. MIT chemist and biologist Ellen Swallow Richards and William Sedgwick's sanitary survey established that water pollution was a public health menace, causing typhoid fever. The regional sewer system was extended to intercept most of the sewers emptying into the rivers around Boston and divert the outflow into the harbor—an enormous improvement for the next half a century, anyway. Finally came the discovery at Lawrence in 1893 of the ability of sand filtration to cleanse drinking water of germs. Many Massachusetts cities constructed filtration plants, greatly reducing typhoid epidemics. This was a public health triumph, although it ironically stalled the momentum that had been building to treat municipal and industrial wastes and actually clean up the rivers themselves, leaving them to grow steadily more polluted through much of the twentieth century.[24]

Once again, Boston looked west for a fresh supply of still more remote, pure water. The Sudbury River (which had been tapped to augment Lake Cochituate) was polluted and inadequate, and Boston wanted to avoid the cost of having to filter the Merrimack. Instead, the city and its suburbs reached deep into the countryside and took the Nashua River above Clinton in 1905, and within a few decades jumped farther west to the Swift and the Ware, on the plateau beyond Worcester. Naturally these reservoirs were christened with Indian names of their own, Wachusetts and Quabbin. They were massive paleotechnic undertakings—each was the largest water supply reservoir in the world at the time of its construction, and each increased the storage capacity of the Boston system four- or fivefold. They demonstrated the unstoppable political clout of the metropolis, which by this time held close to half the state's population. These takings are bitterly remembered in central Massachusetts to this day. Besides flooding several villages for the Quabbin, the Metropolitan District Commission acquired more than fifty thousand acres of surrounding land to be maintained in forest for watershed protection—by far the largest piece of contiguous forestland in the state today and also, interestingly, the most intensively managed.

The Massachusetts countryside was also remade in the paleotechnic era: it was dramatically reforested. Forest cover, which had fallen to a low of about 25 percent of the state between 1850 and 1875, rebounded to about 50 percent by the beginning of the twentieth century. With the rise of the city and the industrial economy, New England farming went downhill—or so it would appear, and so we have often been told. But it would be a mistake to equate this forest recovery with the economic collapse of Massachusetts farming at the hands of cheaper Midwestern production—at least for the period before 1920. Census data tell a different story: between 1880 and 1910 the amount of farmland in Massachusetts was indeed cut in half, but during that same period the value of agricultural production in the state, in constant dollars, doubled. The peak for Massachusetts farming occurred not at the time of the Civil War but of World War I, whether measured in constant dollars or gallons of milk. Plenty of farmland was being abandoned during the paleotechnic era—but not very many farms. Agriculture enjoyed robust growth for half a century after it was supposed to have been mortally wounded, even as the hills were being re-covered by trees.[25]

This anomaly can be explained by one word: railroads. A few farms in remote hill towns surely were given up, but many more across the state were thriving—first and foremost, by reprocessing cheap western feed grain into milk and produce for urban markets. Massachusetts agriculture joined a streamlined national food system, supplying a burgeoning regional market with those products that could not easily be shipped halfway across the continent, and taking advantage of those that could. Urban growth and western grain carried New England farming to its peak production. Suburban towns such as Waltham and Weston, Arlington and Concord, filled with flourishing market gardens that grew fruits and vegetables for the city. These farms absorbed immense flows of recycled urban waste such as horse manure and spent brewery malt. Farmers also provided staggering loads of hay for city liveries and teamsters—often from highly productive hay fields fertilized with the surplus manure from their dairy herds. As the Boston milk shed expanded across the state, farmers within a wagon ride of a rail depot moved toward year-round stall feeding of their cows to increase the flow of milk. They gave up their *pastures,* not their farms. The worn-out pastures grew back, famously, to white pine—a more profitable crop, as many farmers foresaw. This was not an economic collapse but an economic calculation, conditioned by ecological reality: it was difficult and expensive to keep pastures productive and free of pines. By the turn of the century the pasture pine was being cut for boxboards, supplying packaging material for much of the commerce of the day.[26]

While resources poured into the region thanks to the headlong exploitation of the continent's richest farmlands, grasslands, and forests, a more attractive balance was restored over the Massachusetts countryside. Forests recovered half of the landscape, and a more intensive, less degrading agriculture held sway over much of the remainder—beholden though it may have been to the plowup of the tallgrass prairie a thousand miles away. But while the Massachusetts forest was increasing in acreage, as quickly as it could grow back, it was still being heavily cut. An observer might have been forgiven for not seeing the forest for the stumps. The boxboard boom in southern New England coincided with the climax of a wave of clear felling across much of the East. Spruce and fir were cut for pulp, hemlock for tanbark, and with the advent of portable logging railroads, the great hardwood forest fell to provide furniture, ties, pit props, and a thousand other industrial and household products.

Forest fires ravaged the denuded granite hills and sand plains from Minnesota to Mississippi to Maine, raising fears of a coming timber famine. New England's great rivers were jammed with logs from the North that would occasionally perform remarkable feats such as carrying away a railroad bridge over the Connecticut on the spring freshet. Calls to curtail flooding and to safeguard the nation's supply of timber and water did stimulate a public response, a vigorous forest conservation movement that ran alongside the urban campaign for parks and safe drinking water. Today the Commonwealth of Massachusetts owns some five hundred thousand acres of forest, or about a tenth of the state. Most of it was acquired in early decades of the twentieth century in response to the excesses of the raw industrial expansion of the paleotechnic era.[27]

The Neotechnic Remaking of Boston and Massachusetts, 1920–?

The neotechnic era began to emerge with the rise of the automobile in the early twentieth century, but it really took off after World War II. It was, and is, quintessentially the age of oil, and along with it the automobile, the airplane, chemicals, plastics, the expansion to omnipresence of electricity, modern medicine, and telecommunications. With the recent rise of computers, some say we are now in (or at least on the cusp of) a new "information economy"—but whether the Internet will produce another remaking of the urban and rural landscape, or just more of the same, remains to be seen. A true Mumfordian would need to see a new transportation system, new materials, and above all, a new energy base for society. None of that has yet appeared. For now, it seems best to say that we are still in the full flowering of the neotechnic age, and still struggling to enact restraints that are able to protect the public welfare from the mixed blessings brought by our technology, and the economic engine that drives it.

The old industries that had supported the paleotechnic New England economy suffered severe decline in this era, but by the last decades of the twentieth century, metropolitan Boston rebounded as a center of finance, insurance, health, and education. Above all, the region's economy grew as an innovative "knowledge center" tied to the development first of electronic and then of high-tech computer industries. Downtown Boston grew upward with modernist skyscrapers along the "high spine" from Government Center to Back Bay. The big land-making project of this era was in East Boston—giving us a major airport that lies amazingly close to downtown, another mixed blessing.

The essence of the neotechnic remaking of the city, of course, was a stupendous round of highway construction and suburban sprawl. The idea of building graceful parkways was jettisoned in favor of brutally functional expressways punching into, over, and through the city. Charles Eliot's belt of parklands around the city had a massive circumferential highway (Route 128) threaded right through it in the early 1950s, and a second ring road (Route 495) followed a few decades later, another fifteen miles out. Virtually all of the privately held land inside the old metropolis that was still open—and there was quite a lot of it, at midcentury—was filled in with residential and commercial development. Towns within the next ring, between Routes 128 and 495, chose to zone themselves into either low-density affluent residential suburbs or higher-density places of office parks, shopping malls, tract housing, and condominiums. That is the shape of the amorphous neotechnopolis we now inhabit, with outcroppings of several earlier urban and suburban strata poking through it here and there.[28]

In the same tidal rhythms that marked earlier remakings of the city, recent decades have seen renewed efforts to place restraints on both private and public growth of the automotive metropolis. From the late 1960s several major highway proposals have been halted, and a grueling campaign has finally been completed to bury the most egregious one that did get built, the Central Artery, beneath the ground. Persistent attempts have been made to revitalize the inner city's neglected neighborhoods, some more successful than others. Bit by bit, green spaces and walking places have been reclaimed from the highways, and strung along the shores of the harbor and the rivers. Those waters themselves, so central to the beauty of the city but once again overwhelmed by pollution by the middle of the twentieth century, have now been restored to (sometimes) swimmable condition by a massive upgrading of industrial and municipal waste treatment. As for water supply, by the 1970s the metropolis was preparing to reach still farther west beyond Quabbin, to the Connecticut River, to slake its thirst. But this time political resistance was too great for the city to impose its will on the western part of the state, and a successful program of water conservation was implemented instead. Metropolitan water consumption today is only two-thirds of what it was twenty years ago.[29]

The growth of the automotive metropolis continues to affect the countryside, of course. Today the sprawl frontiers stand west of Route 495 in Worcester County, in southeastern Massachusetts between Plymouth and Cape Cod, and in southern New Hampshire. The Cape and the islands are substantially built out. Since midcentury, a wave of residential land-clearing has been spreading across the state in a way reminiscent of the long wave of agricultural land-clearing two centuries earlier—but if that deforestation for farming was soft and reversible, this deforestation for rooftops and pavement could prove hard, and permanent. It poses another enormous challenge for conservation and the reassertion of the public interest in the face of thousands of individually rational private decisions to develop in the easiest way, instead of the best.[30]

The remaking of rural Massachusetts during the neotechnic era has been a two-part affair. The rebound of forest that began in the late nineteenth century has given way to fragmentation and loss in the past few decades. Forest cover reached 75 percent by 1970—an exact inversion of where it stood a century earlier, and a return to where it stood a century before that. Since 1970, though, forest has declined to about 64 percent of the landscape. The past century has seen a severe decline in Massachusetts agriculture. Farmland has been losing ground to the forest on one side and to suburbia on the other. New England farming weathered the paleotechnic opening of the nation's heartland by railroad, but it could not withstand the neotechnic opening of a global market for every crop, made possible by cheap oil, chemical fertilizer, refrigeration, trucks, and airplanes. Real economic decline for Massachusetts agriculture set in after 1920: market gardens were undercut by Sun Belt produce, hay markets vanished along with the horse, milk production steadily consolidated into larger herds on cheaper land elsewhere—in Vermont, or Wisconsin. In the suburbs, land worth far more for growing houses than for strawberries and cream steadily succumbed to its highest and best use. As the decades passed and development moved out across the state, the rate of suburbanization finally overtook the rate of reforestation. Once again, the forest is falling.

Even after the forest stopped expanding and began to shrink in acreage, it continued to increase in volume. As the trees have grown larger and more cut-able, they have been cut less and less. Just as New England agriculture has declined in the face of industrial agriculture, regional forestry has declined in the face of a globalized timber market. The anticipated twentieth-century timber famine never arrived. Instead, the demand for forest products has remained soft in comparison to the vast global supply that has come within reach, and there is growing antipathy on the part of many private and public owners in places like Massachusetts to cut. This puts us in the odd position of an affluent suburban society, mighty consumers of lumber and paper, living in the midst of a forest with which we have almost no working relationship. Massachusetts is among the most urbanized states in the country, and simultaneously among the most heavily forested. Yet the state imports something like 98 percent of the forest products it consumes.[31]

Great strides have been made since the 1960s in pushing back against rampant suburban development. Massachusetts wetlands and waterways have been successfully shielded from further degradation by the wetlands protection act and floodplain zoning. Acquisition of conservation land and easements by federal and state agencies, municipalities and nonprofit organizations has gone steadily forward, so that today some 1 million acres, or about 20 percent of Massachusetts is protected land. All this is to the good, and yet it has done little fundamentally to slow the pace of development. In fact, some argue that the combination of ad hoc land protection and large-lot zoning has only pushed suburbanization more rapidly outward. Meanwhile, less and less is produced from Massachusetts farms and forests, year by year. In the neotechnic era the means by which such basic resources as food, lumber, paper, and energy are extracted from the environment have been pushed far away, out of sight and often out of mind. Leo Marx once labeled this "sentimental pastoralism." Some today have called it the "illusion of preservation."[32]

Lewis Mumford, writing in the 1930s, declared that the neotechnic era he saw unfolding before him was a "cultural pseudomorph." By that he meant that promising new technologies, which might have led to clean, attractive, and harmonious cities and countryside, were instead being driven by the same old paleotechnic forces of heedless growth. In the real neotechnic we got the "townless highways," but whatever became of the "highwayless towns," the garden suburbs that Mumford and others envisioned? One wonders what the next remaking of Boston and of Massachusetts will look like. Will it be just another pseudomorph, where mindless growth drives on in new forms, using new technologies? In part, no doubt. What forces will drive change, and how will the endless tug of war between private enterprise (with its undoubted economic benefits) and the beleaguered public interest play out?[33]

If we follow the Mumfordian paradigm, it seems clear that the next shape of city and countryside will depend on a shift in energy base. A permanent, substantial increase in the price of oil may bring something different in the way of urban and rural development than the sprawl we have been seeing for the past half-century and more. The change will be especially striking if (as some predict) any substitute for transportation fuel proves substantially more expensive than oil was during its heyday. We may also feel the impact of rapid climate change and other unresolved environmental side effects of the fossil fuel era. For example, a rise in temperature of several degrees could make for dramatic changes in the species composition of our forests, while a rise in sea level of several feet would make for arresting changes in the coastline around our city. If such forces dominate our future, one can imagine a remaking of Boston and Massachusetts that would involve renewable sources of energy, more serious reliance on telecommunications in place of (rather than in addition to) frenetic personal travel, a return to denser development of cities and suburbs, and a complete rethinking of what resources we most value in the countryside, and how we can both utilize and protect them.

Without making rash predictions, we can at least imagine a future of enlarged public interest in both city and countryside that is worthy of the likes of Henry Thoreau, Charles Eliot, and Lewis Mumford. In the city this would mean reclaiming green space, walking space, and living space from the automobile. It would mean reviving and expanding a workable system of public transportation that people will actually use. It would mean daylighting our streams, restoring and expanding our riverbanks, our parks, and a healthy urban forest. It would mean the spread of community gardens, urban farms, green walls and rooftops, and everything else that would make the city the fully livable place it ought to be.

In the suburbs such a remaking would mean open space that is not an exclusive retreat from the city, but easily accessible by bicycle trails and public transportation. Thousands of acres of conservation land and hundreds of miles of walking trails already exist in such towns as Weston, Lincoln, and Concord, but are hardly enjoyed. We can envision thousands more acres of community farms providing organic fruits, flowers, and vegetables. Working models are already thriving and engaging people with the land. Farther out, across rural Massachusetts, encompassing a modest but vigorous working farmscape, we can envision "Wildlands and Woodlands": one-half of the state in permanently protected forest. Part of this forest —say one-tenth—could be in large wild reserves, the bulk in sustainably managed woodlands. This would give us a robust natural infrastructure providing biodiversity habitat, watershed protection, recreation, renewable wood products, and carbon sequestration. This vision is not antidevelopment—indeed, many rural towns desperately need more people to restore viable communities, and to care for the land. An affordable "smart growth" approach would embed clustered development in a landscape of working farms and forests.[34]

I would not claim that such a green remaking of Boston, its suburbs, and its hinterlands will come easily, even if the era of cheap fossil fuel gives way to something else. But as a citizen, I can work for that future. As a historian, I do believe in a usable past. Just as we cannot divorce the city from the countryside in understanding the history of Boston and Massachusetts, we also need to connect the city and the countryside in the making of a livable commonwealth today.

7

A City (Only Partly) on a Hill

TERRAIN AND LAND USE IN PRE-TWENTIETH-CENTURY BOSTON

William B. Meyer

IN February of 1847, Edward Everett, then president of Harvard University, climbed a large hill near Boston in the course of a weekend ramble. The scene he found at the top seemed at first to be oddly incongruous. There was a view "of surpassing beauty," but the only houses enjoying it were a few "wretched wooden cottages." Yet the vista and the shacks did go together after all. With the summit "offering no one temptation but Prospect & Air as if this would supply the place of every convenience of life," Everett reasoned, it was no wonder that it attracted nobody who could afford an easier place to live.[1]

Before the twentieth century, hills offered many challenges to those who would live on them that lower ground did not. Today, one does not often see the pattern that Everett noticed in the United States, because these challenges have been met and overcome. High ground in American cities tends now to attract the most affluent of residents and is little obstacle to most kinds of business.[2] The advantages of living on higher ground, chiefly the views and other environmental amenities that it offers, stand out. The inconveniences, having been conquered, have been forgotten. But those inconveniences—

above all, the difficulty of ascent in an era when movement was slow and laborious—carried much weight for most potential occupants before the transportation revolution that culminated with the triumph of the automobile in the twentieth century. Land in the core of the premodern "walking city" was in far more demand than land on the margins, the latter forming a "horizontal urban fringe" that was left mostly to a scattering of uses and users that could not afford more desirable (that is, more accessible) core locations. In the same way, land that was too far elevated above the center formed a "vertical urban fringe" that was used much less intensively than lower ground and by much the same set of mostly marginal occupants—Everett's "wretched wooden cottages" among them—as the horizontal periphery.

The contrast between the earlier pattern and the one that exists today is a significant dimension of Boston's environmental history. It represents environmental change in the sense that the relations of natural features to the human activities that went on around them have changed over time. Social and technological developments, particularly improved modes of transportation, transformed Boston's hills from the liability in the urban landscape they once were to the asset they are today. Their metamorphosis illustrates how "natural resources" and their opposites, physical features that people assess as worthless or worse, are not given natural facts but historical creations. Understanding the fringe character of premodern Boston's uplands, their marginality to the life of the city as a whole, further helps to account for an otherwise puzzling fact. The place famed as the "city upon a hill" not only made quite meager use of its high ground, but during the nineteenth century it seized many opportunities to lower and flatten its terrain. Regarding hills as obstacles rather than resources, Bostonians reduced, reshaped, or removed many of them. This represents environmental change in a second and more obvious sense: the alteration of physical features themselves by human action.

Land-use Patterns in the Premodern City

Several distinctive spatial patterns of land use characterized the premodern American city; they form the background against which to discern any modifications stemming from terrain, and they came from ultimately the same basic sources as the vertical-fringe pattern. Among them, the difference between core and periphery is the best studied and the most striking. A century and a half ago, a contributor to a Boston newspaper made this distinction the basis of an idealized pattern that he suggested was typical of almost all large cities. Business claimed the land in the center, he wrote, "then comes first class residences, then second class, and so on, until you reach the poorest tenements of all on the very outskirts of the city." Residential status, in other words, typically declined from the center outward to the horizontal fringe on the periphery, and so did the presence of most forms of business.[3]

This pattern was beginning to erode visibly even then, but it had been the norm for American cities throughout their history. The chief reason for this lay in the cost and difficulty of transportation in an era dependent for locomotion on walking and horsepower. Businesses, which could outbid all other users for the best locations, needed to be in the center, near the point of easiest access. The richest households, which could outbid all others for the best sites that remained, preferred to live in or near the core. The most socially marginal were thus marginalized in the most literal sense. The paucity of such urban services as water supply, fire and police protection, and street paving and cleaning in the more remote districts added to the fringe's unattractiveness for residence. In accordance with the rule, the spatial core of late-eighteenth-century Boston was of distinctly higher residential status than the outlying areas known as the North, South, and West ends.[4]

A multidimensional mosaic of residential and land-use patterns emerged from the superimposition on the core-periphery gradient of two other patterns. Both of them arose from the same priority given to accessibility over such other considerations as abundant space or environmental amenities. One was the preference of both business and the urban elite for dwellings on the more visible and more easily traveled main streets, with the poor relegated to side streets, back alleys, and courts. The prestige of State Street in late-eighteenth-century Boston illustrates the tendency.[5] So does the preeminence in the nineteenth century of Beacon Street in the Beacon Hill neighborhood, the squalid poverty of many of the Hill's backstreets and courts, the great popularity of the Back Bay avenues with the city's wealthy residents, and the contrasting failure of the South End's more intimate streets and squares. Likewise, the upper floors of buildings—though enjoying views, air, quiet, and seclusion—were so difficult to reach in pre-elevator days that both their commercial and their residential status were much less than that of the lower floors. Only when the elevator replaced the "steep and tiresome stairs" did business expand far upward from the streets of the urban core. In Back Bay mansions, for example, the servants typically lived at the top; in South End lodging houses, rents declined as elevation increased. Newly arrived in Boston, Bartley and Marcia Hubbard, portrayed in William Dean Howells's *A Modern Instance* (published in 1882), end up taking a fourth-floor room at four dollars a week, which is considerably cheaper than the similar accommodations they are shown on lower floors. "'It's pretty high up,'" the landlady tells them with "a warning shake of the head. . . . 'It's high up, and there's no use denying it,'" though she makes as much as she can of the view of the harbor from the windows.[6]

A simple model of an outcast periphery grading into an affluent core cannot completely sum up the margins of pre-twentieth-century American cities, however. Some of the wealthiest citizens, in the Boston area as elsewhere, possessed elegant country residences on extensive grounds in the surrounding towns.[7] Such estates were usually seasonal supplements to central-city residences or the homes of those whose means or occupation did not require them to travel to town on a daily basis. As such, these residences bear out rather than contradict the rule that accessibility was the key to locational choice; only the few who enjoyed unusual freedom in that regard could decide on other grounds where they would live. The scattering of estates and the slums found on the pre-twentieth-century horizontal urban periphery were both drawn there, in the last analysis, by the cheapness of land compared with the crowded core in a period when the "friction of distance" was a serious deterrent to a long daily commute. Equally characteristic of the fringe, and for the same basic reason, were noxious trades that had been expelled from the city—which almost alone among businesses benefited more from seclusion than from accessibility—large residential institutions ranging from colleges to almshouses, cemeteries, and public or undeveloped private open space.[8]

The transition to the modern pattern began several decades into the nineteenth century. New and easier modes of transportation—omnibuses, steam railroads, and horse-drawn streetcars—were its main driving force, coupled with the preference for space and privacy already apparent in the country estates of the premodern elite. Better connections between core and margin gave rise to upper- and middle-class suburbanization on a far more extensive scale than ever before. At the same time, low-income ghettoes peopled largely by immigrants began to appear in central Boston, in areas formerly occupied by the mansions of the affluent.[9] Later in the century the elevator, making vertical movement easy within buildings, began to reverse the relative appeal of lower and upper floors by bringing the superior amenities of the latter within convenient reach.

The prevalent pattern in horizontal space today inverts the premodern center-to-periphery decline in residential status and, within buildings, the upward decline in status. Proximity to the main arteries of traffic too no longer has the appeal that it once had. Most people prefer a more retired setting because of the amenities that go along with it. Many forms of business no longer find a central location indispensable, as witness the decentralization of most industrial, retailing, and service jobs to the suburbs of major American metropolitan areas. At a more basic level the key determinant has shifted from accessibility to amenities. The changes in the relative value of center and periphery, of main and back roads, and of upper and lower floors have been summed up in the saying that technological changes starting around the end of the nineteenth century—electric trolleys, mass car ownership, and the widespread adoption of the elevator—turned American cities "inside out and upside down."[10]

This remark invites application as well to the vertical dimension of urban land. Today's pattern, when the automobile has largely removed the difficulties of life on hills, is clear enough. Considerable evidence suggests that well-to-do residents of American cities tend to seek elevated sites, leaving the lowlands to the poor.[11] Open space, views, and less heavy and noisy traffic are amenities that hills, like the other, now-favored zones of urban areas, tend to afford. Nor do most forms of business, which are also attracted by such site amenities, find terrain a major constraint on location. Were matters different before the twentieth century? High land shared the disadvantages of the premodern horizontal urban periphery. It too was not easy of access, being distant from the center—like the upper floors of buildings, in vertical rather than horizontal space. As a result, it required extra effort to reach or to traverse. (How much, of course, depended on the slope as well as the elevation.) Consequently, one would expect to find that it too, in a time when transportation was laborious, was avoided by the main arteries of movement, by most forms of business and by most residents, and left to users and uses that were poor, highly land-demanding, or both.

The concept of a "vertical urban fringe" extending upward rather than outward from the city center has been proposed as a model of highland occupance in premodern urban settings.[12] It suggests that the same uses and occupants typically found on the horizontal periphery should also have dominated the highest ground of the premodern city and that those typically absent from the former—particularly, most forms of business—should equally have been rare in the latter. Technology conquered the challenges of movement in vertical space later than those of movement in horizontal space.

Only around the close of the nineteenth century did electric trolleys and then automobiles, both capable of scaling slopes with relative ease, become widely available for intraurban movement. Thus a vertical-fringe pattern should have persisted longer than the horizontal-fringe pattern. Pre-1900 Boston bears out these expectations.

An anonymous writer observed in 1830 that Americans seeking a site for a city "give even ground the preference. . . . This is perfectly natural, and in selecting ground to be passed over frequently,—perhaps in after ages to be constantly traversed by dense crowds, it seems proper that the most level should be selected." Where naturally flat land was insufficient, he added, "the surface is generally levelled, often at considerable expense."[13] With hills representing more of an obstacle than an asset in the pre-twentieth-century American city, cutting them down would have increased the value of the acreage that they occupied, possibly enough to return a profit after deducting the costs of the operation, and would generally have seemed an improvement. If a vertical-fringe pattern existed, with poor and marginal residents predominant while the well-to-do and business establishments were few, little effective resistance could have been mounted against projects that their advocates could present as simultaneously ones of commercial redevelopment and of slum clearance for the betterment of the community. A recent article by the historian Matthew Klingle on hill removal and regrading projects in Seattle in the late nineteenth and early twentieth centuries cites the typical arguments that were advanced in their favor: "regrading would eliminate unwanted hills that added costs to shipping and time to travel, open space for building, increase property values, and contain or eliminate unhealthy slums."[14] These were also the standard motives for cutting down hills in nineteenth-century Boston, where such projects did much to alter the city's terrain.

Hills and Land Use in Pre-twentieth-century Central Boston

When the Puritan settlers took possession of the Shawmut Peninsula in 1630, it covered less than five hundred acres and was connected with the mainland only by a narrow neck that the highest tides overflowed. It included three principal areas of high ground rising from the mostly low, flat land of the rest of the peninsula. The most imposing of them by far was the

Trimountain, a large ridge on the northwestern side that took its name from its three separate peaks. Smaller in both height and extent were a rise of land near the water's edge on the south, eventually named Fort Hill, and another (which became known as Copp's Hill) at the end of a northeast-stretching arm of land. The highest point on the peninsula was the middle summit of the Trimountain, which came to be known as Beacon Hill, at 150 feet above sea level. Each of the two lower summits and Fort Hill rose to about eighty feet, and Copp's Hill to about fifty feet.[15]

A notable feature of the early occupance pattern was "the absence of any settlement on Fort Hill, Copp's Hill or the Trimountain." Nearly all of colonial Boston's streets, houses, public institutions, and places of work were on the lower lands of the peninsula, clustering in particular around the harbor. John Hull (1624–1683), one of the few seventeenth-century Bostonians to live on the slope of the Trimountain, summed up its attractions and liabilities when he wrote that "my Habitation is greatly disadvantaged for trade; yet because I alwayes desired a quiet life and not too much business, it was alwayes best for mee."[16] If his experience was typical, the vastly greater importance of centrality than of quiet and seclusion in that period explains why Hull's neighbors on the hill were so few and far between.

War or the threat of war brought Boston's hills fleetingly from the margins of attention to the forefront. The difficult terrain that was their chief drawback at other times became a prime military asset. During the year in which the town and its environs formed the main theater of the American Revolution, the besieged British forces maintained a battery on Copp's Hill and lesser works on Fort and Beacon hills, fought a major battle to win control of Breed's Hill in Charlestown, and finally had to evacuate Boston when the American army moved cannon onto the commanding high ground to the south and west.[17] Of peacetime uses before 1800, a few profited as directly as fortifications did from an elevated site. A tar barrel on a mast for signal fires gave Beacon Hill its name, and a windmill appeared on the summit of Copp's Hill as early as 1632. Large bonfires, cannon salutes, or fireworks displays on the summits, on special occasions of public celebration or protest, enjoyed the advantage of a high pedestal from which the flames could be seen far and wide.[18] But they also benefited from the hills' very emptiness, the abundance of open space with no tightly packed buildings that might be set ablaze by a stray spark.

The other main early uses of the hills likewise illustrate the tendency of the city's vertical fringe, like the horizontal, to attract land-demanding and/or stigmatized activities that had been pushed out of the congested and costly urban center. Those that were forced up onto the largely unoccupied Trimountain included such unpopular neighbors as a gunpowder house and some of the city's ropewalks, which were both nuisances and fire hazards. Cemeteries on the Trimountain and Copp's Hill enjoyed ample space, seclusion, and at the same time some distance from the dense urban populations that had begun to regard crowded burying grounds as a nuisance and a hazard to health. Boston Common occupied much of the gently rising southwestern slope of the Trimountain. It served initially as pasturage and eventually for a wide variety of events requiring plenty of open space, ranging from militia drills to public executions. The upper slope above the Common, from the late seventeenth century until 1800, held the principal municipal institutions for the indigent and the criminal: an almshouse, a jail, and a workhouse, plus the town granary.[19]

The almshouse alone had a resident population by the late eighteenth century approaching three hundred in the summer and sometimes reaching four hundred in the winter.[20] The typical Trimountain dweller of the time was an inmate of one of these public buildings or a resident of the slum on the northern slope known as early as the mid-eighteenth century as Mount Whoredom. The phrase "the Hill," far from being synonymous with wealth and exclusiveness, was a byword for "vice" in Boston into the early nineteenth century.[21] Wealth and exclusiveness, though in a minority, were not absent from the Trimountain. What the historian Walter Muir Whitehill characterized as "a handful of country houses and gardens" served the upper stratum of the hill's population and offered a close parallel in everything but their proximity to town to the estates on Boston's horizontal fringe. The tree-lined "Mall," favored for fashionable promenades, ran along the southern fringe of the Common.[22] All of these occupants together fell very short of filling the Trimountain's expanse. As late as the 1790s, it remained, in the traveler Timothy Dwight's words, "almost absolutely a waste." Mount Vernon, the westernmost peak of the Trimountain, during the same decade was "a repulsive dreary waste, on which only three decent houses were to be seen."[23]

At the turn of the century, the proportions of rich and poor on the Trimountain began to change, but only once it had been made less of a mountain. Between 1803 and 1807, a syndicate of developers, including the wealthy politician Harrison Gray Otis and the architect Charles Bulfinch, cut down Mount Vernon substantially in height and laid out the southwestern slope in lots; the excavated material was used as fill at the base of the slope to claim new land from the flats of the Charles River.[24] The reduction of Mount

Vernon launched the first substantial influx of affluent residents to the Hill. Its evident rationale was that terrain that was more than moderately high and uneven was too difficult to attract more than a bare few residents of wealth and standing and that its successful development for such a clientele necessitated lowering and regrading it to a more convenient elevation. The construction of elegant mansions on Mount Vernon was accompanied by that of a new state capitol on land from the Hancock estate on Beacon Hill, the removal in 1800 of the city buildings from their nearby sites, and, in the early nineteenth century, the razing of the peak of Beacon Hill itself, bringing it level with the State House grounds. The last major reduction of the Trimountain, undertaken for the same reasons, removed its easternmost peak, Pemberton or Cotton Hill, lowering the land by some sixty-five feet in the 1830s and 1840s.[25] As in the Mount Vernon area, concentrated upper-class housing quickly replaced the open space and scattered country estates that had previously occupied the area.

Bostonians of the nineteenth century themselves discerned a clear difference between the land uses that gravitated to high ground and to low. They could see that industry and commerce strongly favored flat central sites and shunned the hills. "For general business and travel, those streets are mostly selected which are nearly level," the *Boston Post* observed in 1870; "the steep grades operate as a barrier to many kinds of traffic," another writer pointed out a few decades later, so that the main thoroughfares and the main streams of business skirted the high ground. Local real-estate experts explained that hills would "interfere with the transaction of business," which depended on walking and on deliveries by railroad cars and by horse-drawn carts and wagons, all of which had great difficulty in climbing steep grades. "Where there is a high hill," one authority testified at a public hearing in 1869, "stores would not be built on account of the difficulty of cartage." On Beacon Hill, he pointed out, "the coal dealers have to supply two extra horses for each load of coal that comes up that hill now." Another witness asserted that even land located "within a stone's throw of deep water" and the busy commercial docks of the port that were concentrated along it would be "used for business purposes" only if it were "dug down to a level." Edward Stanwood wrote in 1872 that "the difficulty that business almost always experiences in ascending a hill" had reserved substantial areas of central Boston, notably most of Beacon Hill, from the commercial and industrial uses that would ordinarily have claimed such prime locations.[26]

As the *Post* remarked in 1870, "the hill-sides are left for private residences or public institutions." The remark especially holds true if one understands the term "public institutions" broadly to include such features as cemeteries, parks, and undeveloped open space as well as hospitals, colleges, and asylums: some of the most typical occupants of the horizontal urban fringe. The same two populations that characterized the horizontal fringe stood out on the vertical; the residents of this zone (the inmates of the institutions among them) were no representative sample of the city's population overall. On the hills, and especially on their higher reaches, certain classes—the poor (particularly immigrants and minorities) and the quite affluent—predominated. The well-to-do who settled on elevated land within or near the city core possessed such resources as horses, carriages, and servants to cope with the challenges of these sites. In return, they could enjoy the amenities of "elevation, retirement, and proximity to business" that one Pemberton Hill promoter advertised in the 1830s, and also of views and cleaner air (especially in a period when much disease was blamed on the miasma associated with low-lying, badly drained land).[27] The very poor, for their part, had their own incentive to settle on high ground; the need to live on cheap land within reach of their workplaces and accept in return the difficulties of the site that made it affordable.

It became a matter of common remark throughout the century that the reshaped Beacon Hill (as what was left of the Trimountain now came to be called) was home at once to many of the richest and many of the poorest residents of the city, the latter on the northern summit and slope and the former on the southern. In 1830 the Hill held both "the most elegant houses and eligible residences in the city" and a district of "the most miserable huts, tenanted by some of the most wretched inhabitants," portions of the city's African-American population in particular. The northern slope was the central city's principal black neighborhood of the nineteenth century. Portions of it, especially in the enclosed courts and back streets, often drew attention for their cramped and unsanitary conditions; in the 1890s newspapers continued to refer jocularly to the northern slope in general as "the shady side of the hill," "the back, black side of Beacon Hill."[28]

Residences shared Beacon Hill chiefly with institutional buildings and the open space of the Common, Pemberton Square, and the cemeteries. The Sanborn fire insurance maps of central Boston through the end of the nineteenth century document the entire absence of industry and of most forms of business from the upper hill despite its proximity (in horizontal space) to the intensively occupied city core. Among the establishments that rose the highest up the gentle grades of the southeastern slope (the side closest to downtown), hotels, museums, and theaters predominated, all of them businesses that would have been less inconvenienced than most by an elevated site. A high-service reservoir was installed on the hill's summit next to the State House as part of the city water system in the late 1840s. Other public institutions on Beacon Hill included the State House, the Boston Athenaeum, numerous churches, and, from the 1870s onward, Boston University—the only institution of higher learning then located in the central city.[29]

Even some public institutions would have found a site on the hilltop inconveniently difficult of access. In the early 1870s, repeated efforts were made to have the site of the reservoir on the summit of Beacon Hill (about to become redundant, thanks to a larger high-service reservoir on the mainland) ceded for the site of a new Suffolk County courthouse. Lawyers who would have had to visit the courthouse several times a day objected that the climbing would "wear out" those who had to do business there regularly, that it would be feasible only "if mankind had wings."[30] Not until the following decade was a new courthouse built, midway up the slope of the hill in Pemberton Square.

Nineteenth-century Beacon Hill thus displayed a marked vertical-fringe character, as did even the lowest of the core's three principal hills. The slope of Copp's Hill was home to one large industry whose character as a nuisance, like that of the Beacon Hill ropewalks, helps account for its otherwise unlikely location on high ground. The city gasworks was established in the 1820s on a site stretching from the water's edge up the hill's slope. As it grew, complaints about danger and air pollution, voiced when the gasworks was first established, became more frequent, and it not only reflected but reinforced the undesirable character of the area for upper-class residence. So did the cemetery on the hill's summit, dating back to the mid-seventeenth century, which like the city itself had begun to feel the pressures of overcrowding. By the 1850s the "noxious effluvia from the burial-ground," a Boston physician recorded, were "a frequent source of annoyance, if not of sickness, to the block of houses immediately adjoining," which were also exposed to surface drainage that was both unsanitary and offensive. In 1873, during an outbreak of smallpox, the city selected a site on top of Copp's Hill for an isolation hospital for victims after angry neighbors destroyed a building elsewhere that had been previously chosen for the purpose.[31] The Sanborn insurance maps through the end of the century depict no businesses on the hill's slopes or summit apart from the gasworks. The predominantly Irish residents by the 1870s had won the area the name "Connemora Hill."[32]

The Demolition of Fort Hill

The scanty early residential occupation of central Boston's third principal hill took in the two extremes of social status. Fort Hill possessed both lower-class residents and mansions. The latter began to multiply after the city opened lands on the hill for development in the early nineteenth century, and an elegant pedestrian mall (later converted, with the space it enclosed, to a public park under the name Washington Square) was built on the summit.[33] As late as the mid-1840s, part of the hill remained an address of status and prestige. By 1850, however, its name had become a synonym for poverty and lawlessness. It was one of the chief areas the influx of post-famine Irish refugees settled in the mid- and late 1840s. The historian Oscar Handlin's census-based map of Boston's Irish-born population in 1850 marks out Fort Hill as its single densest area of settlement, just as his map of the African-American population in the same year shows its continued heavy concentration on the north side of Beacon Hill. By the time of the Civil War, murders, frequent small riots and affrays, and numerous instances of petty crime, plus the drunkenness and disorder associated with a reported hundred-plus saloons, made "Fort Hill rowdyism" a byword and consolidated the Hill's reputation as one of the worst slums in Boston.[34]

Grades like those of Fort Hill, as a Boston newspaper remarked in 1865, were "too steep for any purpose but residences." Indeed there were no factories or warehouses on its upper slopes higher than thirty feet above sea level, despite its prime location near the docks and downtown, and even though "by the end of the Civil War, warehouses had completely encircled the Hill." The open space of Washington Square occupied the summit, and institutional buildings were not absent: a city schoolhouse, some charitable homes, and, during the cholera epidemic of 1849, an emergency hospital that the Board of Health had established in an old city gun house on the hilltop.[35]

The devaluation of some fifteen to twenty acres of prime real estate by the character of its terrain, and its seemingly entrenched occupation by a marginal population, soon suggested to some observers that razing the hill might erase a notorious slum and pay for the work by making a large tract of land available for the needs of business. The first serious public suggestion to this effect appeared in the summer of 1852. A large fire on Fort Hill in mid-July caused extensive damage. Within a few days an anonymous writer proposed in a city newspaper not to replace the buildings the blaze had destroyed and to treat the conflagration instead as an opportunity to transform the whole district. In 1854 the legislature chartered a Fort Hill Corporation to undertake the work on a private basis. Lacking the essential power of eminent domain, however, it quickly stalled and made no impression on the hill or its patterns of occupance.[36]

The legislature revived the project in early 1865 with the passage of a bill authorizing the city to undertake the work, to condemn property as needed, and to assess betterment taxes on landowners. Newspaper comment treated the act as a measure equally of slum clearance and of economic development. It would eradicate a place of "abject and squalid poverty," of "filth, which would do honor to the dirtiest portion of London or Liverpool," "this mountain of sin." Observers had little doubt that the area's appalling condition and uselessness for business purposes were things for which "the physical features of the locality are solely responsible." "The only apparent remedy for this evil," the *Boston Post* concluded, "consists in *razing* Fort Hill with its superstructure of untidy dwellings, and substituting exclusively in their stead a class of wholesale stores and warehouses adapted to the growing wants of commerce in that neighborhood."[37]

The city did not launch the first stage of the Fort Hill removal until the late summer of 1866, after the failure of several legal challenges to the project. This step involved the reduction to local grade level of a single thoroughfare, Oliver Street, traversing the hill's summit. September and October saw the pavement torn up and the buildings on both sides of the street vacated and razed. These months also saw the first of what would be a number of casualties in the hill's removal. A fifty-year-old widow, Mary Quinn, recently evicted from her Fort Hill tenement, was scavenging for firewood in a half-demolished building she had formerly lived in when a wall fell and killed her. "During the demolishing of the houses in the vicinity," the *Boston Post* reported, "large numbers of women and children have been engaged in piling up what wood and rubbish they could find." It was one symptom of the widespread distress that the project caused to those evicted, who "besieged the Committee [in charge of the project] with petitions to delay the work till the winter months had passed." *The Pilot,* the weekly organ of the Boston Catholic Archdiocese, several times deplored the removal of the hill as a measure to benefit business at the expense of the mostly Irish poor who lived there, turning the latter out of their homes at a time when housing was already costly and difficult to find. Digging had not progressed far beyond the excavation of the deep Oliver Street trench before it became apparent that the hill was mostly composed not of the high-grade gravel that had been expected but of "a hard earth" that was both more trouble to extract and of much less value for other use. Work soon came to a halt from the difficulties both of removing and of disposing of it, and laborers threw footbridges across the trench to reunite the twin summits.[38]

In 1869 another shorefront expansion project, the creation of Atlantic Avenue, afforded a new place for the debris from the hill and the digging resumed. The city cleared the previously undisturbed side streets on the slopes of their inhabitants and buildings. The use of steam shovels accelerated the work, as did the construction of an inclined-plane railroad to carry the dirt down to the vicinity of the Atlantic Avenue improvement. By May of 1870, a newspaper could report that "Fort Hill is being rapidly reduced and will soon be down to a respectable and profitable level." Although the great Boston fire of 1872 and then the national financial panic of the next year delayed its redevelopment, the flattened Fort Hill area over the course of a decade turned into precisely the commercial-warehouse-industrial district that the advocates of removal had envisioned.[39]

Hills and Land Use in Three Outer Wards

Further insight into the use of the hills comes from an examination of land use on the highest ground in the city's most urbanized outlying wards at the close of the nineteenth century: Charlestown (an independent city until annexed to Boston in 1873), South Boston, and East Boston. Each ward possessed one area above one hundred feet in elevation, as high as or higher than the top of Beacon Hill. Here, too, there was a distinct pattern to the use of high land, marked by the near-absence of business and the predominance of residences, institutions, and open space. There was no such uniform tendency as appears today for the affluent to claim the highest ground, and, as in central Boston, projects to lower the hills to facilitate more intensive use of the ground that they occupied were numerous. One modern writer has described late-nineteenth-century Charlestown as increasingly divided between "the natives entrenched on the heights" and "the Irish on the slopes."[40] Monument Square on the summit of Breed's Hill was indeed an elite (mostly Yankee) residential area during the latter part of the nineteenth century. As every student of Revolutionary history knows, however, Bunker Hill is higher than Breed's Hill, and the top of Bunker Hill, at more than a hundred feet above sea level much the highest land in Charlestown, was the site in that period not of the mansions of the well-to-do but rather of a collection of classic urban fringe uses.

Bunker Hill's earliest important occupants in the 1820s had been the Charlestown Almshouse and a Catholic burying ground. Two grading projects to make access easier and promote development lowered the hill's summit by some fourteen feet in the 1840s. As late as 1871, Charlestowners were still discussing a much more drastic further reduction, in tandem with a land-making project on the Mystic River flats, that would have left Bunker Hill lower than Breed's Hill. A city commission assigned to study the project reported favorably, arguing that the reduction would increase the land's value and open it to more desirable uses than currently occupied it. The Almshouse had moved at midcentury to a new site on the Mystic's banks, but a cluster of Catholic institutions—a church (St. Francis de Sales), a convent, and a school—developed adjacent to the cemetery. The Charlestown Heights city park, running from the summit down a slope that was too steep for ordinary development, opened in the 1890s. Together, these institutional uses covered approximately half of Bunker Hill's crest at the turn of the century.[41] The rest of it, far from being or having ever been an affluent Yankee enclave, was home to a heavily immigrant (mostly Irish) and working-class population. "Day laborer" was the single most common occupation recorded among the hilltop's sixty-four households in the 1900 Census. The area's health status matched its socioeconomic character. Of the six sanitary districts into which late-nineteenth-century Charlestown was divided, the one centered on the summit of Bunker Hill, with the highest average altitude, had the highest average annual death rate.[42]

The original ridge forming the highest ground in South Boston has gone by several, partly interchangeable names: Dorchester Heights, Mount Washington, and Telegraph Hill. Its peak as late as 1850 extended 130 feet above sea level. Extensive lowering and grading operations during the 1850s, both to increase accessibility and to obtain material for landfill, left it little more than a hundred feet in elevation, about the same height as Beacon or Bunker Hill. These operations eradicated the relics, undisturbed until then, of fortifications and batteries on the Heights that had played an important role in the Revolutionary siege of Boston and had been refurbished for defense in the War of 1812.

A resort hotel on the hill's upper slopes, the Mount Washington House, had failed to prosper, and in 1839 it became the property and new home of the Perkins Institution for the Blind (previously housed on Fort Hill). A visiting Charles Dickens noted the sad irony that the inmates had no way of enjoying the site's magnificent view—which indeed suggests that that advantage, often cited as a reason for hilltop or hillside sites for hospitals and public institutions, really had very little to do with the choice of location. As Perkins's annual reports made clear, land so elevated was richer in drawbacks than in advantages. The site did offer "room and air," but its chief benefit was that it was the only kind of land that the institution could afford in close proximity to the resources of a large city. The grading and lowering of the streets to lessen the obstacles to traffic in the 1850s "left the building at an undue and very inconvenient elevation, and accessible only by a long flight of steps which were unsightly to look upon, toilsome to ascend, and badly exposed to wind and rain." The piped city water for many years after the completion of the Cochituate system would not rise higher than Perkins's basement and often failed even to reach that far. In general, the way such establishments were sited led to such problems: "Location and structure should be made with a view to the happiness and the advantages of the special class of unfortunates for whom it is erected. But alas! these are often overlooked; and the establishment is located to secure some gift, or to please some neighbor; and it is set upon a hill." It is not surprising that the young reformer Charles Sumner, a close friend of the Perkins's director, looked forward in 1845 to an enlightened future when "hospitals for the sick, and other retreats for the unfortunate children of the world, for all who suffer in any way, in mind, body or estate, shall nestle in every valley."[43]

Another large charitable facility, the Carney Hospital, a Catholic institution offering care to indigent patients, opened in the 1860s on the hilltop, where it had been preceded by a large reservoir surrounded by open city land constructed as part of the Boston's municipal water system. By 1900 the public open space of Thomas Park (including the now disused reservoir) and the Carney's buildings and grounds occupied well over half of the summit of Telegraph Hill (the Perkins Institution stood just below the hundred-foot line). The only large business establishment on the upper slopes was the plant of the Suffolk Brewing Company, an apparent exception that proves the general rule: nineteenth-century breweries found hills advantageous because they could easily excavate cellars to store their product at cool temperatures during the summer. On Telegraph, as on Beacon Hill, the residents included the affluent and upper class, mostly on the circular avenue surrounding Thomas Park, and the working class, who predominated on the side streets and in the hospital's wards. The summit's extensive open space was not an unmixed blessing. A South Boston newspaper reported in 1892 that the disorderly behavior that it attracted was one of the factors making the Thomas Park area "undesirable for residence," despite the "fine view" and the pure, cool air.[44]

The highest land originally found on Noddle's Island, the heart of the "island ward" of East Boston, stood on Eagle Hill at nearly a hundred feet above sea level. Grading projects lowered it considerably, however, ending with the construction in 1850 of a reservoir for the new Boston water system that occupied the hill's entire summit. The only land elsewhere in East Boston that exceeded a hundred feet in elevation was the large drumlin called Orient Heights on Breed's Island, which landfill eventually attached to Noddle's Island. The summit of Orient Heights, with more than twenty acres lying above the hundred-foot contour, had even by 1900 failed to attract more than the scantiest population, although it had been subdivided for sale years earlier by the Boston Land Company. Hardly more than a dozen households, none of them upper class, lived on the summit, scattered amid scores of still-empty lots.[45]

The Hills and the City

While Boston grew more crowded, its upland area shrank in both absolute and relative terms. Some of the slopes and summits were lowered or removed altogether while land-making projects (many using earth taken from the hills themselves) expanded the area of low and flat land. The hills that remained acquired more residents and uses over time than they had had before, but they remained a distinct zone within Boston throughout the nineteenth century and marginal to the main thrust of intensive urban development. In 1891 the landscape planner Charles Eliot observed that in the Boston area, "Population, which has everywhere avoided the heights, has, like the waters, settled in the valleys."[46] Business, except in a few specialized forms, did not appear above the hundred-foot contour or even close to it. That same elevation appears to have been a limit to affluent residential development on any substantial scale, as it was in central Boston, although it was not an absolute limit to working-class residents. Much more acreage than one would expect near the center of a large and crowded city was left open or occupied by uses characteristic of the urban fringe, such as public institutions, cemeteries, and parks.

The geographers Michael P. Conzen and George K. Lewis noted in 1976 how the hills "were avoided by early settlers because of their very steep slopes," leaving them open for occupation in the nineteenth century by such uses as hospitals, parks, and reservoirs demanded by a growing urban population on the flat land below. In 1930 the Harvard geographer Derwent Whittlesey likewise observed that high lands in the greater Boston area were favored locations for certain classes of use, for "reservoirs, monuments, schools and hospitals," for "parks and reservations." He noted also that the hills of the longer-settled "inner core" showed a pattern of occupance somewhat different from those of the later-developed "outer ring." "High-grade residential sections on high or hilly sites" characterized the high ground of the outer ring. That of the inner core had a much more heterogeneous housing stock, combining the extremes of wealth and poverty, a mix, in Whittlesey's words, of "slums and mansions."[47]

The one pattern was a legacy of the nineteenth century and earlier, when difficulty of access made hills unattractive for residence to all but the very well-off and the very poor. The other was a no less characteristic product of the trolley- and automobile-aided development of the twentieth century, when access had virtually ceased to be a problem and the amenities of hills had become their salient feature. Today the most striking differences from the earlier pattern lie in the more intensive use of high ground, the absence of large low-income populations except where they represent a legacy of past development, and the comparative indifference of business to elevations and slopes that once represented serious barriers. Relics of the earlier pattern, though, are still strong and not only in the central core of Boston. Public institutions, cemeteries, and parks appropriated many of the hilltops in the greater metropolitan area in the nineteenth century, forestalling subsequent residential development or other forms of use that transportation changes had made newly feasible, and leaving them even today occupied by such typical fringe uses of the earlier period.

Two Other Fringes: Waterfronts and Islands

As well as its horizontal and vertical fringes, pre-twentieth-century Boston possessed two other zones characterized by the same kinds of occupance and owing them to the same general considerations. Waterfront lands are peripheral within a city, being on the margin of the settled area. The sole exception in a port city, like Boston, is the land fronting on navigable waters, particularly the deep-water main harbor, and central and convenient to the activities concentrated there. Islands are the most remote of all parts of a city in that land-based forms of transportation cannot reach them without a bridge, a causeway, or landfill. One could expand the model of the horizontal and vertical fringes, then, to embrace waterfronts and islands, whose premodern residents should have been a mix of the rich and the poor and among whose other uses business should have been rare and open space, public institutions, and noxious trades should have predominated.

The shoreline of nineteenth-century central Boston along the Charles River, the inner harbor, and the Fort Point channel indeed housed many of the same urban fringe uses as the hills did. The eighteenth-century ropewalks found havens for a time both on the hills and on the shore. Large public institutions sited along the land's edge included the state prison in Charlestown, the Massachusetts General Hospital, the Boston City Hospital, and, in the early years of the nineteenth century, the city almshouse after its removal from Beacon Hill, first on the Charles River in the West End and then on the hilly waterfront city lands in South Boston. Waterside land, however agreeable in some ways, was not regarded as necessarily healthy in nineteenth-century Boston. The low-lying site of the City Hospital, near the miasmatic waters of the Roxbury Canal, struck observers as particularly unhealthy. In the years after the Civil War, when it was proposed that the city build an asylum for its insane poor on an elevated waterfront site in Winthrop, critics of the idea thought that such a location would be neither healthy nor agreeable for the inmates.[48]

The waterfront, like the hills, was the home of both notably rich and notably poor Bostonians, among whom the latter again predominated. The other important black neighborhood besides Beacon Hill in antebellum Boston was the waterfront North End around Ann Street. Charles Eliot in the late nineteenth century spoke of the zone in the city where land met water as containing "the cheapest lands near Boston" and as the place to which "the housing of the lowest poor and the nastiest trades" naturally gravitated. Urban planner Karl Haglund's *Inventing the Charles River* provides ample detail bearing out the characterization.[49] At the same time the city's most prestigious postbellum addresses were on the waterfront land of the newly developed Back Bay, "the water side of Beacon Street." Even that location had notable disadvantages to offset its agreeable aspects—above all, the stench of the tidal flats in summer (albeit a time when most wealthy residents lived out of town).

The islands of nineteenth-century Boston harbor, accessible only by boat, were much more spatially marginal even than the horizontal fringe, the shoreline, or the hills, and they displayed the purest collection of distinctive fringe occupants and land uses.[50] Before the middle of the nineteenth century, they were largely devoted to such periurban uses as livestock grazing, farming, and the quarrying of gravel and stone—activities that served the city but used land far more extensively than they could have used any ordinary city land. The extensive tracts of open space, a rarity in a city, attracted large military encampments and training exercises during the Civil War to Long, Lovell's, Gallop's, and George's islands. Wealthy Bostonians converted several of the islands, in whole or part, into summer or year-round country estates analogous to those on the horizontal fringe and on the eighteenth- and early-nineteenth-century Trimountain.

During the colonial era the internment of Indians on Deer Island during King Philip's War in the 1670s and the establishment of a quarantine hospital on Rainsford Island in the early eighteenth century prefigured another urban-fringe use for the islands. Toward the mid-nineteenth century these remote and sparsely populated bits of land became favored sites for public institutions, especially those whose marginalized or stigmatized occupants made them unwelcome elsewhere as neighbors. State and city poorhouses found refuge on Deer, Rainsford, and Long islands; quarantine stations and isolation hospitals on Deer, Gallop's, and Rainsford; and the Suffolk County prison on Deer Island. The islands' isolation offered the added advantage of making escape more difficult than it would have been on the mainland. Advocates for the better treatment of the residents of these institutions sometimes bitterly criticized a harbor-island location as inhumanely bleak and isolated.[51]

Other unpopular public and private urban activities sought or were pushed to the islands on the same principle: gunnery testing, refuse and sewage disposal, and the rendering of horse carcasses. Some also became fringe zones of crime and illicit amusements such as prizefighting. A few of the islands were newly developed in this era for opulent summer residences and resort hotels. Here, too, the striking juxtaposition of wealth with poverty so typical of the premodern city fringe appeared, for the chief population of this zone in 1900 consisted of the residents of the public institutions, chiefly the Almshouse and Hospital on Long Island and the House of Correction on Deer Island.[52]

An enduring stereotype contrasts the city, especially the large and densely populated city, with "nature": an entirely artificial environment with one unaltered by human actions. There is no city, though, that does not exist and operate on and within a natural setting, and the crowding together of human activities in the city may even increase the significance of any natural feature that helps or hinders them. One cannot understand the pattern of location of many uses of land in Boston, both past and present, without taking features of terrain—hills, shorelines, islands—into account. At the same time one can by no means determine what those features have meant for human activities from their physical properties alone. Changes in the ways of life with which the hills, shorelines, and islands interact have altered the roles that they play. The urban development that terrain has helped to form, moreover, has reshaped that terrain in its turn. Nineteenth-century Bostonians lowered hills, remade shorelines, and joined islands to the mainland to make them more useful in an era that put a premium on easy accessibility. Zones whose location or physiography made them difficult of access represented fringe areas of cheap land that had a striking tendency to house Boston's poor and very poor, its public institutions, and very little of its business. The same zones, or what remains of them, are regarded today as more than ordinarily desirable for residence and for many business purposes as well because their wealth of environmental amenities more than outweighs their relative (and now minor) difficulty of access. It would be anachronistic to suppose that nineteenth-century Bostonians saw them as we do.

8

Reforestation in Norfolk County, Massachusetts, 1850–1910

David Soll

FOR those familiar with the traditional narrative of agriculture's inexorable decline in heavily industrial nineteenth-century New England, the contents of the *New England Farmer* in the early months of 1883 may come as a bit of a surprise. By the early 1880s many of the region's farmers had embraced innovative practices, generally grouped under the rubric "intensive agriculture," and expressed guarded optimism about the prospects for agriculture in the region. Although the editors opened the 1883 inaugural issue with the oft-repeated lament that soil depletion had led to "a constant tide of emigration," they also observed that "the means are within reach of every intelligent farmer to make his own fields as fertile, as productive and as profitable as are the fields of the pioneer."[1] Speaking to Vermont dairymen two weeks later, A. W. Cheever, a farmer from Norfolk County, Massachusetts, southwest of Boston, noted: "The days of soil robbery in New England are nearly passed. Honest agriculture is beginning to pay a small profit."[2]

A series of articles by local farmers detailing their newfangled agricultural practices betokened the promise of New England agriculture. "Ensilage at Florence Farm," which described the attempt by a pair of Norfolk County dairymen to determine the effects on milk yield of giving their cows silage (fermented feed that farmers produced by storing green crops in silos), reflected the experimental bent of the region's farmers.[3] The matter-of-fact heading of another piece, "Another Silo," testified to the widespread adoption of ensilage by the region's husbandmen.[4] The contributions by agricultural improvers generally focused on methods to increase milk yields by modifying feeding techniques for milch (dairy) cows.

The editors of the *New England Farmer* highlighted another unmistakable trend: the reforestation of metropolitan Boston. Touring western Norfolk County in early 1883, they found that "thousands upon thousands of acres of what were once considered fair farming lands are now covered by heavy growths of white pine, and other forest trees."[5] In the early 1880s few observers drew a connection between changes in agricultural practices, the most prominent of which was the tremendous expansion in milk production, and incipient reforestation. Even today, with the benefit of more than a hundred years' hindsight, historians have generally failed to appreciate the ways in which the new agricultural practices associated with dairy farming altered the appearance and landscape composition of the cities and towns near Boston. This chapter remedies the oversight by exploring the connection between the boom in milk production in Norfolk County and the reforestation of the county's farms that occurred in the second half of the nineteenth century and the early years of the twentieth century.

Scholars have long recognized that, beginning around 1850 or so, Northeastern farmers adopted new crops to compete with their Midwestern counterparts. With the advantage of flat, seemingly endless expanses of fertile land, Midwesterners produced grains at prices Northeastern farmers could not match. Forced to sell their farms or to innovate, farmers along the eastern seaboard shifted to producing perishable foods (such as fruits, vegetables, and milk) to sustain their farms and their families. By 1899 agricultural journalists described this shift in the dispassionate tone generally reserved for distant events: "When railroads were extended to those new lands, as they soon were, serious competition of the east in grain growing was at an end. Land did not decline even then, for farmers turned their attention to crops in which competition was less scarce and which generally required more hand labor."[6]

Despite the recognition that many Northeastern farmers, especially those near major population centers, turned to production of perishable commodities in the face of Midwestern competition, the dominant historical narrative has emphasized not agricultural innovation, but farm abandonment, as in this account by two forest historians: "As farm mechanization increased, these small farms could no longer be worked profitably in competition with the rich, stone-free farmlands of the Midwest, whose products had been made more accessible to eastern markets by the construction of the Erie Canal and the railroads."[7] According to this interpretation, Midwestern competition simply overwhelmed Northeastern farmers, forcing them to abandon their fields, which over time naturally converted back to woodlands. Scholars have also stressed the decline of sheep raising and the lure of urban living as primary causes behind farm abandonment and the ensuing reforestation of New England.[8]

The farm abandonment thesis holds true for some hill towns in western Massachusetts, where many families left their land for industrial jobs in the cities, but it reveals little about the ways in which most Bay State farmers responded to changing agricultural markets. In 1989 geographer Michael Bell persuasively challenged the dominant narrative of the nineteenth-century downturn in New England agriculture, observing that it grossly overstates the extent of the decline and misrepresents its timing. It was only around the turn of the century, noted Bell, that farming in Massachusetts began to atrophy.[9] Not surprisingly, farm abandonment fails to capture the complexity of developments in the rural areas surrounding Boston. Contemporary descriptions of changes in rural life often drew a distinction between conditions in the agricultural hinterlands and those in areas closer to major urban centers.

An 1887 account in Boston's *Daily Advertiser* sounded a typically wistful note in its depiction of an abandoned farm: "The stillness of a solitude haunts the place . . . what labor the reddened and weary hands once expended here, and yet the sons have not stayed to reap any profit from the father's toil." But, lest city readers get the impression that the entire countryside had become a blighted wasteland, the editors concluded by noting, "We have been referring only to the farming locations remote from the cities and from the large towns. There are farmers who acquire wealth."[10] The residents of Norfolk County were among those suburban farmers who adopted new agriculture techniques, often profiting handsomely in the process.

Contrary to the agricultural abandonment theory, the overall population of Norfolk County increased in the second half of the nineteenth century, and the number of farms held steady or decreased only slightly throughout the county. From 1875 to 1895 milk production in Norfolk County more than doubled, rising from 1.9 million gallons to almost 5.4 million gallons.[11]

Farmers did not achieve this dramatic increase by breeding or purchasing more cows; instead, they adopted new feeding methods that greatly increased the amount of milk each cow could produce. Much of the reforestation that took hold in this period stemmed from changes in dairy cattle feeding practices. Although farmers continued to pasture their cows and feed them hay over the winter, they supplemented these traditional staples with grain and meal—mostly from the Midwest—reducing their reliance on degraded pastureland. Cheap grain allowed farmers to obtain increased milk yields from smaller parcels of land. Pastureland that had formerly supported cattle was no longer needed to maintain milk production and soon began to sprout trees.

As the popularity of grain feeding spread, suburban Boston farmers repeated this cycle many times over, resulting in a significant reduction in pastureland and a concomitant increase in woodland. When demand for wood for use in the construction of boxes and crates spiked in the early 1900s—what historians now refer to as the "boxboard boom"—many farmers elected to harvest the white pine trees that had sprouted in their former pastures. However, the boom represented the last hurrah for extensive logging in the county. As rural areas outside of Boston developed, landowners allowed these cut-over woodlots to mature. They now provide the backdrop for the sylvan towns of suburban Boston.

Before exploring these changes in more detail, it is worth highlighting a crucial component of this story: the importation of grains from the Midwest for use as cattle feed. Intelligent Norfolk County farmers were not content to sit around and lament Midwestern dominance in grain production. Instead, they put the cheap grains of the Midwest to good use by feeding them to their cows. Thus the interregional dynamic between the Midwest and the Northeast went far beyond the simple tale of displacement at the heart of the farm abandonment thesis—the claim that Northeasterners, pummeled by competition from Midwesterners, either migrated west or flocked to cities for industrial jobs, allowing their former farms to revert to woodlands. Instead, in areas like Norfolk County, where dairymen enjoyed excellent access to expanding urban markets, Midwestern grains proved to be the crucial missing ingredient in the recipe for agricultural success.

The role of Midwestern grains in Northeastern landscape change also complicates the city-hinterland dynamic that the environmental historian William Cronon limns so thoughtfully in *Nature's Metropolis*. Cronon focuses on the ways in which Chicago's insatiable appetite for raw materials radically altered the economy and landscapes of rural areas, ultimately transforming much of the Midwest.[12] Boston's growth also changed the face of the surrounding countryside, but in a somewhat more complex fashion. Northeastern cities such as Boston drew heavily on the natural resources of other regions of the country to fuel their economic development. This reliance on the crops and timber of other regions freed Boston's rural suburbs to assume different functions than the prairies and forests of the Midwest. The city's suburbs became a site of second homes for urban elites and produced perishable food items for its growing population. In the case of dairy farmers, their embrace of the resources of another region—Midwestern grains—resulted in significant changes to local landscapes. These changes reflected not simply the dynamic of a hinterland producing for its central city, but also the impact of regional economic exchange, which transformed the Norfolk County landscape in very particular ways.

Massachusetts Agriculture at Midcentury

If competition from Midwesterners in traditional staple products such as corn and beef made specialization in perishable products an obvious path to economic survival for Norfolk County farmers, an important question remained unanswered: How could they secure adequate incomes from fields that suffered from deteriorating fertility? In the first half of the nineteenth century, eastern Massachusetts dairymen adopted butter and beef as their primary market crops and fed their cows grass and hay—itself a major market crop—to ensure a steady supply of both. Most farmers opted to clear new land rather than take the trouble and expense (labor costs were high) of adequately fertilizing their hayfields and pastures. Although such practices may have made short-term economic sense, they led to a disordered landscape chockablock with unsightly and unproductive overgrazed pastures. As pastures wore out, cattlemen often converted them to hayfields, many of which became degraded in turn, convincing their owners to allow them to return to woods.[13]

As Bay State farmers struggled to feed growing populations in the first half of the nineteenth century, Massachusetts began to lose much of its forest cover. George B. Emerson, widely perceived as the foremost expert on the state's forests, lamented this deforestation in his well-known study, *The Trees of Massachusetts*: "A few generations ago, an almost unbroken forest covered the Continent. . . . Now, these old woods are everywhere falling. The axe has made, and is making, wanton and terrible havoc. The cunning foresight of the Yankee seems to desert him when he takes the axe in hand. The new settler clears in a year more acres than he can cultivate in ten, and destroys at a single burning many a winter's fuel, which would best be kept in reserve for his grandchildren."[14] Emerson accurately described the changes in forest cover that had occurred since the American Revolution, but by the time he published these words in 1846, deforestation as a result of farm clearance was nearing an end. As early as 1853, a Middlesex County farmer remarked that a good deal of his former pasturage and tillage was "under a promising growth of young wood, which is the most profitable crop we can raise here; it will grow without our paying laborers nine shillings a day to hoe up weeds."[15]

Most agricultural observers sounded a more pessimistic note, focusing on what they viewed as the inexorable degradation of the state's pastures. Charles Flint, the secretary of the state agricultural board, emphasized the widespread nature of pasture deterioration: "In very many towns the number of acres in pasturage, also, is decreasing, many old pastures having become so poor as to be abandoned to bushes, or converted into woodland." Another participant at the board's annual meeting criticized farmers for failing to ensure the fertility of the pastures on which they based their livelihoods: "Many thousand acres have furnished all the milk, butter, and cheese sold from the farm, without receiving so much as a pennyworth in return."[16]

At midcentury most Bay State landowners struggled to make the transition from an extensive style of agriculture (which was itself actually a fairly new phenomenon) in which they cleared new land rather than renew the fertility of existing plots, to a newer, more intensive mode of agriculture in which their ability to manage their land and animals more efficiently would determine their livelihoods.[17] Farmers recognized the economic opportunity in the increasing urban demand for pure milk (milk produced in urban areas was widely and correctly recognized as impure) but were constrained in their ability to seize this opportunity by the degraded conditions of their farms. To produce consistently high milk yields throughout the year, cattle required abundant grass and high-quality hay and other feeds, but growing such feeds required heavy applications of manure, which could only come from cows that enjoyed rich diets. More visionary farmers recognized that the way out of this conundrum was staring them in the face: They could increase their cows' productivity, and therefore the overall productivity of their farms, by feeding them Midwestern grain.

Changes in Feeding and Agricultural Practices

By the 1850s cattle feeding methods had become a favorite theme of agricultural journals and at state agriculture meetings. Not content to guess about the effects of changes in feed on the quantity and quality of the milk their cows produced, farmers undertook their own experiments to determine optimal feeding practices. At the 1853 meeting of the state board of agriculture, William Porter, who farmed north of Boston, reported that his cows produced an average of three fewer pints a day when he removed shorts—a form of grain—from their diets. Porter experimented with a variety of feeds, including shorts, oil meal, and rye meal, but he concluded that his cows produced the most milk on corn.[18]

By the 1870s a consensus had begun to emerge concerning the benefits of grain feeding. T. Whitaker, one of Norfolk County's most respected farmers, strongly advocated grain feeding. Whitaker resided in Needham, fast emerging as one of the county's major milk-producing towns. His reputation for agricultural innovation earned him a position as an occasional contributor to the *New England Farmer,* where in 1873 he published an article describing some of the changes he had recently undertaken on his farm. In a revealing metaphor that suggested the growing influence of the industrial mind-set among rural types, Whitaker called cows "machines for the manufacture of milk and manure" and fed them accordingly on rye, oats, barley, and fodder corn that he raised himself. He recognized, however, that to maximize his overall productivity, he needed to look beyond the farm. Whitaker observed, "To increase the amount of manure, I keep more animals on the place than I produce feed for; so I levy upon the West for shorts, linseed meal and corn meal." Careful collection and application of manure, which had increased in both quantity and quality with the introduction of Midwestern grains to his herd's diet, allowed Whitaker to reclaim portions of his farm for hay production that previously had yielded "nothing but moss and hardhack."[19]

Whitaker's observations point to the development of a more intensive style of agriculture, one that would result in significant changes in farm management. As he noted, high-quality feed lay at the center of this agricultural transformation. Supplementing locally raised crops with more nutrient-rich grain imports from the Midwest increased the productivity of both dairy cows and the crops they fed on. The cycle was simple: High-quality Midwestern grains led to high-quality manure, which in turn led to higher crop yields. Higher crop yields meant farmers could raise enough food to sustain both their animals and their families without having to rely on degraded pasturelands. As Whitaker suggested, applying the enriched manure to these tired lands often replenished them, resulting in welcome crops of hay. However, farmers often opted to "shrink" their farms, allowing thousands of acres of eroded pasturelands to return to woods.

In retrospect, the benefits of adopting intensive agriculture may seem obvious, but in altering their agricultural practices the Bay State's suburban farmers underwent what amounted to a sea change in their thinking about cattle. Maximizing milk production entailed more than simply putting grain in a cow's trough; it signaled a revolution in farmers' prevailing attitudes toward their most valuable possessions. Speeches given at the Massachusetts State Board of Agriculture meetings attest to the substandard level of husbandry practices. Henry Colman, who as commissioner of the state's surveys of agriculture in the late 1830s and early 1840s traveled throughout the region, observed in 1838 that "the general treatment of the cows in New England would not be an inapt subject of presentment by a grand jury."[20] Fifteen years later, even as some farmers described their experiments with new feeding techniques, Peter Lawson of Dracut, a town near the New Hampshire border, lamented: "It seems to be the farmer's boast that he can show with how little food and shelter he can enable his cow or cows to keep life (if not soul) and body together, and yet produce milk."[21] Members of the agricultural press routinely criticized farmers for such practices well into the 1860s. The period from 1850 to 1870 thus represents a transitional phase in the animal husbandry of eastern Massachusetts in which some farmers, especially those with ready access to urban milk markets, improved their practices, while others hewed to the old ways.

Improvements in the care of cattle were often closely linked to changes in feeding practice. The trend toward feeding more grains and fodder crops meant that cows now enjoyed ample nourishment even after the grazing season ended. As a result, farmers found that their herds produced high-quality milk for most of the year, a stark contrast to the traditional annual cycle, in which milk production dropped off significantly once the grass began to wither. Not surprisingly, farmers showed more interest in the welfare of their herds once they revealed their capacity to generate income through the winter months. Writing in 1914, Frank Smith, a longtime resident and unofficial scribe of the Norfolk County town of Dover, located approximately twenty miles southwest of Boston, linked the change in farmers' milking practices to the availability of high-quality winter feed: "Our cows were not expected to give milk during the winter months, but were turned out to browse in the woods . . . the ensilage of forage makes it possible to furnish succulent food to animals at all seasons of the year."[22]

George Fletcher, a dairy farmer from Milton, just to the southeast of Boston, typified this shift to a more aggressive milking regimen. In 1885 he reported that six of his ten cows were in milk for more than 300 days, with the average length of milking 315 days. The lure of high profits in the winter milk market—when supplies fell but demand remained steady—convinced Fletcher to extend his cows' productive season. Whereas farmers like Fletcher had previously supplied cows with enough feed—generally hay—to make it through the winter without starving, they now added other grains and fodder to the mix to ensure continued high levels of milk production long after the year's pastures had been exhausted. Fletcher sustained yields by feeding his herd a mixture of the hay he harvested and brewers' grain and cottonseed meal that he purchased. Without the addition of these purchased feeds, Fletcher could not have extended milk production into the winter months.[23]

Markets and Railroads

The crucial factor underlying the growth of dairy production in suburban Boston was the ever-increasing thirst of city dwellers for milk. This growing market reflected both urban population increases and rising per capita consumption. From 1850 to 1900 the city's population boomed, increasing from just over 135,000 to 560,000; adding the populations of inner-ring suburbs brings the 1900 total to approximately 900,000. During this period milk became a dietary staple of many Bostonians. In the mid-nineteenth century the market remained quite small because "few people drank milk, children drank most of it, and most of it was not safely drinkable."[24] Cows fed on the remnants of the urban food supply or brewery slops supplied most urban dwellers. Many health experts attributed the high death rates among urban babies to their consumption of unsanitary milk. In response to such problems, in 1856, Massachusetts passed the country's first ordinance regulating the quality of milk. A combination of regulations, claims by reformers touting the health benefits of dairy products, and the widespread perception that the newly abundant supply from rural areas was far superior in quality to city milk, led Bostonians of all ages and classes to make milk consumption a daily habit. By the 1890s the government estimated per capita consumption in the Boston area at almost a pint a day.[25]

Milk's relative affordability also played a key role in boosting urban consumption. An 1883 *New England Farmer* editorial, "The Low Price of Milk," lamented that the market "is constantly crowded with a flow of milk."[26] Fifteen years later, plentiful supplies continued to depress prices. A federal government report on the region's dairy market identified a major cause: More dairy farmers were producing milk on a year-round basis. In the mid-1870s families generally paid eight cents more for milk in winter than they did in summer. This differential reflected the relative paucity of winter supplies (cows fed primarily on hay struggled to produce any milk at all in the winter) in a period before the vast majority of farmers had made grain a core component of their cows' diets. By 1891 the spread had been reduced to four cents, and by 1896 the difference was down to two cents.[27]

The final core component of the fluid milk boom—along with Midwestern grains and growing urban markets—was the development of an extremely dense network of railroad lines in the Boston area. In a report for the U.S. Department of Agriculture, George Whitaker, who served on the Massachusetts Dairy Bureau for many years, observed, "Boston seems to have been the pioneer city of the United States in the transportation of milk by railroad."[28] Agricultural data for individual towns in Norfolk County strongly suggest that farmers eagerly embraced milk production only when assured of a reliable means of transporting the highly perishable commodity. Statistics from the Norfolk County towns of Dedham and Medfield reveal a similar pattern: Output increased substantially following introduction of rail service. In 1845, when most towns in the county still lacked rail connections to Boston, only a handful of Norfolk County towns produced fluid milk for sale, but farmers in all of the county's cities and towns sent butter and cheese—slightly less perishable items that did not need to be shipped within a day of being produced—to market. In Dedham, which established rail connections with Boston in 1834, milk accounted for 25 percent of the cash value of all crops by 1845, a pattern that the county's outlying towns would replicate when they established rail connections to Boston.[29]

By 1870 rails linked all towns in the county to Boston. In their annual report for 1870 the selectmen in Medfield observed, "We now have railroad facilities, second only to comparatively few towns in the Commonwealth."[30] In 1865, four years after the extension of rail lines to Medfield, the town's farmers sold only twenty-five hundred gallons of milk; ten years later farmers had turned their attention to the Boston market and produced more than fifty-six thousand gallons.[31] By the 1890s the milk production and delivery system in Boston had become a decidedly corporate and efficient business. Five wholesalers, known as contractors, dominated the market and contracted with dairymen to transport their milk to the city; one man—George Whiting—effectively controlled three of the five firms. The contractors rented railroad cars, built loading platforms at railroad stations, erected ice houses in key rural locations to ensure that milk did not spoil before a train arrived, and boasted large warehouses and offices near the railroad terminus in Boston. The firms purchased all the milk brought to market, but, much to the chagrin of farmers, paid lower prices for surplus supplies that had to be converted to butter or cheese to prevent spoilage.[32]

A dairy farmer marked his day not only by his milking regimen, but in many cases also by the trip to the railroad depot to deliver his milk. Contractors assigned each dairyman an identification number printed on a sticky piece of paper that the farmer affixed to the wooden stopper on his milk cans. They then simply tallied the number of cans a farm delivered to the depot to compute the payment owed. In some cases groups of farmers contracted with a neighbor—usually a former dairyman who had decided to become a full-time transporter—to convey their milk to the rail station, but most lived close enough to a station to avoid this extra expense. In a subsequent report on the milk business, George Whitaker captured the nonchalance of a group of Bay State dairy farmers whose livelihoods depended on the railroad, but whose movements were not strictly bound by a train's arrival or departure time: "The farmers have loaded their empty cans before the train arrived and some begin to drive away before the train leaves, reading their morning papers as their horses jog slowly along."[33]

Big Changes Rooted in Small Changes

The image of vast amounts of grain shipped from the Midwest to the Northeast lacks the romantic appeal of small-scale dairying for local markets, but imported grain underpinned the boom in milk production. State census figures clearly reflect the decline in the production of grain crops that could be grown much less expensively in the Midwest. Massachusetts had long procured a large percentage of its wheat from other states, but in 1845 it still produced 47,986 bushels; by 1885, Bay State farmers harvested just over 7,000 bushels.[34] We must turn to federal data to appreciate the seismic shift in grain production to the Midwest. In 1849 central states produced 36 percent of the nation's corn crop; by 1879 these states accounted for almost 69 percent of the nation's corn. By 1879, Massachusetts produced less than one bushel of corn per state resident, while Midwestern states produced fifty-five bushels per person.[35]

In an address entitled "Harvesting Corn," delivered at the annual state board of agriculture meeting in 1881, Professor Levi Stockbridge offered a clear explanation of the economic rationale behind the shift from sowing cereals to dairying and market gardening: "We could not grow this crop with profit, either because of the cheaply produced and transported western product, or because of the demand of our markets for supplies, in producing which the West could not compete: therefore it was a stroke of good policy to produce for this market demand, and purchase the corn needed for home consumption."[36] The reliance of the Bay State on grains from outside the region manifested itself both directly and indirectly. The increasing number of grain dealers dotting the streets of Boston and surrounding towns represented perhaps the most obvious sign of this shift. In 1867 grain dealer listings ran to four pages in the *Massachusetts Register and Business Directory*; by 1878 they occupied almost six pages. As more farmers turned to imported grain, dealers began to open up shops in smaller towns. By 1899 most towns in Norfolk County had a local grain dealer. Farmers' diaries, like that of George Wadsworth of Franklin, record periodic trips to the grain dealers, especially in cold-weather months.[37]

An increase in overall farm productivity provided a more subtle indication of the widespread feeding of grain crops to dairy cattle. Although the situation of every farmer differed depending on the land he worked, eastern Massachusetts agriculture reached a tipping point in the 1870s. A substantial number of farmers concluded that profitable agriculture required an infusion of fertility from outside the farm in the form of grain fed to their cattle. Agricultural papers began to run stories with titles such as "Concentrated Farming," which urged farmers to make good use of manure and commercial fertilizer to increase crop yields, thereby reducing the acreage devoted to tillage.[38] In 1875 state statistician Carroll Wright identified increased crop yields as one of the more significant agricultural trends of recent years: "An examination of the relation of quantities raised to acres tilled convinces one that the farmers of Massachusetts have learned the important lesson of tilling less land and tilling it better, for we find that even in the crops where loss in the aggregate occurs there has been a great improvement in results."[39]

Perhaps because the state census did not track amounts of grain purchased from outside the state, Wright failed to appreciate that higher crop yields reflected more assiduous application of manure, which had become more abundant and higher in quality because of the widespread introduction of Midwestern grains into the diet of dairy cattle. In the words of environmental historian Brian Donahue, grain "effectively replaced fresh land as the source of the critical increment of fertility that kept the system running."[40] The massive grain farms of the Midwest thus facilitated the intensification of agriculture in Massachusetts. As early as the 1850s, Bay State farmers began to retire their deteriorated pastures, content to let them sprout saplings that would mature when they became old men. Of course, grain did not fall like manna from heaven; individual farmers had to incorporate it into their cattle-feeding regimens and decide what products to take to market with the increased yields the grain afforded. In the case of some farmers, we must infer their decision to incorporate grains into their herds' diets from other data; in other cases, farmers left firsthand accounts of their embrace of grain feeding. Federal census data allow us to track the changes in crops that individual farmers planted.

We can draw some conclusions about the adoption of grain feeding by examining data for two farmers—Benjamin Shumway and Caleb Ellis—from the Norfolk County town of Medfield. In 1860, Shumway sold a modest amount of butter and harvested a variety of crops. Although he continued to grow crops, a portion of which he fed to his cows, Shumway shifted his focus to milk production. Between 1860 and 1870 he expanded his herd of cows from two to eleven, abandoned butter production, and became one of Medfield's largest milk producers. Caleb Ellis followed a similar path. By 1870 his herds produced thirty-seven hundred gallons a year, more than any other farmer in town. Whether out of habit, nostalgia, or as a partial income hedge, he continued to produce a small amount of butter—one hundred pounds.

From 1870 to 1880 both men reported significant increases in milk yields, likely tied to their adoption of grain feeding. Without adding a single cow to his herd, Shumway increased his milk production fourfold, from eighteen hundred gallons to seventy-two hundred gallons. Ellis actually trimmed his herd from seven cows to four, but boosted overall production due to a doubling in milk yield per cow. Absent the introduction of feed from outside the farm and a corresponding extension of the milking season, it is virtually impossible to account for such significant increases. Although federal census data do not indicate the amount or value of feed bought, they do reveal that Shumway purchased commercial fertilizer in 1879. Given his willingness to buy fertilizer, he probably considered cattle feed a prudent investment as well.[41]

The data available on Shumway and Ellis allow us to draw reasonable conclusions regarding their likely embrace of grain feeding, but A. W. Cheever's description of his efforts to wring a livelihood from his family farm in the Norfolk County town of Wrentham provides us with concrete evidence of the impact of grain feeding on the reforestation of metropolitan Boston. Cheever's success with the practice led him to advocate it in the pages of the *New England Farmer* when he became the newspaper's agricultural editor in 1884. When he inherited the family farm in 1859, Cheever faced a problem common to many Massachusetts farmers: His land seemed poorly suited for most forms of agriculture. Concluding that his grasses were too sour for sheep and that his wet, cold soil and uneven land would not lend itself to either Indian corn production or orchards, Cheever opted to become a butter farmer. Cheever's farm was not located near railroad lines, so although he produced some milk, he wisely decided to concentrate on butter production, for which swift transport was less crucial.[42]

A model of the ecologically astute businessman, Cheever recognized the physical and fiscal connections that united his operation. He fed his eight cows hay, grain, and, when in season, green crops, keeping them stabled except in summer, when they spent part of the day in pasture. Atypical of most area farmers in that he grew virtually no grain, Cheever still valued and made use of the large amounts of manure that stall-feeding generated. He also observed that by reducing his reliance on pasture, stall-feeding allowed him to trim expenses related to maintaining fences.[43] His practices may have had a strong internal coherence, but Cheever's livelihood depended on forces outside the farm. Relying on stall-feeding meant that his net income reflected both the cost of his inputs as well as the payments he received for butter and the other commodities he produced. Cheever reported to the Norfolk County Agricultural Society's supervisory committee, "As I buy all my grain, I endeavor to do so at such times and in such quantities as will be most advantageous." Cheever's reliance on the market meant that his profits fluctuated significantly from year to year. In 1865 he expected to realize a significant gain over the previous year because grain prices had declined by almost half.[44]

Cheever recognized that by adopting stall-feeding, he set in motion a series of physical changes on his farm, changes he believed would increase his bottom line. He noted of a large parcel of pasture: "It is now gradually coming up to pines, birches and other kinds of forest trees. By the old management it was worthless land, as it would not pay expenses. Now it is slowly, and without any expense, becoming valuable. One lot of three acres on which I rode horse to cultivate a field of potatoes twenty-five years ago, is now covered by a thick growth of white and pitch pines that are about thirty feet high and a foot through at the butt." Approximately twenty-five of the farm's ninety acres consisted of old pasture on its way to resprouting. For Cheever letting his pasture go to woods made economic sense because he could harvest the trees once they attained sufficient size. Hundreds of farmers in Norfolk County and thousands throughout the state came to the same conclusion, and through their individual economic calculations they effected a significant shift in the composition of the landscape in the greater Boston area.[45]

Reforestation and Cutting: Farms Shrink and Boxboard Booms

In the late 1880s and early 1890s, Norfolk County began to change. While the milk boom continued, rising real-estate prices made residential development an attractive alternative to farming, especially in towns located near Boston. By the late 1880s eighteen trains a day ran from Milton's three train stations to Boston, transporting the estimated one-sixth of the town's working population that commuted to jobs in the city. The process worked the other way as well, as affluent Bostonians purchased homes in Milton to serve as summer retreats, a key factor in the near doubling of real-estate values in the town between 1885 and 1895. Farm acreage tumbled as the growing nonagricultural population consumed more land for its residences and businesses.[46]

Residential development may have proceeded more slowly in such towns as Medfield and Franklin, but many well-off Bostonians spent their summers in these outlying communities soaking up the agricultural ambience. By the 1890s regular summer visitors to communities in southern and western Norfolk County would have noticed a number of significant changes, most notably the silos that had sprouted up on many farms. Ensilage had two main benefits relative to hay: It was cheaper to produce, and it increased milk yields. From 1895 to 1905 milk yields increased an impressive 22 percent, owing in large part to the widespread adoption of ensilage by the county's farmers. Bostonians accustomed to the sight of grazing herds might have wondered where all the cows had gone. Farmers had moved them into the barn to feed on silage and grain rather than keep them on pasturelands that had often dried out by late summer.[47] By 1898 the trend toward feeding grain and silage had gained such momentum that dairy farmers routinely asked, "Shall we grain our cows the year 'round, or without grain feed while at pasture?"[48]

As grain feeding gradually became the norm among Bay State farmers from 1870 to 1900, observers began to note the changing appearance of local landscapes. Standing on the summit of the Blue Hills in 1884, Milton town historian Albert Teele admired the "immense tracts of waving forests" as he looked south toward the towns of Canton and Sharon.[49] Farmers on the ground generally eschewed Teele's romantic perspective, preferring to view their woodlots as sources of additional income. The year before Teele surveyed the county from his perch atop the Blue Hills, the *New England Farmer* remarked on the vigorous reforestation occurring in Bellingham, in southern Norfolk County along the Rhode Island border: "A great many acres are now covered by heavy growths of pine lumber, which finds a ready market in the neighboring villages of Milford, Medway and Franklin, Mass., and in Woonsocket, R.I."[50]

Although small-scale forestry provided a livelihood for some county residents, few farmers realized significant income from their woodlots in the 1880s. Some cut cordwood from their lands in the winter to earn extra cash, but the woodlots of most county farmers boasted very young growth not suitable for cutting. This would change, however, in the twentieth century. By 1905 census takers classified almost a third of Norfolk County farms as woodlots, a designation signifying that their owners derived at least half of their income from the sale of wood products. County farmers sold more than $250,000 of wood products, an increase of over 50 percent from 1885. However, it is worth emphasizing that even those farmers whose woodlots provided them with significant cash income generally continued to practice dairy farming. From 1895 to 1905 milk production in the county grew by almost a third, due primarily to increases in milk yield per cow as more farmers adopted the feeding of grain to cattle.[51] Despite these increases in overall production, Norfolk and surrounding counties were becoming bit players in the Boston milk market. By the 1890s the extension of railroad lines allowed the Boston "milkshed" to expand into Vermont and New Hampshire; suburban towns now accounted for only a small percentage of the total volume shipped into the city. One farmer observed that Bay State dairy farmers struggled to earn a living due to competition from "milk that is produced and shipped from one hundred to nearly three hundred miles away from our markets."[52]

Farmers who could spare a moment between tending their cows and cutting their trees recognized the tremendous impact that intensive farming methods had on their lands. The 1905 state census provides a clear picture of the landscape changes that had occurred in Norfolk County and throughout the state. Since 1865, the percentage of land in Norfolk County forests had been steadily on the rise, increasing from 35 percent to 46 percent in 1895. But the decision of census enumerators to designate some land "unclassified" masked the significant reforestation that was well under way by the mid-1880s. As saplings became trees around the turn of the century, enumerators reduced the amount of land they considered unclassified, deeming it woodlands instead. As a result, forest volume in the county skyrocketed, increasing from roughly sixty-eight thousand acres in 1895 to more than ninety thousand acres in 1905.[53]

From the perspective of turn-of-the-century Norfolk County farmers looking to diversify their income sources, two facts about their woodlands loomed large: They would soon be of marketable size, and they consisted largely of white pine. Although scholars have suggested different dates for the beginning of the boxboard boom, the mass cutting of New England's forests for box and crate construction, all sources point to an acceleration of cutting between 1905 and 1910. New England's timber production in 1909, considered the peak of the boxboard boom, was threefold the amount cut in the region in a typical year in the early 1990s.[54] Box makers relied heavily on white pine to carry the products churned out by the expanding economy. Farmers who had embraced grain feeding in the 1870s, allowing their fields and pastures to reseed themselves in forest of white pine, now prepared to reap the indirect benefits of their innovative agricultural practices.

The boxboard boom offers a textbook example of the conjunction of supply and demand. Just at the time that many stands of white pine in New England were attaining merchantable status (which generally took thirty to thirty-five years of growth), industry began to demand a seemingly inexhaustible supply of boxes. A 1907 Forest Service circular highlighted New England's strong industrial base and its corresponding need for packing materials: "It would be hard to name an article made of cotton, wood, wool, leather, or metal which, to some extent at least, is not manufactured within its borders. To ship the great bulk of these products there is needed each year an enormous number of packing boxes."[55] White pine proved the ideal wood for packing materials. Its newfound abundance in New England made it an obvious candidate, but boxmakers prized it for its combination of unique qualities. Its light weight kept freight charges to a minimum; moreover, it was odorless, strong, and light in color. One forest expert observed that "for the box maker, it is an ideal wood and the New England manufacturer would be hard pressed to find a suitable substitute."[56] Unfortunately, less valuable species such as gray birch and red maple were more likely than white pine to establish themselves on these woodlots, leading to calls from foresters for a deliberate policy of white pine planting.

In the heyday of the boxboard boom, however, farmers did not have to practice sophisticated arboriculture to realize gains from their woodlots; they only needed to prevent their timber from going up in flames. Farmers who protected their woodlands stood to earn substantial amounts of cash when they harvested them, in large part because of the shortage of box material. One forest expert commented that despite the market shortage, "it is probable that through natural extension on abandoned fields more land in New England supports a young growth of white pine today than for many years past."[57] A 1910 report on the Bay State's wood-producing industries clearly conveyed the insatiable demand of industry for New England's white pine. Although most of the wood products made in Massachusetts consisted almost entirely of trees grown outside the state, more than a third of the wood used in box construction contained Massachusetts white pine. Manufacturers used it to make pails, refrigerators, musical instruments, and other products, but almost 90 percent of it ended up in the form of boxes and crates.[58] With no industrial plantations to supply them, New England's box manufacturers had to "depend for their supply of timber mainly upon farmers' woodlots."[59]

Ironically, the tendency of farmers to abandon their pastures only after they had been thoroughly depleted of most of their nutrients proved a boon for them in later years. One forester noted of the land on which most second-growth white pines stood, "After its moderate fertility was exhausted by agriculture it became better suited to the growth of pine than to more exacting species."[60] Second-growth white pine generally developed in "pure" stands, which contained no more than 20 percent of other species. Because most seeds germinated at the same time, farmers enjoyed even-aged stands of white pine. When a stand became large enough to profitably harvest, farmers generally clear-cut it, eager to reap maximum profits. Compared with the increasingly meager returns available from dairying, forestry offered Norfolk County landowners a substantial income relative to the minimal amount of labor they invested in their trees.

Around the turn of the century, the forestry community began to call for a more disciplined approach to woodlot forestry, exhorting farmers to plant particular species and to leave enough trees standing to reseed harvested plots. However, as Allen Chamberlain, secretary of the Massachusetts Forestry Association, recognized, most farmers were content to reap the benefits of nature without lending a helping hand: "It is common enough to see an old pasture, for instance, growing up thickly to white pines. For the most part they grow up as best they may, and at the end of forty years, say, they are cut and sold for cheap box boards."[61]

Henry Bullard, a dairyman whose land straddled the western edge of the county in Medway, embraced this new, more aggressive mode of forestry. Bullard actively managed his woodlots, removing birch and other small trees to make room for more profitable species. He also ran a larger operation than most neighboring dairymen; his sixty cows made his herd one of the largest in the county. However, in many ways he typified the changes that had swept through the county over the previous thirty-five years. He began his agricultural career in 1869 as a butter farmer, but ultimately he decided to concentrate entirely on fluid milk production. He fed his cows grain year-round, "in order to maintain regular and uniform shipments and price." For all his success as a dairy farmer, Bullard found that "the woodland has been one of the best investments, counting taxes, interest and all expenses as an offset." In the years leading up to 1905, Bullard sold more than five hundred thousand feet of pine and hardwood timber, netting himself a hefty cash supplement to his milk proceeds.[62]

Henry Bullard's land likely remained devoted to agriculture well past the turn of the century. However, around the turn of the century, Norfolk County began to turn away from agriculture, a shift that had significant economic and ecological repercussions. Its eastern portions were quickly becoming Boston's residential suburbs, and affluent city dwellers began to buy up large properties in outlying towns for second homes. A 1906 article in an agricultural newspaper titled simply "Farm Property Is Good Property" reflected the increased demand for farm property among urbanites.[63] Meanwhile, Boston contractors had begun to purchase the bulk of their milk from farmers in northern New England, leading to a market glut and rock-bottom prices. Many of Norfolk County's dairy farms shut down by World War I. Their owners had presided over an important shift in the composition and character of the landscape, as white pines grew up in the county's abandoned pastures. These men reaped the benefits of these trees in the years of the boxboard boom, leaving parcels throughout the county almost completely barren. But new trees soon replaced the ones that had been turned into boxes, creating the heavily wooded suburban landscape so familiar to Boston-area residents.

THE period from 1850 to roughly 1910 represented a transitional interlude in the landscape of suburban Boston. As extensive farming gave way to a more intensive style of agriculture based largely on production of milk for the Boston market, farmers adopted new practices, most importantly the feeding of grains to dairy cattle. Grain feeding allowed for the "shrinking" of farms, as degraded pasturelands returned to woods. Norfolk County farmers harvested many of these incipient forests in the boxboard boom of the early twentieth century, but these cut-over lands sprouted trees again, constituting the core of the post–World War II sylvan landscape. Stone walls may run through forests in suburban Boston, but the original reforestation of the region reflected not a turn away from agriculture, but rather the embrace of a much more intensive mode of farming. Recovering this story should remind us that the environmental history of suburbs did not commence with the mass arrival of cars and subdivisions in the post–World War II era. Instead, we can only understand the landscape of such places as Norfolk County in the context of a particular agricultural history, a history that differs from that of more rural regions. Far from being abandoned in the late nineteenth century, Norfolk County farms were reengineered. Dairymen moved their herds inside, letting the forest reclaim the land that their ancestors had worked so hard to clear.

9

How Metropolitan Parks Shaped Greater Boston, 1893–1945

James C. O'Connell

BOSTON has been called a "city of firsts," and among its innovations it was the first American city to create a metropolitan park system and the first to undertake regional planning. Established in 1893, the Metropolitan Park System was one of the most influential civic achievements of Progressive Era Boston. Boston took Frederick Law Olmsted's concept of networked parks and applied it to the rapidly growing metropolitan region. The parks were the first regional effort to protect environmentally significant areas and mitigate the most damaging consequences of the rampant real-estate development of the latter nineteenth century. For half a century the metropolitan parks and parkways provided a physical framework for Boston's suburban growth, with extensive preserved open spaces, recreational facilities, and advanced motorways.

The same environmental and social issues that galvanized Progressive reformers around the country between 1890 and World War I engaged Boston's park advocates. Members of the Yankee elite who were dismayed at chaotic urban growth sought to create a metropolitan park system that would preserve wilderness areas surrounding the city and clean up the riverbanks and beachfronts. This effort was the urban counterpart to the Progressive campaign of Theodore Roosevelt, Gifford Pinchot, and others to preserve natural wonders and rationally manage natural resources by creating national parks and national forests.

The Boston park reformers, led by Charles Eliot and Sylvester Baxter, believed a new level of metropolitan government was necessary to plan the region and carry out major public works projects. The effort of Boston's Metropolitan Park Commission reflected an emerging reform faith that comprehensive planning could create a rational spatial and infrastructure framework for metropolitan development, influencing other cities around the country. The park-planning efforts would succeed because of the creation of an appointed metropolitan commission guided by enlightened civic leaders, professional landscape architects, and engineers acting in the "public interest" and avoiding the rough-and-tumble of urban machine politics.[1] Boston's park reformers also reflected Progressive concern about uplifting the condition of the urban masses by providing exposure to nature and fresh air, playgrounds, and recreational facilities. According to the historian Jon Peterson, the planning associated with the City Beautiful movement "must be understood as a popular cause, largely cultural in expression, within the context of Progressive Era urban reform."[2]

Boston initiated its regional planning efforts after its attempts to annex suburban communities into a large central city stalled in the 1870s. Following the Civil War, Boston absorbed the surrounding municipalities of Roxbury, Dorchester, Charlestown, Brighton, and West Roxbury. In 1874, Brookline voted down annexation, as its citizens sought to preserve their political autonomy. With the exception of Hyde Park in 1912, Boston added no more surrounding towns. Yet rapidly growing suburbs were experiencing difficulty in delivering public services and considered metropolitan solutions. Sewage disposal was a persistent problem, since effluent polluted rivers and ponds, ruining local water supplies and spreading disease. Waste and clean water were quintessential regional issues because polluted ponds, streams, and aquifers did not respect municipal borders. In 1889 the Massachusetts legislature established the Metropolitan Sewerage Commission to build and manage waste systems for Greater Boston, creating the first regional public works authority in the country. So successful was the commission that the state created a Metropolitan Water Board in 1895 to supply water to the Boston region (the sewerage and water authorities merged in 1901). The metropolitan commission model also seemed to be appropriate for developing a regional system of parks.

The concept of a regional Boston park system had been part of the civic conversation for half a century. Scotsman Robert Gourlay's "General Plan for Enlarging and Improving the City of Boston" (1844) recommended preserving regional beauty spots, such as Fresh Pond, Spy Pond, and the highlands that ringed Boston, and connecting the scenic areas with paths and drives. In 1869 landscape gardener Robert Morris Copeland proposed a regional system of parks and connecting parkways, calling for a grand boulevard to circle Boston, starting at Squantum Point in Quincy and passing around the city to Revere Beach.[3] Central Park designer Frederick Law Olmsted set in motion Boston's efforts to create such a system with his proposal to build a network of preserves linked by parkways. In 1875 the Boston Park Commission initiated creation of the "Emerald Necklace" system, which included the Charles River embankment, the Fenway, Jamaica Pond, Arnold Arboretum, Franklin Park, and City Point in South Boston. In contrast to building a single large park like New York's Central Park, Olmsted sought to distribute reservations throughout Boston's urban fabric, establishing the foundation for a larger metropolitan system.[4]

By the early 1890s there was a strong appetite for new parks in Greater Boston. According to the journal *Garden and Forest,* civic leaders and the public agreed that parks were "essential not only to the pleasure and comfort, but to the physical health and the mental and moral growth of the people."[5] A number of suburban communities had established municipal commissions in accordance with the Massachusetts Park Act of 1882. Unfortunately, they found that expenditures for schools, roads, sewers, and waterworks were taking precedence. This led them to seek a regional solution. The preeminent advocates for creating a metropolitan park system for Boston were landscape architect Charles Eliot and journalist Sylvester Baxter. Eliot was the son of Harvard president Charles W. Eliot and Olmsted's protégé. Baxter, a Malden resident, was a supporter of Progressive causes, including metropolitan government. As the preeminent disciple of Utopian novelist Edward Bellamy, author of *Looking Backward* (1888), Baxter organized a Nationalist Club in Boston, which promoted nationalization of railroad, telephone, and telegraph systems. He advocated central planning to solve social problems. Eliot represented the environmentalist perspective, and Baxter exemplified the social reformer behind the metropolitan parks movement.

Eliot wanted to preserve scenic spots on Boston's outskirts in the face of rapid, indiscriminate development and reckless clear-cutting of forests. He had personally tramped through the undeveloped woodlands and open areas surrounding Boston and, with the aid of an 1890 Appalachian Mountain Club "Map of the Country about Boston," he had identified places worth preserving. An adherent of Olmsted's belief in the restorative properties of nature, Eliot thought that a Boston metropolitan park system would expose the urban populace to the aesthetic, psychological, and physical experiences of nature. He called park reservations "the cathedrals of the modern world."[6] Concerns about mitigating the unhealthy crowding of industrial slums dovetailed with the parks movement that swept late-nineteenth-century America. Frederick Law Olmsted promoted open spaces as breathing spaces for overcrowded cities. Parks could diminish actual density while transforming perceptions of density. Picturesque reservations were supposed to be "pieces of country" that provided a substitute for rural summers, with fresh air, meadows, and lakes, which many city dwellers would not normally enjoy.[7] This concept suited the educated elite, who celebrated the pastoral myth and regarded it as a way to acculturate immigrants. Eliot wrote: "For crowded populations to live in health and happiness, they must have space for air, for light, for exercise, for rest, and for the enjoyment of the peaceful harmony of nature which, because it is the opposite of the noisy ugliness of the town, is so refreshing to the tired souls of the townspeople."[8]

Baxter argued that if open spaces were not preserved, Greater Boston would be "in danger of becoming a vast desert of houses, factories, and stores, spreading over and overwhelming the natural features of the landscape . . . relieved by hardly an oasis."[9] For him, metropolitan parks could conserve natural beauty and provide recreational opportunities for congested neighborhoods in an era of industrial pollution and blighted development. The park system could create an orderly framework for suburban development, whether for the wealthy, the middle class, or working people. The development issue was acute in such communities north of the Charles River as Cambridge, Malden, and Somerville. They had filled up so rapidly with factories and congested worker housing after the Civil War that they sorely lacked open space and visual appeal. In 1888 the *Somerville Journal* lamented the unplanned growth: "Had it been possible to foresee how great the growth of the city would be and to make a general plan by which its growth might be regulated to the best advantage, Somerville today would be a much more attractive city than it is."[10]

In 1891, Eliot and Baxter helped to establish a private nonprofit trust to preserve scenic open spaces. According to Eliot, the Trustees of Reservations, the world's first such conservation organization, would hold "small and well-distributed parcels of land . . . just as the Public Library holds books and the Art Museum pictures—for the use and enjoyment of the public."[11] It soon became apparent that the Trustees of Reservations would be unable to move quickly enough to acquire the most valuable open spaces when a complicated pattern of ownership blocked the organization from acquiring Belmont's Waverley Oaks. The Trustees realized the need for the powers of eminent domain of a government agency. Eliot and Baxter organized a coalition of "gentleman reformers" and suburban park commissioners, which convinced the state legislature to establish a Metropolitan Park Commission (MPC) to develop a regional parks blueprint.[12] Baxter became secretary of the commission, Eliot was its landscape architect, and Charles Francis Adams Jr. was its chairman. Adams, the grandson and great grandson of presidents and a personal friend of Olmsted, hated "development fiends" and the destruction that industrialization was wreaking on his beloved Quincy.[13]

The resulting 1893 *Report of the Metropolitan Park Commissioners* became the blueprint for preserving Greater Boston's endangered nature spots and was the nation's first regional plan. This brilliant program guided the Metropolitan Park Commission in its acquisition of properties.[14] Eliot wanted Greater Boston to take advantage of its varied natural features, situated between the sea and the rim of hills that surrounded the city. He proposed preserving and improving several types of open spaces in the metropolitan area:

1. *Outer rim wild forest.* Large forest preserves were located on the edge of urbanized Boston along the ridge of the Boston Basin. They were as yet undeveloped because of their craggy terrain. The Middlesex Fells, which Eliot compared with Boston's Arnold Arboretum for scenic beauty, encompassed more than two thousand acres in Malden, Medford, Melrose, Stoneham, and Winchester. To the south of Boston, the Blue Hills, with seven thousand acres located primarily in Milton, were the highest point on the East Coast south of Maine. Eliot remarked that the Blue Hills had "vastly finer scenery than any of the great public woods of Paris."[15] Other wooded reservations he proposed were Stony Brook (475 acres) in West Roxbury and Hyde Park, and the Waverly Oaks in Beaver Brook Reservation (59 acres) in Belmont and Waltham.

2. *Riverbanks.* Three major rivers flow west to east through Greater Boston into the harbor: the Charles, the Mystic, and the Neponset. The rivers were polluted and industrial detritus fouled their banks. Eliot argued in favor of clean-

ing up the watercourses and beautifying the banks. This was a radical proposal for an era that turned rivers into open sewers and their banks into ugly dumping grounds. The Metropolitan Park Commission developed reservations along the three rivers, while the Metropolitan Sewerage Board diverted untreated effluent away from the rivers. Along the Charles River, the MPC created the Charles River Basin in Boston and Cambridge, the Middle Charles River at Newton and Weston, and Hemlock Gorge between Newton and Wellesley. The Mystic River Reservation included the Mystic Lakes, which were the source of that river. One of its tributaries, Alewife Brook, also became an MPC reservation.

3. *Oceanfront beaches.* Boston was the first major city to create public oceanfront reservations. Olmsted's Emerald Necklace plan of the 1880s proposed the Strandway along the South Boston waterfront and linked it with Franklin Park. Eliot observed: "Here is a rapidly growing metropolis planted by the sea, and yet possessed of no portion of the sea-front except what Boston has provided at City Point [Strandway]."[16] He set out to create parkways and public beaches along much of the metropolitan Boston oceanfront. The two great waterside parks were Revere Beach (the first in the United States acquired by a public authority) and Nantasket Beach in Hull. The Metropolitan Park Commission also developed public beaches in Lynn, Nahant, Quincy, Swampscott, and Winthrop. Eliot even proposed preserving the Boston Harbor Islands for parkland, a concept that achieved fruition in 1996, when Congress established the Boston Harbor Islands National Recreation Area.

4. *Parkways.* To make the metropolitan preserves accessible to the public, Eliot proposed a network of parkways between the city and the reservations. They were originally laid out as roads for carriages and bicycles, often with streetcar rights-of-way. The Metropolitan Park Commission parkways emulated those that Olmsted had designed in Boston—namely, the Arborway, Fenway, Jamaicaway, and Riverway. The Blue Hills Parkway and the Fellsway connected the large forest reservations with the urban core. The Charles River, Mystic Valley, and Neponset Valley parkways turned riverbanks into landscaped transportation corridors. The Revere Beach Parkway brought thousands of beachgoers to the shore, and Winthrop Shore Drive, the Lynnway, and Quincy Shore Drive provided attractive boulevards along the oceanfront.

The comprehensive 1893 metropolitan parks report persuaded the state legislature to establish a permanent Metropolitan Park Commission that very year. Eliot, Baxter, and Adams continued in their leadership roles, with Adams calling the legislation establishing the MPC the "Magna Charta for parks."[17] Reflecting the gradual approach to reform pursued in Boston in the

1880s and 1890s, the MPC was a pragmatic response to the need to provide an orderly setting for rapid suburbanization. Regional planning provided an infrastructure framework for hyperactive, market-driven real-estate development. Under the state's umbrella the MPC utilized eminent domain powers to acquire vast tracts of endangered open spaces. The MPC achieved favorable bonding terms, avoided excessive taxation of communities, and forged a fiscal arrangement whereby Boston paid half the bond interest and the other communities paid the rest.

The MPC started as an elite institution administered by Brahmin reformers interested in nature conservation and moral uplift. From this point on, state and metropolitan authorities were increasingly involved in managing environmental problems and building physical infrastructure. Relying on such technical specialists as landscape architects and engineers, the MPC sought to impose a rational, efficient order onto the metropolis. The MPC prefigured Theodore Roosevelt's conservation of national parks and forests by a decade. The Metropolitan Park Commission bridged the Olmsted era, which regarded parks as works of art, and the Progressive era, which championed rational environmental management.

In reviewing Eliot's 1893 metropolitan parks plan, *Garden and Forest* observed that park design issues could only be dealt with by a "true artist."[18] As a landscape architect, Eliot recognized the importance of artistic design and his approach to environmentalism emphasized scenic beauty. Yet he also wanted to achieve a "scientific 'park system'" that included ocean beaches, the Boston Harbor Islands, the major riverfronts and tidelands, large forest areas in the rocky hills, and neighborhood parks and playgrounds.[19] It was Eliot's comprehensive regional approach to the environment that differentiated Boston's park-planning efforts from those of other cities in this era. Nevertheless, this environmental pioneer did not pay attention to the ecological issues of plants and wildlife, habitat, and the biotic health of water bodies that were considered essential by the 1960s.

Metropolitan Boston had the first park system to be fitted comprehensively into the regional environment. Most cities, according to *Garden and Forest,* came "into possession of a certain number of acres of land by some chance" and had to adapt it to turn it into a park without an overarching plan.[20] Rational planning required coordinated government action, which Eliot strongly advocated. He wrote that the profit-making imperative of private land ownership inevitably led to "cheap makeshift structures of small taxable value; whereas public ownership will so enhance values that the whole community will reap a profit in the end."[21] The Progressive reform ideal of regional planning constrained private-sector environmental abuse and prefigured the late-twentieth-century environmental movement. Cities across the country introduced planning initiatives to provide a level of services that the marketplace on its own could not deliver.

The major achievements of the MPC under Charles Eliot's supervision until his sudden death at thirty-seven in 1897 were the Blue Hills and the Middlesex Fells reservations. These woodland parks provided vast open space adjacent to rapidly urbanizing communities, comparable to state parks that were created in rural parts of Massachusetts during the twentieth century. Because of their rugged topography, the land was inexpensive to purchase. As with several metropolitan reservations, the Middlesex Fells were intended to protect Greater Boston's water supply—in this case, Spot Pond, Middlesex Fells Reservoir, and three smaller Winchester reservoirs. The wild and hilly Middlesex Fells also provided valuable green space to counterbalance development in surrounding Malden, Medford, Melrose, and Stoneham. To the south of Boston, in Canton, Braintree, Milton, and Quincy, the Blue Hills Reservation formed a physical barrier that slowed southward suburbanization.[22]

Eliot initiated the preservation of the Charles River Basin, which became the true "central park" for Boston and Cambridge. He proposed damming the Charles River to turn the tidal estuary into a lake-like basin and develop the banks as a series of reservations and parkways. Although the damming of the Charles River was not effected until 1910, work on a comprehensive riverfront development plan began in the 1890s. Eliot also spurred preservation of the Middle Charles River between Newton and Weston when he designed the parklike setting at Hemlock Gorge. The Echo Bridge, a stone-arched aqueduct over the Charles River, added to the site's picturesque qualities. The Metropolitan Park Commission subsequently established the Middle Charles River Reservation downstream, absorbing the Riverside Recreational Grounds, which were privately managed between 1898 and 1914. This park was a place for active recreation, with four thousand canoes and a swimming pool with a sand beach. The Charles River was kept relatively clean until World War II by sewer lines built by the Metropolitan Sewerage Commission, which also maintained it.

The most popular metropolitan reservation was at Revere Beach, which was three miles long. The MPC moved a railroad line back from the crest of the beach and cleared three hundred buildings, many of which were shabby shanties.[23] The commission built a broad boulevard and elaborate pavilions and a bathhouse along the edge of the sand. Eliot initially envisioned Revere Beach as a genteel residential neighborhood, lauding "its capabilities as a place of residence, equipped with a broad esplanade and drive, and lined with houses and hotels facing the southeast and the sea."[24] But as soon as Revere Beach opened in 1897, it attracted the masses. On one Sunday that summer, forty-five thousand people came to Revere Beach, and the number grew steadily for decades.

After the turn of the century, swimming exploded in popularity. No longer did fully dressed adults limit their enjoyment to a stroll along the sand. Both men and women answered the call of Progressive reformers led by Theodore Roosevelt for the "strenuous life" and took the plunge, learning how to swim the crawl and donning form-fitting bathing suits.[25] The middle-class and working people who frequented Revere Beach were not as interested in contemplating nature as in active recreation and amusement parks. While earlier city dwellers were often overworked laborers simply seeking restful contact with nature on a Sunday afternoon, many turn-of-the-century urbanites worked in offices or stores and sought more active exercise. South of Boston, the MPC established Nantasket Beach (1899) in Hull and Quincy Shores Reservation (1907). Quincy Shores approximated more of what Charles Eliot had anticipated for Revere Beach. Middle-class homes lined it. The views to the Boston Harbor Islands were superb. Nantasket Beach, however, was a South Shore counterpart of Revere Beach, with its crowd-pleasing rides and amusements. By 1905 the world's largest roller coaster opened at Paragon Park across the street from the MPC reservation.

The metropolitan parks were such a success because of propitious timing. Frederick Law Olmsted, designer of the Emerald Necklace, recognized that the MPC plan was "the finest opportunity . . . of dealing with lands especially fitted by nature for public recreative purposes."[26] The thoughtfully crafted MPC plans were a relative bargain to execute. Public officials of the 1890s saw the value of public improvements and utilized the creative financing mechanism of state bonding to pay for them. And a public seeking new leisure activities was ripe to enjoy a park system spread across the metropolitan area. Remarkably, the metropolitan vision that Eliot and Baxter had championed in 1893 was realized within less than a decade. By 1900 the Metropolitan Park Commission had acquired 9,177 acres of reservations, thirteen miles of oceanfront, fifty-six miles of riverbanks, and built seven parkways. The $6.5 million price tag was a bargain.

Boston's Metropolitan Park System represented a blend of Progressive ideas related to solving urban problems associated with massive population growth, overcrowding, poverty, indiscriminate development, destruction of wilderness areas, and water and air pollution. The reformers' vision combined planned growth and the provision of scenic open space for breathing wholesome air. They held the utopian belief that woodlands, scenic lookouts, and bucolic stream banks were cathedrals of the modern world. The 1893 metropolitan park report and the commission that it engendered became the Progressive reformers' instrument for re-creating urban space. The metropolitan preserves furthered the reform belief in parks as a means of social control, using them to foster a healthy democracy and inculcate the masses with a middle-class appreciation for nature and beauty.

The Heyday of Boston's Metropolitan Parks

Boston's metropolitan parks became an international model, as they were the first and largest such system in America. At the Paris Exposition of 1900 the MPC displayed an enormous plaster topographical scale model (thirty-foot circumference) of the five-hundred-square-mile Metropolitan District, depicting the parks and parkways and the communities that encompassed them. The model was so popular that it traveled to the 1901 Pan-American Buffalo Exhibition in Buffalo, the 1904 Louisiana Purchase Exposition in St. Louis, and the 1905 Lewis and Clark Centennial Exposition in Portland, Oregon.[27]

Sylvester Baxter claimed that Boston's metropolitan parks had done "more to advertise Boston all over the world than any other modern factor."[28] The historian Jon Peterson has ranked Boston's metropolitan park system with Central Park and the Chicago Sanitary and Ship Canal as "urban triumphs" of American comprehensive planning.[29] The MPC's 1893 plan, America's first regional effort, preceded by sixteen years architect Daniel Burnham's celebrated regional scheme for Chicago, which featured preservation of the lakefront, a network of concentric parkways around the city, and a forest preserve encircling the city. Ironically, after World War II Boston fell behind many metropolitan areas in regional planning and the regional provision of public services.

The park-and-parkway system developed in Boston was a fundamental element of City Beautiful planning between 1890 and the beginning of

World War I. Peterson has argued that "park systems figured far more prominently than civic centers in the outburst of great civic schemes" undertaken under the banner of the City Beautiful movement.[30] Such cities as Atlanta, Buffalo, Chicago, Cleveland, Denver, Kansas City, Minneapolis, Seattle, and Washington, D.C., created parkway systems within their city limits, but they were not as extensive or as integrated into the suburbs as Greater Boston's. Although the topography of each city was different, the parks and parkways followed comparable patterns, particularly since a small circle of landscape architects was responsible for them.

The City Beautiful era was a high water mark for American urbanism. With unprecedented energy, Americans beautified their cities and developed urban amenities in the suburbs. The Greater Boston park-and-parkway system helped to create leafy suburbs such as Melrose, Milton, and Winchester. The early-twentieth-century garden suburb landscapes of other cities with park-and-parkway systems, such as Philadelphia and Chicago, resembled Boston's. The early-twentieth-century Boston suburban landscape shaped by the metropolitan park system is clearly delineated today as the area inside Route 128. The corridor for Route 128, most of which was built after World War II, passed through farmland located just beyond the developed areas of suburban Boston.

Plans did not always turn out as expected, however. Revere Beach sprouted honky-tonk amusements, the Alewife and Revere Beach parkways became commercial strips, and the Charles River Basin was transformed by the construction of Storrow Drive in the 1950s. Such developments underscored the role of planning in Greater Boston, as they directly contradicted the intentions of the first planners. The early success of Boston's metropolitan parks demonstrated that the Metropolitan District could be a useful instrument for regional planning. In 1907 the state legislature established the Metropolitan Improvements Commission and charged it with recommending transportation and economic development improvements for the Metropolitan District. Sylvester Baxter was named secretary, and Arthur Shurtleff (he changed his surname to Shurcliff in 1930), a landscape architect who had worked with the firm of Olmsted, Olmsted & Eliot, was the lead planner.

The Metropolitan Improvements Commission report was the first American urban plan to move beyond the civic art concerns of the City Beautiful era and focus on public works infrastructure and the efficient delivery of services that became known as the City Functional movement.[31] The plan covered railroads and terminals, the waterfront and docks, and waterways, but the most important element was Shurtleff's proposal for a circumferential parkway system to connect the radial roads that emanated from downtown Boston and facilitate traffic circulation around the suburban periphery. At the time most motor vehicles crossing the region had to travel through the twisting and congested streets of downtown Boston. Shurtleff's circumferential parkway system would enable motorists to bypass downtown.

The Metropolitan Improvements Commission report spurred a movement to create a full-scale metropolitan government to carry out the recommendations. Edward A. Filene, Louis Brandeis, and James Jackson Storrow organized the "Boston—1915" campaign to promote a metropolitan planning board. According to the historian Mel Scott, Boston's proposal was the country's most ambitious when Progressive planning efforts were at a peak.[32] When the metropolitan planning board finally came before the state legislature in 1912, however, suburban representatives voted it down. They feared the threat centralized power posed to local autonomy and believed that the public improvements could be realized locally.

By rejecting the metropolitan planning board, the legislature decided against regional control over growth and left authority over land use in municipal hands. In 1913 the state legislature passed the first Zoning Enabling Act, which delegated zoning powers to municipalities. Within eight years twenty-eight Greater Boston municipalities had adopted zoning, putting the region on a highly fragmented planning path, which still persists today. By rejecting regional planning for local land-use regulation, the state decoupled infrastructure development from growth management in Greater Boston. Nevertheless, the Massachusetts state government continued to search for a more comprehensive approach to infrastructure planning for Greater Boston. The legislature merged the metropolitan commissions for parks, water, and sewerage into the Metropolitan District Commission (MDC), which began operation in late 1919. The MDC served nineteen municipalities with water service, twenty-six with sewer service, and thirty-eight (one joined later) with parks and parkways.

The managerial approach of the MDC was more bureaucratic than the patrician reformer impulse that had characterized the Metropolitan Park Commissioners. The creation of the MDC marked the end of Progressive park reform and the emergence of a growing bureaucracy and an engineering culture that focused more on construction projects than on environmental preservation or civic beautification. Combining the metropolitan parks with the region's water supply and sewerage services, which were essentially engineering endeavors, reinforced this tendency.

An emerging urban population seeking to enjoy sports, concerts, and personal relaxation brought about the "golden age for recreation" in the park system.[33] Gradually, interest in recreation surpassed the environmental concern for preserving nature. The early-twentieth-century park embodied the Progressive Era ideal that leisure activities should be organized for the urban masses.[34] In 1912 the *Christian Science Monitor* editorialized in "Parks for the People": "It is not at all surprising that the demand for more playgrounds and greater facilities for sports in the metropolitan park area should have come. Boston's own experience with city parks . . . [is] forcing social workers, educators and progressive parents to favor the playground program."[35]

Nothing indicates the new type of recreational park better than Revere and Nantasket beaches, which swarmed with thousands of visitors on summer weekends and attracted amusement rides and penny arcades to their periphery. The MPC's Riverside Park, on the Middle Charles River between Newton and Weston, offered rowing, swimming, tennis, track, baseball, and football facilities. Riverside was a public-sector counterpart to the nearby commercially managed Norumbega Park, which featured a zoo, amusement rides, a theater, and the popular Totem Pole Ballroom. The MPC commissioners worried that the public was not taking advantage of the woodland reservations, such as Blue Hills and Middlesex Fells, in the same numbers that were flocking to Revere and Nantasket beaches. To attract people to the Middlesex Fells, in the Sheepfold the MPC placed "see-saws, swings, coasters, baseball diamonds, sand boxes, shelters, tables with settees for picnicing parties, and a drinking water supply." The MPC started operating a zoo there in 1900, when Superintendent Charles P. Price exhibited his collection of birds and later added a bear cub from Maine. At Houghton's Pond in the Blue Hills Reservation, the MPC built a cement bathhouse to encourage swimming and added waste receptacles and drinking fountains to "induce families to eat their luncheons there."[36]

Children became an important audience for parks. Play directors were hired and team sports organized to create alternatives to street play. Boston opened the nation's first dedicated playground, with special equipment and activities, in 1890. Early on, the MDC built ten playgrounds in its reservations, and municipalities built dozens more. The MDC commissioners reflected the Progressive view of playgrounds, declaring that children "must be started correctly for the good of their future citizenship and thus kept from the dangers of the street."[37] These spaces were an expression of the Progressive "fresh air" reform movement, which regarded outdoor recreation as essential for the health and assimilation of immigrant urban children.

The Metropolitan Park Commissioners reflected the view that government should guard public morals and provide for public needs, maintaining that privately owned facilities would not provide adequate services unless government established standards. At Nantasket Beach the Metropolitan Park Commissioners advocated replacing "unwholesome" amusements with "wholesome" refreshment stands and bathhouses.[38] The metropolitan parks offered free and respectable entertainment. In 1914 the MPC spent $25,000 on 322 concerts at metropolitan parks, with Revere and Nantasket beaches presenting free concerts six days a week during the summer.[39] Band concerts attracted thousands, even to such places as the Mystic Valley, Furnace Brook, and Woburn parkways. In 1929 the MDC initiated the Boston Pops concerts on the Charles River Esplanade.

Recreational use of the metropolitan parks grew in the 1930s, when unemployment increased people's free time and New Deal funds poured in. Park administrators abandoned efforts at social reform and focused on simply providing a broad range of recreational opportunities. Parks were seen as community amenities, which no longer required moral claims for justification.[40] By the 1930s the MDC had developed golf courses at Riverside Park (Leo J. Martin Memorial Golf Course) in Weston and at the Ponkapoag Golf Course in Canton. The agency created ski trails, a toboggan run, and a ski jump at the Blue Hills Reservation. The 1935 St. Moritz Ice Carnival at Blue Hills attracted thirty-five thousand people. Beaches at Lynn, Nahant, Revere, Swampscott, and Winthrop set attendance records in 1938 with a total of twenty million visitors.[41] Along the Charles River the MDC operated swimming facilities at Magazine Beach and Gerry's Landing in Cambridge, Faneuil in Brighton, the Newton-Boston line, and Havey Beach in West Roxbury. During the 1930s the commission even maintained a beach on the Mystic River in Medford. After World War II, however, these beaches were closed because of the increasing volume of "dirty water" produced by pollution and sewerage runoff.

The infusion of New Deal funds into the agency to boost local employment in the Boston area produced an impressive array of activities. In 1938, for example, the WPA-MDC Recreation Project hired supervisors to organize nature walks (106,860 attended), baseball games (104,144), horseshoe matches (45,854), ski outings (13,850), moonlight walks (3,090), birdhouse-making (2,500), and weenie roasts (2,500).[42] To provide facilities for these activities, the commission used federal and state funds for ball fields, sidewalks, plantings, signage, fences, and water bubblers. Hundreds of relief workers maintained the MDC park system at an exceptional level, reinforcing the legacy of the pioneering metropolitan parks of the 1890s. One of the leading New Deal projects was the Civilian Conservation Corps (CCC) camp at Blue Hills Reservation. Under the CCC two hundred unemployed men lived there and built roads, picnic areas, and parking lots, cleared forests, and controlled gypsy moths.

Starting in 1935, the MDC used CCC workers to build trails, picnic grounds, and swimming areas at the newly established Breakheart Reservation in Saugus. In 1939 the commission used federal funds to develop an expansive picnic area at Nahant Beach Parkway, with stone tables and seats and large fireplaces. During this era picnic grounds were popular excursion destinations for motorists. Although the Boston region suffered economic stagnation during the 1930s, New Deal funds allowed the MDC to modernize its facilities. The most notable MDC development of this era was the Storrow Memorial Embankment along the Charles River Esplanade. Dedicated in 1936, the Storrow Memorial Embankment doubled the amount of parkland along the river between the Charles River Dam and the Massachusetts Avenue Bridge, creating a new playground and facilities for boating and swimming. The expansion included the Music Oval, which eventually accommodated the Hatch Shell.

The MDC had come a long way in providing visitor amenities and recreational facilities from Charles Eliot's vision of creating natural woodland parks for communing with nature. The MDC parks had become more urbanized, with an extensive system of well-designed and maintained park spaces. The construction of recreational facilities grew out of the engineering orientation of the MDC. The tally of the agency's facilities at the close of the 1930s was impressive: twenty-seven refreshment stands, thirty-five comfort stations, five boathouses, five outdoor skating rinks, one ski jump, two golf courses, ten playgrounds, thirteen bandstands, fourteen baseball fields, twenty-eight picnic areas, and twenty tennis courts.[43]

The increasing emphasis on recreation during the 1920s and 1930s marked a shift away from the nature worship of the Progressive reformers to a more populist conception of parks and the rise of organized athletics.

The eclipse of Progressive reform had started with the defeat of the "Boston—1915" campaign for a metropolitan planning board and the merging of the metropolitan parks with water and sewer service under the Metropolitan District Commission. The MDC was little interested in moral uplift, concerned instead with functional management of Greater Boston's park-and-parkway system. It increasingly regarded itself as an infrastructure agency creating the physical framework for metropolitan growth.

Parkways and Parks Shape the Suburban Landscape

The parkway network produced the Metropolitan Park System's most significant consequences. It created a development framework for Boston's first-generation automobile suburbs (1900–45), which allowed lower residential density, created well-landscaped neighborhoods, and spurred automobile commuting. When Charles Eliot and Sylvester Baxter were working on the blueprint for Boston's Metropolitan Park Commission, they originally viewed parkways as connections between land preserves and concentrations of population. Their idea was to landscape traffic corridors so that they seemed like parks. Eliot explained that "a parkway is simply a highway made as agreeable as may be possible."[44] An urban counterpart to these avenues was the boulevard.

The parkways were planned to be 125 feet wide and included lanes for pleasure driving and ordinary traffic as well as a right-of-way for streetcars. The center lanes were for racing horse-drawn carriages and followed the prototype Olmsted developed in 1868, when he designed Brooklyn's Ocean Parkway, which connected Prospect Park to Coney Island. In 1886, Henry C. Whitney, president of the West End Street Railway Company, hired Olmsted to draft a similar boulevard plan for Beacon Street in Brookline. It would accommodate streetcars and horse-drawn vehicles, thus opening up the abutting property for development. According to City Beautiful planner Charles Mulford Robinson, Olmsted's Beacon Street was the first American boulevard "frankly to accept the car tracks as a permanent feature of the avenue and one that could and should be turned to decorative purpose."[45]

The Metropolitan Park Commission used the Boulevard Act, which allocated five hundred thousand dollars to build boulevards and parkways and provide work relief for those left unemployed by the Panic of 1893, to build the Blue Hills Parkway and the Fellsway, as well as to construct and widen such major Boston arterials as Blue Hill, Columbus, Dorchester, Huntington avenues, and Washington Street to Forest Hills.[46] This initiative extended Commonwealth Avenue through Brighton and Newton to the Middle Charles River Reservation and added streetcar lanes on Blue Hill, Huntington, and Commonwealth avenues. Much of Boston's current arterial system dates from this period.

Some of the most significant parkway planning took place north of the Charles River in communities that were poorly served by surface arterials. Given license to create a system of northern parkways, Charles Eliot surveyed routes connecting Mount Auburn Street in Cambridge to Revere Beach and Lynn. These included Fresh Pond, Mystic Valley, Lynn Fells, Revere Beach parkways, Winthrop Shore Drive, Winthrop Parkway, and the Lynnway. Construction began in Eliot's lifetime, but completion took twenty years. As parkway planning progressed, Eliot realized that he was doing more than connecting parks. He was creating a traffic circulation network for metropolitan Boston, something desperately needed in a city made up of disconnected neighborhoods and one of the few U.S. cities not built on a grid. Parkways formed the stems of street systems and provided the circulation framework for developing new neighborhoods.[47] The new transportation arteries changed land uses on the urban periphery, making abandoned farms and neglected open land available for suburban growth. Eliot argued for the economic benefits of parkways: "That both parkways [Blue Hills Parkway and the Fellsway] will develop building lands at present remote from all means of rapid transit is not to be counted against them; since it is from the increasing taxable value of these very lands that a huge share of the cost of constructing the new ways may be derived."[48]

Eliot also understood that riverbanks could be ideal locations for parkways because of their scenic qualities and the ability to circumvent developed areas. He wrote: "There is really nothing to prevent the reservation of the banks [of the Mystic River] for public use, and ultimately, though perhaps many years hence, the construction of a river road which would provide the pleasantest possible route to Boston from Medford, Arlington, and Winchester and all the towns beyond."[49] The parkways along the Charles, Mystic, and Neponset rivers brought the public into pleasing contact with the riverfronts unlike the highways of the 1950s and 1960s, which walled off rivers from local communities.

No one originally anticipated that motor vehicles would use the metropolitan parkway system. The roadways were intended for country-like pleasure drives for carriages, horses, and bicycles.[50] Ironically, however, the year the Metropolitan Park Commission was established—1893—was the same year that inventors Frank and Charles Duryea built America's first gasoline-powered automobile in Springfield, Massachusetts. At first automobiles were a rich man's hobby. Yet by 1905 motorcars were riding on parkways in such numbers that the gravel paving surfaces started to deteriorate and the MPC had to lay more durable asphalt. Traffic counts for suburban arteries grew by 131 percent between 1909 and 1912, and by another 121 percent between 1913 and 1914.[51]

Increased attention was given to highway planning because of an astronomical rise in Massachusetts automobile registrations (mostly in Greater Boston). Registrations soared from 3,772 cars in 1904 to 102,633 in 1915 to 1,019,460 motor vehicles in 1929. Thoroughfares between downtown Boston and the North and South shores were constantly jammed. Sylvester Baxter observed that accommodating automobiles was overwhelming the recreational and conservation concerns of the Metropolitan District Commission.[52] The *Boston Globe,* in 1916, reported: "In 10 years, especially in the last five, automobiles have changed the character of the [Metropolitan] district and the use of the parkways and boulevards has made the whole district an open space which has become of great, if not greater, importance than the reservations." Baxter observed in 1913: "Metropolitan Boston could hardly exist without these new elements [parkways] in the general plan that originated in quite another purpose." He forecast that, by 1950, the growth the parkways encouraged would reach out as far as Framingham, Wilmington, Canton, and Beverly—and he was correct.[53]

The development of metropolitan Boston's parkways reflected the planning paradigm of the late nineteenth and early twentieth centuries, which promoted traffic mobility along broad roadways to accommodate expansion of the city, even before automobiles became prevalent. The metropolitan parkways exponentially expanded Boston's handsome boulevard network, which included Commonwealth and Huntington avenues, into the suburban hinterland. The park-and-boulevard system that Olmsted had pioneered and Eliot expanded became the "design mainstay of the City Beautiful" by the turn of the century, according to the historian William H. Wilson.[54] Boston's curving parkways followed the contours of the landscape, reinforcing the physical form of the metropolitan area and the identity of individual communities. The parkways' objectives were distinctly aesthetic and ecological. Shady oaks, elms, maples, and beeches lined their medians and borders. The authorities prohibited commercial vehicles in order to screen out much of the heaviest and most noxious traffic. The early-twentieth-century parkways blended into and defined neighborhoods, unlike the concrete highway corridors that fragmented urban and suburban communities after World War II.

The metropolitan parks and parkways helped shape a distinctive early-twentieth-century suburban landscape that was an extension of the city, as opposed to the suburban rejection of urbanism that followed World War II. The Metropolitan Park System was a reaction to the dense, disorderly, teeming cities of the late nineteenth century. The Progressive park planners set aside significant swathes of natural landscapes for public use that had more natural beauty and open areas for recreation. The Metropolitan Park Commissioners observed, in 1911, that the timing of the parks had been propitious for assembling open space and creating attractive suburban landscapes: "It was a fortunate beginning because the District was rapidly changing from one in which its population was gathered into detached cities and towns separated by intervening strips of land with sparse population, to one in which, by the pushing out of buildings to occupy these intervening and hitherto unoccupied lands, reasonable lines of intercommunication were becoming more necessary each year and increasingly more difficult to provide."[55]

The metropolitan park system promoted real-estate development, attractive neighborhoods, and increased property values. Metropolitan Parks Commission Chairman William B. de las Casas pointed out that parks and parkways gave "definiteness of character and development" to their neighborhoods.[56] Public parks buffered neighborhoods from industry and slums and enabled middle-class and even working-class families to live with natural amenities that upper-class residents enjoyed in such suburbs as Brookline, Newton, and Wellesley. The reservations resulted from the conscious effort of Progressive reformers to raise the standard of living of the urban masses by exposing them to woods, hills, meadows, streams, and the seafront. The new parks ensured that marginal yet scenic properties, whether rocky uplands, riverbanks, or beachfronts, did not become blighted dumping grounds.

The suburban communities to which the Metropolitan Park System gave form filled in rapidly as electric streetcar service was introduced in the 1890s. Henry C. Whitney, who had developed Beacon Street in Brookline with Olmsted, converted Boston's streetcars from horse power to electric power in 1889. In the horsecar era commuters could comfortably live within two and a half miles of downtown, but by the turn of the century, the effective commuting distance was nine miles.[57] The new suburbs were a less dense version of the "streetcar suburbs" that historian Sam Bass Warner described in *Streetcar Suburbs: The Process of Growth in Boston, 1870–1900.* Where multifamily houses predominated in the post–Civil War "streetcar suburbs" of Dorchester, Roxbury, and West Roxbury because of the need for density near the urban core, the suburbs the streetcars and parkways developed in the early twentieth century tended to have a larger percentage of single-family homes and larger yards. The parkways and the autos that used them enabled even lower residential densities, which began to appear after 1920 and predominated in the post–World War II era.

The suburbanization that took place during the development of the Metropolitan Park System made Boston the fourth largest metropolitan area in 1910, trailing only New York, Chicago, and Philadelphia. Metropolitan Boston had a population of 1,520,470 in thirty-six communities. In 1917, *A Handbook of New England* observed: "This great concentrated population, equipped with the intensive transportation facilities of a huge metropolis, is what invariably astonishes the stranger who, with census figures in mind, expects to find Boston a city of the St. Louis, Cleveland, or Baltimore type, rather than one comparing with Chicago and Philadelphia."[58] Some of Greater Boston's most desirable neighborhoods were located near metropolitan parks and parkways. In the era before auto traffic became prevalent, many of the wealthiest families situated their homes along parkways. Boston's Jamaicaway, for example, was the setting for Mayor James Michael Curley's shamrock-shuttered twenty-one-room palazzo overlooking Jamaica Pond. Commonwealth Avenue in Newton attracted wealthy families to build prestigious homes. According to *A Handbook of New England,* Commonwealth Avenue offered "the best entrance and exit to and from Boston."[59] Only after 1920 did wealthy families abandon their parkway homes and move to more secluded locations in rural suburbs, leading the trend toward privatism in suburbia that held sway after World War II.

At the coastal communities of Hull, Lynn, Quincy, Revere, Swampscott, and Winthrop, beach reservations and parkways oriented these communities to the oceanfront. The industrial communities of Everett, Malden, Medford, and Stoneham benefited from their proximity to the Middlesex Fells Reservation, which provided a nearby wilderness atmosphere and recreational opportunities. The Metropolitan Park Commission preserved the views from the Fells toward Boston by preventing houses from being built at the edge of the tableland. The MPC planted thousands of trees to protect the water quality of the reservoirs and grazed sheep on the Sheepfold to maintain a pastoral scene. According to the *Boston Globe*, the Fellsway promoted "artistic development and physical growth" in Medford and Malden, and "vast tracts of territory heretofore of no particular value" were brought into development.[60] Running north from Somerville to Melrose, the Fellsway had originally been constructed to bring visitors to the Middlesex Fells, but gradually it became an automobile commuter corridor. It encouraged suburbanization in Stoneham and other communities north of the Middlesex Fells. The Mystic Valley Parkway spurred development in Arlington, Medford, and Winchester. Revere Beach Parkway opened up development opportunities in Everett, Malden, and Revere. The Lynnway facilitated traffic flow between the North Shore and Boston.

One can appreciate the impact of the MPC on the emerging suburban fabric by examining a working-class community such as Somerville, which experienced a major industrial growth spurt after the Civil War, before the introduction of parks and parkways into the metropolitan landscape. Slaughterhouses and brickyards lined Somerville Avenue and abutted the railroad tracks. Developers filled in wetlands for railroad tracks and factories. There were extensive tracts of dense working-class tenements. Somerville reserved only fifty-two of its twenty-four hundred acres for parks and playgrounds.[61] City fathers avoided building parks because they wanted to keep taxes low, and builders wanted the maximum amount of land for development. By the time the *Somerville Journal* ran a story headlined "Beauty or Business: Shall We Have Parks or Factories" (on March 13, 1896), the decision had been made in favor of industry. The only MPC presence in Somerville was at the edge of the community—at the Alewife Reservation and Parkway. The city's main open space was a postage stamp-sized Broadway Park (Saxton Foss Park) and Nathan Tufts Park in West Somerville.[62] This densely developed industrial suburb was the type of community that Eliot and Baxter had wanted to avoid creating in the future.

In contrast, upper- and middle-class enclaves took advantage of the Metropolitan Park System. The Mystic Lakes Reservation, between Arlington and Winchester, attracted prestigious residences. The larger estates were on the Winchester shore because the MPC protected the entire east side of the lakes as parkland. One real-estate advertisement praised the location: "The Beauty Spot of Greater Boston—overlooking the Mystic Lakes on the Winchester car line—large lots—restricted to singles houses only."[63] The MPC did not own the Arlington side of the lakes, and the lakeside neighborhood of that town had subdivisions with smaller lots and less expensive homes than in Winchester.

Winchester led the way in using the Metropolitan Park Commission to secure its status as a prestigious suburb. Businessmen Forrest Manchester and Edward Ginn persuaded the MPC to extend the Mystic Valley Parkway from the Mystic Lakes along the Aberjona River into the heart of town. Near the center, a tannery, coal yard, lumberyard, railroad freight yard, and worker tenements marred the community's appearance. Manchester and Ginn wanted to replace them with a park and a playground. The town's wealthy residents supported the new park, while workers opposed it from fear of losing their jobs. After a heated campaign, Manchester and Ginn obtained support from the town and the MPC for the park, partly by donating their own funds. When Manchester died a few years later, Winchester named the park in his honor.

Parkways Between the Wars

One of the foremost examples of a parkway that formed the landscaped spine of an attractive neighborhood was Fresh Pond Parkway in Cambridge. The original Fresh Pond Parkway, opened in 1900, stretched from Mount Auburn Street to Huron Avenue (it later was extended to link Memorial Drive and the Alewife Brook Parkway). The Metropolitan Park Commission created Lowell Park at the head of the parkway, next to the former house of writer James Russell Lowell. Designed by the firm of Olmsted, Olmsted, & Eliot, Lowell Park served as a handsome centerpiece to a prestigious neighborhood. Unfortunately, the oak-shaded parkway did not turn out as originally intended—the automobile traffic became so heavy that many homes erected barriers along the roadway to screen out the noise.

Fresh Pond Parkway vividly conveys how the MPC parkway system shaped suburban development in the early twentieth century. The attractiveness of the parkway originally made the neighborhood popular with Harvard faculty, just as parkways spurred the development of desirable residential districts from Winthrop and Quincy to Winchester and West Roxbury. The great unanticipated consequence of the routes was that automobiles started clogging them with traffic in the 1920s. They evolved from purely recreational drives into the Greater Boston region's first motorway system, knitting together the region's fragmented communities and spurring suburbanization. After World War II automobiles so overwhelmed the thoroughfares that Greater Boston required an entirely new system of limited-access highways.

During the 1920s the Metropolitan District Commission expanded its planning and engineering activities, as it became the lead agency for transportation planning in Greater Boston. Individual municipalities were incapable of building a coordinated regional roadway system, and the State Department of Public Works (DPW) was still mainly concerned with building rural roads.[64] In 1923 the MDC established the Division of Metropolitan Planning to address transportation planning, including highways, transit, and shipping. Arthur Shurtleff became its chief consultant. The MDC extended or built fourteen parkways between 1922 and 1932. It constructed a South Shore connection with Old Colony Boulevard and built the Northern Artery from the Charles River to the Fellsway, a route later christened the Monsignor O'Brien/Monsignor McGrath Highway. During the 1920s the agency extended parkways along the Charles River to Newton and Watertown. Building these links created an integrated system, achieving the inner circumferential route Shurtleff had proposed in 1909. A motorist could start on the North Shore at Lynn Shore Drive and follow MDC parkways west to Watertown and Newton, south through Boston to the Blue Hills and east to the Quincy Shore. This route was a precursor of Route 128, which the State Department of Public Works started constructing as a limited-access highway out at the edge of the Metropolitan District in the late 1930s.

One of the biggest challenges to the MDC was adapting the sinuous parkways to the growing crush of traffic. The historian August Heckscher has argued that the City Beautiful boulevards and parkways were "ripe for the automobile to take over and deform."[65] By the mid-1920s the Metropolitan District Commissioners stated that "one of the most difficult problems presented to the Commission has been the adaptation of roadways intended for the use of horse-drawn vehicles to enable them to function under the stress of intensive motor vehicle use without destruction of marginal plantations of trees, shrubbery and grass, and without bringing noise and danger into the parks themselves."[66] The MPC/MDC constantly upgraded and beautified the parkways before World War II. The Metropolitan Park Commissioners reported that "much has been accomplished in improving the shade trees and adding the shrubbery" along the Revere Beach, Mystic Valley, Middlesex Fells, and Charles River parkways.[67] When unemployment was high during the Depression years, the MDC employed thousands on work relief planting bushes and trees, cutting brush, and erecting directional signage.

The MDC resurfaced the parkways, added lanes, erected light fixtures, and planted trees. After World War I it no longer included streetcar lanes in new parkways, as automobile use surged and streetcar ridership declined. Some boulevards, such as Newton's Commonwealth Avenue in 1930, removed streetcar tracks. They excluded trucks and buses and separated through from local traffic. Parkways limited automobile access and restricted curb cuts for abutters. The Trustees of Reservations, in 1937, noted how parkways had beautified Boston: "The approaches to the City have been made open, accessible, and attractive, which otherwise would have been congested and ugly."[68]

The struggle to maintain the pastoral character of the parkways intensified during the 1920s. The MDC waged a strenuous battle to maintain its ban on commercial enterprise abutting the routes. Municipalities were often lax in zoning adjacent properties for nonobtrusive uses. The MDC commissioners complained in 1936: "The limited areas of land owned by the Commission . . . and the disregard of protests filed by the Commission to the changing of the zoning . . . makes it impossible in some areas to maintain the high type of parkways originally planned for the district."[69] The MDC requested jurisdiction from the state to control all land use within five hundred feet of a reservation or parkway, but the legislature did not grant it, taking the side of local "home rule." The lack of cooperation from municipalities in preserving the attractiveness of parks and parkways and state's recalcitrance at providing the MDC adequate authority to implement planning set the stage for the neglect they suffered after World War II.

The growing amount of traffic on parkways diminished their residential appeal compared with the pre-1920 era, when they were showcases for imposing homes. The West Roxbury Parkway became lined with middle-class Tudors and Colonials during the late 1920s and early 1930s, while the homes on VFW Parkway from the late 1930s and 1940s were even more modest. Yet despite increasing traffic on parkways, Harvard University landscape architects John Nolen and Henry V. Hubbard argued as late as 1937 in *Parkways and Land Uses* that the routes remained desirable places to live. Parkways still increased property values. Nolen and Hubbard argued that "the community *as a business* will be better off financially on account of the parkway because it will ultimately be receiving annually in added taxes."[70] Ironically, the landscaped parkway was nearing its end as an organizing element of the suburban landscape.

To experience the difference between the MDC parkway design and the functionalist highway designs of the federal government after World War II, drive out of Boston along the Jamaicaway and the VFW Parkway to the Route 1 corridor in Dedham. Rows of trees on the median and each side of the road beautifully shade the Boston stretch. Signs read: "Pleasure Vehicles Only." A mix of wooded areas, houses, churches, hospitals, schools, and minimal retail lines the roadway. The parkway ends at Dedham, however, and Route 1 becomes a treeless concrete stretch of retail sprawl. Even where the banks of the Charles River appear on the west side of the highway, concrete abutments and retail stores obliterate the scenery.

Postwar Changes in Parks and Parkways

When World War II broke out, Boston's metropolitan parks were at a height of popularity they would never see again. The physical improvements and recreational activities the New Deal sponsored were greater than the MDC's founders ever anticipated. Parkways were the backbone of the region's automobile transportation system and regional infrastructure planning continued to set a national standard. After the war the metropolitan parks and beaches entered a period of decline. The wartime arms buildup reduced resources for parks. The slide into bureaucratic routine at the MDC was evident in the *Annual Reports* for 1940 through 1947, which were published in a single, thin volume and which described no major initiatives. As residents left Boston and inner-ring suburbs, some parks became neglected. Suburbanites in the communities outside of Route 128 and the Metropolitan District shied away from creating sites comparable to the metropolitan parks. Meanwhile, rampant highway construction and private development paved over swathes of parkland.

When the MDC developed its 1956 master plan, the agency focused on the rapid growth and changing recreational needs of the metropolitan area. The children of the "baby boom," as well as workers and retirees with increased leisure, were seeking expanded athletic facilities such as swimming pools, skating rinks, and ballfields. The swimming pools were an alternative to polluted rivers and ponds, while enclosed artificial skating rinks permitted skating year-round. Overall, the MDC built twenty-seven skating rinks and nineteen swimming pools between 1950 and 1975, as state legislators found these facilities an effective means of bringing home the bacon to their districts.

In line with developments elsewhere in the country, during the 1950s and 1960s, the MDC created no scenic parks. Although central city residents continued to use large urban parks, including Boston's Franklin Park and the Emerald Necklace, amenities beyond the city attracted increasing numbers of metropolitan residents. Revere and Nantasket beaches continued to draw large crowds, while forest preserves such as Middlesex Fells, Blue Hills, and Stony Brook lost visitors. It seemed more exciting to drive to the White Mountains or Cape Cod. Leisure had changed with greater affluence, the embrace of the automobile, and paid vacations for workers.[71] The Metropolitan District Commission treated its parkways in a utilitarian manner. The experience of driving through a bucolic setting lost favor, as speed, horsepower, and easy access became the primary objectives of motorists. Americans gave up cycling and walking on parkways, instead preferring motoring, a dominating obsession of the era. Fresh Pond, Alewife, Mystic Valley, and Revere Beach parkways lost almost any semblance of the pastoral due to the loss of trees, medians, and scenic views, proliferating curb cuts, and encroaching strips of commercial structures.[72]

Policy changes at the federal and state levels rendered obsolete the MDC's traditional parkway design. The Federal Highway Act of 1956, which created the interstate highway system to facilitate military transportation, placed a premium on functionality over landscape beautification. In Massachusetts the 1948 *State Highway Master Plan* proposed an ambitious statewide highway construction program that shifted the leadership for highway construction to the State Department of Public Works. The state DPW's capacity to obtain federal highway funds, which the MDC could not receive, further strengthened its position. The decline of the parkways was part of a broader trend of parkland neglect. Growth pressures forced the MDC to turn over acreage for development. Incompetence and outright corruption sometimes influenced sales. The state cut highways through the Blue Hills, Middlesex Fells, Neponset River Reservation, and the Mystic River Reservation. In East Boston, Massport's Logan Airport gradually swallowed up Wood Island Park, which had been designed by Frederick Law Olmsted. The most controversial loss of MDC parkland occurred when Storrow Drive was carved into the Charles River Reservation in 1951.

The MDC commissioners, in their 1956 plan, recognized that "the changing patterns of land use, and in particular the almost fantastic increases in land values in some localities, have created enormous pressures for the release of parklands to private ownership or their diversion to highways, to military use or for other public purposes."[73] After a decade and a half of Depression and wartime stagnation, public opinion "favored development over open space."[74] Harvard landscape architecture professor Norman T. Newton lamented: "All too often the reservations have been gouged out for the sake of highways, but this is only one manifestation of a still larger fault: utter obeisance to the automobile in administration of what was born as a park system. . . . If Charles Eliot were to see the condition of his Metropolitan Park system today it is doubtful that he would think it much of a memorial. Surely this is one of the supreme ironies—that something so far ahead of its times when he conceived and brought it into being should be so far behind today."[75] Instead of regarding parks as public amenities that raised property values, many people believed that they were a drain on taxes and that using the land for development would expand the tax base.[76] Neither the political leadership nor the public held a vision that strongly valued the role of parks in metropolitan Boston.

The concept of metropolitan parks had little appeal to the booming postwar suburbs outside the Metropolitan District and the Route 128 beltway. The MDC's 1956 plan contained a section on "Metropolitan Sprawl" that discussed the problem of unplanned construction, arguing that suburbs outside the Metropolitan District were incapable of preserving adequate open space in the face of rampant development. The report cited Eliot and Baxter's 1893 study for its admonition that preserving open space was the key to maintaining "suburban charms."[77] The agency proposed creating new reservations in Braintree and Weymouth, but housing subdivisions sprang up on the targeted open space before it could be purchased. The MDC's acquisition wish list was ignored in the 1950s, although several recommendations were later followed. At the same time water supply and sewerage functions absorbed much attention at the commission, especially as Boston's harbor and rivers became increasingly polluted. Water pollution, in turn, forced the closing of beaches and caused the public to shun walks and other activities along the edge of rivers.

Whenever there was discussion about expanding the MDC to include more distant suburbs, there was resistance. In the early 1940s the Trustees of Reservations proposed enlarging the metropolitan park system to include eighty-six new communities within a twenty-five-mile radius of downtown Boston (approximately as far out as today's I-495 beltway). The MDC, City of Boston, State Planning Board, and a governor-appointed committee approved the idea. But representatives of many of the candidate communities came out strongly against the proposal. They wanted no new tax assessments and no interference from the MDC or Boston interests.[78] Similar efforts in the 1960s to expand the metropolitan park system were also resisted.

Not only were growing outer suburbs jealous of local control, they conceived of their landscapes differently from Metropolitan District suburbs. Boston's outer suburbs thought of themselves as more rural than "suburban." These country towns preferred "natural" open space to the refinements of public gathering spaces, lookouts, and recreational fields. The rural suburban ideal defined itself in opposition to the city, not integrated with it, as Metropolitan District suburbs had been in the prewar years. Postwar Boston suburbs abandoned the Olmsted-Eliot–style mix of planned parks, parkways, and sidewalks that had long formed the physical framework for metropolitan development. Middle-class homeowners focused on their private yards, which tended to be larger than those in the older Metropolitan District. With this privatized sense of open space, suburbanites seldom tried to create large municipal parks, building instead school athletic fields and playgrounds. Recreation came to be the province of organized sports teams rather than individuals hiking or cycling on their own. Meanwhile, the sprawl development patterns dictated by the highway, the strip, and the residential subdivision undercut nature conservation and produced a fragmented landscape.

The post–World War II era in Boston's metropolitan parks and parkways signified a dramatic reversal of the remarkable achievements of the earlier twentieth century. Although the MDC built numerous recreational facilities such as skating rinks and swimming pools, the agency neglected to maintain natural areas of parks and even sold off acreage for development. The commission conserved no new natural areas, and emerging suburbs on the metropolitan periphery took no interest in having it create parks in their communities. The parkways were despoiled as trees were cut down and commercial sprawl proliferated. The management of the MDC was dominated by a combination of construction-oriented engineers and political hacks rather than by the idealistic reformers who had founded the agency. Bostonians neglected the scenic beauty and natural attractions of parks. With the decline in interest, budgets shrank and general maintenance deteriorated, further reducing their appeal.

Tracing the Arc of Urban Planning

The development of Boston's Metropolitan Park System from the early 1890s until the post–World War II era traces the arc of urban planning across the country. When Charles Eliot and Sylvester Baxter wrote the blueprint for the country's first metropolitan park-and-parkway system, they were marking the transition from the germinal park planning of Frederick Law Olmsted to the City Beautiful movement (which Olmsted helped propagate with his landscape design for the 1893 Chicago World's Columbian Exposition). The Boston Metropolitan Park System plan was predicated on Olmsted's desire to create "natural" parks in exploding industrial cities, but it also contained elements of City Beautiful, which initiated a comprehensive approach to urban planning and introduced aesthetic design as an essential consideration in developing the urban landscape. The result was a metropolitan landscape that integrated the central city with the suburbs, creating a pattern of development that was hailed as a national model.

By the 1920s and 1930s the metropolitan parks were well established and heavily used, as the Metropolitan District Commission added a wide array of ballfields, golf courses, tennis courts, swimming areas, and even a ski run on the Blue Hills. Regional and park planning was evolving from an emphasis on beautification and preservation of nature to providing a broad array of recreational opportunities. During the 1910s and 1920s the MDC also developed the first regional highway system, converting the sylvan parkways into a network of four-lane motorways. In this era of the City Functional, elite reform and moral uplift no longer drove park planning. The overriding goal of public policy was to create an efficiently functioning metropolitan infrastructure. Thus the state created the MDC in 1919 to combine the region's sewerage and water supply services with the parks and parkways. During this era the suburban landscape of the Metropolitan District, which the Metropolitan Park System was intended to shape, reached maturity.

After World War II the tenor of regional and park planning changed dramatically. The highest priority was construction of a high-capacity, limited-access highway system to accommodate burgeoning auto use and to facilitate development of the outer suburbs. In the city this entailed widespread demolition, while in suburbia the highways plowed through natural areas. Planners gave little thought to environmental preservation or beautification, as utilitarianism trumped aesthetic considerations. Postwar suburban flight weakened urban communities, leaving them less concerned about maintaining parks. According to analysis by the Metropolitan Area Planning Council, "open space for active and passive recreation, conservation and related purposes has been considered a land use of secondary importance, a luxury that need not be considered until other forms of land use are established."[79] This era spawned chaotic suburban sprawl that communities have long been struggling to repair.

Not until the 1970s and the rise of the environmental movement did the public once again begin to appreciate Boston's metropolitan parks. This sentiment encouraged the rehabilitation of the metropolitan parks, restoration of the beaches, and creation of new bike and walking trails. The parks revival, despite periodic funding shortfalls, helped reintegrate the city and suburbs into more of a metropolitan whole. After years of neglect, planners, politicians, and citizens rediscovered Charles Eliot's system of parks and parkways, regarding it as an urban icon, a soothing escape from the stresses of city life, and a building block for community redevelopment.[80]

10 Reclaiming the Middle Charles River Reservation

Daniel Driscoll and Karl Haglund

IN 1987, near the economic high point of what was then called the "Massachusetts Miracle," the state legislature passed a five-hundred-million-dollar bond bill for the purchase of open space. It allocated fifty million dollars to the Metropolitan District Commission (MDC), the successor agency to the Metropolitan Park Commission.[1] Acclaimed at the beginning of the twentieth century as the first regional organization of public land in the world, the MDC had shifted its focus to active recreation facilities in the 1950s.[2] The 1987 bond bill followed the appointment as commissioner of William Geary, a Boston native who had grown up near the parkways and beaches that the MDC managed. Geary was determined to renew the open lands under the agency's stewardship. He hired eight planners, and under the banner of "Bringing Back the Green," the agency produced a video presentation and held eleven widely publicized meetings at public venues throughout Greater Boston to solicit comments on the most valuable remaining open space that it might add to the twenty-thousand-acre park system.

The agency acquired almost two thousand acres of land between 1988 and 1995, but by 1990 the Massachusetts economy had deteriorated precipitously. The MDC park management budget declined by 31 percent and full-time staffing fell by 38 percent during those years.[3] Several of the planners hired under the bond bill began work on some of the neglected properties already under the agency's stewardship. One of those projects looked at land in the middle section of the Charles River Reservation from Watertown to Newton. Over the next twenty years a significant habitat restoration project reclaimed dozens of acres the commonwealth had acquired a hundred years before, and the skepticism and opposition of the abutting landowners largely transformed into vocal support for a new riverfront greenway. Beyond the conflicts over access, vandalism, and maintenance that such corridor projects have faced across the country, the agency would ask property owners to remove illegal incursions on state property—lawns, gardens, commercial parking lots, and equipment storage—and to pay for the restoration of the land in lieu of back rent.[4] The construction of the greenway reflects a successful effort to strike a new balance among conflicting interests and to transform encroachers into supportive partners in the stewardship of public lands.

The Vision

In their 1893 plan for the Metropolitan Park system of Greater Boston, Charles Eliot and Sylvester Baxter created a vision that called for "reserving" —acquiring as a public trust—the "rock hills, the stream banks, and the bay and the sea shores" surrounding the city. The Charles River Reservation, extending inland in a meandering course for almost eighty miles, was the centerpiece of this regional vision. The best-known stretch of the river is the Charles River Basin. In 1910 the construction of a large earthen dam below the Longfellow Bridge transformed the once tidal estuary, established a fixed water level for the easternmost nine miles of the river, and covered forever the wide expanse of noisome mudflats that previously had been exposed at every low tide. Harvard University and the Massachusetts Institute of Technology began construction of their riverfront campuses, and the Charles River Basin and its esplanades became the foreground for the long views of Beacon Hill and the Back Bay, and for the more intimate views where the river narrows along the parkways abutting Harvard's campuses in Cambridge and Allston.[5]

Upstream of the university, the public margins of the Charles in Watertown, Newton, and Waltham were managed more intermittently, and long stretches reverted to scrub growth. Nearby urban development restricted access to the river and obscured its central place in the region's environment. As the planners in the 1990s saw it, the reclamation of this almost unknown portion of the river would connect the now densely developed cities west of Boston and Cambridge with the basin and restore a missing link in the regional vision of the park system.

The work of the last two decades of the twentieth century realized many of the objectives articulated in the four-volume *Open Space and Recreation Program for Metropolitan Boston* of 1969, which had argued explicitly for a renewal and expansion of Baxter and Eliot's 1893 plan. The four volumes provided detailed recommendations for land acquisition and landscape development, and one volume was devoted exclusively to the margins of the three major metropolitan rivers—the Mystic, the Charles, and the Neponset. The open land along the rivers offered the possibility of "linear open space within easy walking and driving time of every inner community."[6] But of the 224 miles of waterfront along the three rivers, only 86 miles was in public ownership, and the fencing off of private property prevented what the report called "'linear' public use"—what we would now describe as greenways.[7]

For the Charles just upstream of the Watertown dam, the report notes that the "early plans for acquisition and control of the banks were never completed as proposed," and this section of the Charles was "being squeezed by residential, commercial and industrial development as well as by incompatible public uses."[8] In promoting the report's recommendations, the staff of the Metropolitan Area Planning Council and their consultants met with town planning and conservation commissions and showed photographs of future opportunities—parking lots extending to the edges of rivers, missing links in the open space network. Other groups were enlisted in the cause, in particular a number of chapters of the League of Women Voters. Out of these efforts new organizations developed; among the most important were the Conservation Law Foundation and the Charles River Watershed Association.[9] As in the 1890s, the discussion of open space was the starting point in the promotion of a regional vision.

To lay the groundwork for action in the communities along the middle Charles, citizens and city staff in each affected community undertook detailed studies of their portion of the river. The work reflected in four reports completed in the 1970s was of particular importance. The Boston Conservation Commission published *Charles-to-Charles: A Conservation and Recreation Corridor for Boston, Brookline, and Newton.* William Giezentanner, writing for the Newton Conservation Commission, confirmed that the Charles "is perhaps the most important natural recreational and educational resource in the City." He mapped a continuous footpath along the river and documented that the path already existed along eleven miles of the thirteen-mile route. John Wacker and Associates completed studies for both the Watertown and Waltham Conservation commissions.[10]

The First Steps: Reclaiming the Riverfront

For the first three miles above the Watertown dam, residential and commercial land uses alternated in Watertown and Newton. Backyard gardeners, small-scale businesses, and the declining remains of nineteenth-century brick factories lined the river, sometimes extending down to the water's edge. The river changes character again in the center of Waltham, where in 1814 the Boston Manufacturing Company had taken advantage of a twelve-foot waterfall to build the first power looms in America. Damming the river at Moody Street to drive the wheels and shafts of the new factory, the company also created the "Lakes District" upstream—a series of meandering coves and inlets along the riverfront in Waltham and Newton. Although the river was much wider here, by 1970 the same jumble of land uses dotted the water's edge.

Between Moody and Prospect streets was a large commercial property that included a narrow strip of open land along the Charles, extending thirty-one-hundred feet between the river and the railroad, the longest stretch of privately owned riverfront between the Watertown dam and the town of Weston. Adjacent to a portion of the property were the commuter rail tracks that the Massachusetts Bay Transportation Authority (MBTA) managed. When the property went up for sale in 1985, the MDC was interested in the riverfront portion of the property but could not afford to purchase the whole parcel. Ron Vokey, planner for the City of Waltham, Charles Stewart at the MBTA, and Julia O'Brien, the director of planning for the MDC, discussed the possibility of a three-way transfer of parcels. Without waiting for the land transfers to be resolved, the city hired Carol R. Johnson Associates, a firm of landscape architects and planners, to develop a preliminary design, and the MDC agreed to pay for construction if the land could be transferred to the agency.[11]

In the final agreement, following extensive negotiations with city and state agencies, Nova Biomedical Corporation agreed to donate almost five acres along the river, most of it then part of an existing parking lot.[12] Nova also paid $65,148 to the MBTA for an easement along the tracks for use as replacement parking. The MDC paid the MBTA $61,386 for the riverfront parcel between Nova's donated parcel and Moody Street. Permitting this short stretch of riverfront for a public pathway required the approval of the City of Cambridge Water Board (since the city's reservoir along Route 128 supplied all of its water through a thirty-six-inch pipe that passed under the property), the conservation commissions of Newton and Waltham, the U.S. Army Corps of Engineers, the Massachusetts Environmental Policy Act Office (MEPA), the Massachusetts Water Resources Authority (MWRA), the Massachusetts Historical Commission, and the Massachusetts Natural Heritage Program. After seven years of property negotiations, permit review and approval, and construction, the pathway was finally dedicated in October 1992.

As in Waltham, the Watertown Conservation Commission had been discussing improvements along the river for several years before commissioning the 1975 study of the river. And as in Waltham, the first reclamation of riverfront land followed from a major building project with an adjoining parking lot along the Charles. The largest building in Watertown Square at the start of the twentieth century was Lewando's, a massive, rectangular, three-story brick structure with the company's name and business—"Cleansers Dyers Launderers"—spelled out on top of the building in large letters visible from the square and along the riverbank.[13] Riverbend Office Park, the new owners of the building, inherited a paved parking lot built by the previous owner that extended from the building to the river, part of it on public land. The city's conservation commission brought up the possibility of funding a four-hundred-foot walkway, and the partnership agreed to pay for the construction and provide an easement to the MDC. "The river walk," as one of the company managers said at the time this section of the greenway was completed, "doesn't go anywhere. It ends with a cyclone fence. But it definitely is used."[14] The paving was torn up, wild flowers were planted along the half-acre parcel, and the park opened in June 1992, a few months before Riverwalk in Waltham.[15]

Once the construction of the new walkways in Waltham and Watertown had begun, the staff of the MDC Planning Office turned to the larger vision for the three miles of the Charles between Riverbend and Riverwalk. The 1890s annual reports of the Metropolitan Park Commission (the MDC's predecessor agency) described acquisitions by the commonwealth of most of the riverfront land in Watertown, Waltham, and Newton, and most of the taking plans were still in the agency's archives. But all the boundary markers in the field had disappeared. To realize the hundred-year-old metropolitan vision for the Charles would require a new survey. Even though encroachments were a familiar problem for the managers of the commonwealth's extensive parks and forests, in 1990 there was no formal program to identify or resolve conflicts over the use of state lands.

Recognizing the potential for unhappy neighbors if surveyors appeared unannounced along the property lines of residents and businesses, the MDC distributed a letter in August 1991 over the signature of the MDC commissioner, Ilyas Bhatti, announcing to abutters and local political officials the start of the survey. Its contents were quite general. The scope of the work was to conduct "a boundary survey of selected stretches of the Charles River shoreline in Watertown, Waltham and Newton"; the "ongoing objective" of the agency was "to expand and improve river corridor open space for the enjoyment of local business employees, community residents and passive boaters. . . . Clarification of MDC park boundaries will allow the MDC, working with private abutters, to better assess potential park land improvements."[16] The words "greenway" and "riverwalk" were not to be found in the notice, and there was no mention of any interest in building continuous paths along the river's edges. Originally scheduled for completion in October, the survey took place the following spring. The *Boston Globe* reported that the project "left a trail of bright orange stakes and stone markers along both sides of the river and a list of 150 businesses and residences encroaching on public land" (the final number was 92).[17] Concerned that abutters might be upset by the now-visible boundaries and pull up the survey stakes, the MDC made a video recording of the entire survey line.[18]

In June 1992 the agency organized a public meeting to present the results of the survey. The *Boston Globe*'s advance coverage of the meeting included the first widely circulated statements of a much more specific objective for the Charles: "a continuous river walkway linking Watertown Square to Needham's Cutler Park." If that were successful, it would be then possible to walk along the river from the Hatch Shell in Boston to Route 128 in Newton. Commissioner Bhatti tacitly acknowledged that the agency had neglected its stewardship of this stretch of the river for decades by saying he hoped the project would "demonstrate to the community a reaffirmation of our commitment to open space."[19]

Several abutters shared their opinions with the press before the meeting. The vice president of a construction equipment company wondered if new public access would lead to security problems.[20] A past president of the

Boston Area Bicycle Coalition raised several specific safety issues—a fire hydrant was in the path along Riverwalk Park, a curve along the Nova Biomedical property seemed too tight, and as the trees along the path grew they might block the views for cyclists. "Here at last," he said, "is a place where a 5-year old can come down and ride, but it should be reasonably safe."[21]

Francis Boudrot, a sixty-eight-year-old resident of Pleasant Street in Watertown, whose vegetable garden extended onto MDC riverfront property, said before the meeting, "We're against it, naturally." He had had a garden there for thirty years and did not, according to the *Globe*'s reporter, "like the idea of a new footpath so close to his yard."[22] In an interview five years later, Boudrot's wife, Theresa, observed that the other abutters along the river were not owners: "A lot of people probably weren't interested in the meetings. They probably thought that it would never become a reality." At first she and her husband did not think it would happen either.[23] Other neighbors spoke in favor of the plans. Another Pleasant Street homeowner suggested that there already was an issue of teenagers loitering along the river. She said that "if it's cleaned up and well-lit and more open, teen-agers are less likely to hang around here." And the manager of a business on the Newton side of the river said the company knew it was encroaching and would lose space but still favored the park development.[24]

A few months later, after Riverwalk Park had opened in Waltham, the *Globe* wrote again about the middle Charles development and described a "grand plan to extend riverside paths another 13 or 14 miles . . . from Watertown Dam . . . to Brook Farm at the West Roxbury-Needham line." The story compared the new projects in Waltham and Watertown with River Street Park along the Charles in Wellesley, created from an area used by the town to dump snow. A manager for Spaulding Investment Company said that "it was an eyesore before, overgrown; now it's really a plus. Our tenants who look over the river through it, they love it."[25]

Resolving Encroachments

Why were there so many encroachments on this short stretch of river? Kevin Hollenbeck, a site supervisor for the MDC in the south region of metropolitan Boston, suggested that a large majority of the residential abutters to the state reservations encroached one way or another. It might be from ignorance of the boundaries, or because the agencies did not maintain the property lines. Some people were well aware of what they were doing, but suggested that they were improving the adjacent property: "You will stir up a hornet's nest when you tell people to get off the land. They may threaten to go to their [state] representative." Another problem was the difficulty of funding surveys. Although the commonwealth owned 450,000 acres of forests and reservations, at the start of the greenway projects along the Charles it employed no registered surveyors. Surveys are not a priority for the park agencies: "No one is pressuring the agency to get the encroachments off MDC property," Hollenbeck suggested. "If you have an ice skating rink roof fall in, you will instantly have an organized group of people protesting to fix it. No one writes letters to the Commissioner complaining that their neighbors are on MDC property. The people say, 'There are 6,000 acres here. Why are you kicking me off my 100 feet?'" Commercial abutters, however, are more likely to cross over the line for their own benefit, to the detriment of the public landscape.[26]

Along the Charles the MDC had ignored the encroachments for decades and had made little effort to use the land or to notify abutters of the state's interests. In the two decades after World War II, the agency focused on active recreation and leased or sold large tracts of open land for commercial and institutional real-estate development. The parcel that Nova donated to the MDC for Riverwalk Park, for example, had been sold by the agency for next to nothing almost forty years before. Although the survey contract for the Charles was only twenty thousand dollars, it was nevertheless a major effort to secure approval for the project. Once the surveyors began working, the MDC also engaged the state attorney general's office in a lengthy discussion to determine alternatives for resolving the encroachments. The staff of the MDC Planning Office had hammered out long-term planning and management objectives for the state's land before the survey work began, and those goals were crucial in persuading James Milkey, an assistant attorney general for the commonwealth, that his office should invest in a major effort to settle the state's property rights.

In addition to identifying encroachments, the legal review also included current and expired uses granted to some abutting property owners. Milkey would later say that the legal work along the middle Charles may have constituted the most encroachments ever resolved in a single area in the commonwealth.[27] The illegal uses included well-tended gardens such as the Boudrots', abandoned and overgrown plots, and active industrial and commercial uses such as the Boston Edison site, where an array of telephone poles and utility boxes were regularly used to train field service technicians. Boston Edison's was one of only a few encroachments by large corporations; most were by small businesses and residential properties.

After extensive discussions of the legal issues, the MDC and the attorney general's office determined that the development of a consistent, objective formula would reduce the possibility that the abutting owners would find the damage assessments to be arbitrary and would minimize the likelihood of litigation. Instead of charging back rent, which the law allows, the final formula was based on the cost of removing the "semi-permanent" encroachments and restoring the land by removing one foot of substrate and replacing it with one foot of loam, then seeding the area: eight dollars per square yard for existing gravel surfaces and twelve dollars per square yard for paved surfaces. To guarantee that the damages collected would be used to carry out the restoration of the land (and not be transferred to the state's general fund), an "Encroachment Reclamation Trust" was established to hold the collected damages.[28] This was the agency's first trust fund for maintenance, and it required a lengthy set of approvals, including the state's Executive Office of Administration and Finance as well as the governor's office. The trust was instrumental in persuading abutters to pay damages.

Every encroaching abutter received a written notice of the requested assessment and a form for "Acknowledgement of State Ownership/Agreement to Quit." For "semi-permanent" encroachments, there was a memorandum of agreement to detail the circumstances of each such property. Months before abutters received these forms, they had all had several formal or informal meetings with MDC staff to discuss the basis of the damage claim and to learn how the commonwealth would use the funds. The approach that the MDC and the attorney general's office developed was remarkably successful, in no small measure because of the three years of door-to-door, property-by-property negotiation between 1991 and 1994. Despite the initial fears of some abutters and the warnings that state legislators would prevent the construction of the greenway, all ninety-one of the encroaching property owners signed agreements with the commonwealth, and a maintenance trust fund of $120,000 resulted from the collected damages.

As the *Boston Globe* noted, "What is surprising is not that the MDC finally retook its own land, but that it has resolved the vast majority of the encroachments without a single lawsuit."[29] Fifteen of the encroachments were resolved immediately, twenty-four were relatively minor, and forty required payment of damages to the MDC. Of the forty requiring payments, twenty-nine were commercial properties and eleven were multifamily residential. The largest payment by a commercial owner, Boston Edison, was $30,000; the average was $6,200. The average payment by residential property owners was $2,200. Eleven encroachments on scenic easements, with no public access rights, remain to be resolved.[30]

Even the opponents offered their grudging respect for the lengthy negotiations. The owner of a dry-cleaning business, who employed twenty people at a processing plant on the river, noted that "the best thing I can say is this could have been a disaster but wasn't. I think they're doing a great thing in restoring the banks of the river, but I think they should have been a little more sensitive to businesses. . . . They've owned this land for a long time and never done any improvements or upkeep and now they are charging me to remove asphalt I didn't put down."[31]

The Master Plan and the First Construction Projects

The funding for the first phase of planning was set at $350,000, and the MDC hired Carol R. Johnson Associates to conduct an inventory of the reservation, to analyze in detail the issues identified, to develop alternatives, and to prepare detailed recommendations for the corridor.[32] Near the end of the resolution of the property issues, in November 1994, the agency convened a public meeting to announce the beginning of a "Restoration Master Plan" for this portion of the river. Commissioner Bhatti opened the meeting by emphasizing the opportunity to make the Charles "the people's river." Julia O'Brien, the director of planning, confirmed the importance of citizen participation in designing the greenway to meet the communities' aspirations. A slide show reviewed the river's decline between 1930 and 1970, as well as the revival that Riverwalk and Riverbend parks represented. Images of the riverfront confirmed that despite the decades of residential and commercial encroachments, the reservation nonetheless supported a thriving and diverse wildlife habitat. Residents, local politicians, and city staff were invited to sign up for the Citizens Advisory Committee and commit to working together to resolve the competing interests that had already been identified in the discussions with encroaching abutters—security and privacy, preserving and restoring wetlands while creating public access, enhancing wildlife habitat and encouraging the active presence of walkers, joggers, and bikers.[33]

Since the riverfront had been ignored for decades, the participants in the advisory committee meetings had few preconceived notions of what should be done. O'Brien later said that "this section of the River had been unavailable for easy public use for so long that it didn't have the sort of constituency that the Charles Basin, for example, has . . . people are going to be skeptical that the plan is going to reach an implementation phase in a short enough term to make it worth their while to be involved. The demands on good people's time are extensive." In her view the project was "for the most part, noncontroversial—we were taking a bad situation and making it better."[34] The relatively small group of participants in the meetings were receptive and enthusiastic about the proposed designs.

Published four years later, the master plan identified ten goals for the project, outlined in general terms:

1. Improve access to the river and the greenway.
2. Improve circulation and open-space connections along the river corridor.
3. Eliminate gaps in public ownership.
4. Reclaim all MDC property.
5. Protect and enhance the character of open space and the shoreline along the river.
6. Protect and improve the visual and scenic quality.
7. Promote sustainable environmental quality.
8. Maximize educational and interpretive opportunities.
9. Limit potential conflicts between reservation activities.
10. Maximize the safety of reservation users and the privacy of abutters.

Some of these goals were clearly in conflict with others. Improving access, for example, would require cutting hard-surfaced paths through neglected but often lush natural growth. Could access for people along such narrow corridors be significantly increased while at the same time enhancing wildlife habitat? Abutters would realize the greatest privacy by requiring opaque barriers, natural or constructed—but that would interfere with the principle of "eyes on the park" that by this time was a hallmark of park planning. The plan addressed these issues by providing sketches and photographs to illustrate how to apply general design principles in diverse riverine landscapes. In several stretches there would be elevated wooden boardwalks over narrow, hard-packed dirt paths, allowing the reestablishment of wetland plants below the boardwalks. Pathways would be introduced into some natural areas, but the removal of extensive zones of gravel and asphalt, to be replaced with grasses and indigenous vegetation, would compensate for the loss of this habitat.

An unusual aspect of the master plan contract was that it also included the preparation of construction documents for three segments of the pathway. In the past the usual approach at the agency had been to complete master plans as the first stage of developing or expanding the reservations, and then to wait to prepare final design drawings when the legislature had appropriated funding for construction. With the blessing of the agency's contract administrator, the scope of the contract with Carol Johnson included the final design and permitting of three segments of the riverwalk that had been identified in the master plan and approved by the advisory committee and the agency. Because construction documents were complete and permitted, the MDC was able to move quickly when a major new source of funding appeared: the 1991 federal transportation funding bill, the Intermodal Surface Transportation Efficiency Act (ISTEA). The federal appropriations, approved every six years, allocated gas tax funds to state highway projects according to a congressionally mandated formula. The 1991 ISTEA legislation marked the first time that federal gas taxes were available to fund "enhancement activities," including pedestrian and bicycle projects.[35] A grant from the Massachusetts appropriation paid for most of the construction of the first segment of the master plan, which was chosen because of the support of the Watertown community.

A major, unanticipated difficulty in the final design of the three segments was the siting of the paths. The landscape architects developed the alignment from data on aerial photographs, and the surveyors were directed to stake out the path's center line in the field. When the MDC planning staff reviewed the proposed alignment, however, they were startled by the number of mature trees that were slated to be removed. The project manager for the MDC determined to revise the alignment in the field, a hundred feet at a time, by taking a garden hose to mark the modified center line of the pathway and then driving stakes along the new route. The revised layout avoided most of the big trees.

One of the most important and most successful of the design elements was the introduction of granite entry markers, four feet high and etched with the blue heron logo of the project. These markers would signify the entrances to the river walkway, which was especially important along the busy roadways where the path had to cross the traffic lanes. Although the newly built walkways created fifteen-foot breaks in the dense vegetation along the river, the entry points would be easy to miss without the granite markers.

And where breaks in the path required park users to follow city sidewalks, blue painted "heron tracks" marked the routes to the connecting sections of the riverfront walkways.[36]

Another planning issue that evoked strong opinions was the physical form of new access to the water's edge. Where the grades were flat, the design called for narrow paths that would cut through the existing vegetation. In wetland zones wooden boardwalks would be built over the legally protected areas. The landscape architects and some MDC staff originally argued for locating the pathways as close to the water's edge as possible, to maximize river views. But this seemed to some to be an unnecessary conflict with the goal of habitat restoration. In the final design the paths were located as far away from the river as possible, with access to the riverfront provided by short paths or boardwalks perpendicular to the main walkway. These off-path overlooks were also a subject of spirited discussion, since some feared that they would be vandalized and become hangouts for illicit activities—perfect spots for the activities that pathway opponents feared. Looking back on almost a decade of experience, however, the overlooks have become favorite spots for people who want to step out of the way of walkers, joggers, and bikers—and sit quietly on the edge of the water, watching the movement of the current, waiting patiently for a glimpse of the river's wildlife.

The singular design landmark of the whole project is the Blue Heron Bridge, completed in 2006, which links Newton and Watertown in a densely wooded section of the river. Designed by T. Y. Lin International, a well-known engineering firm, it has "quickly become a revered local landmark. In the summer, fathers and kids throw fishing lines into the water off of it, angling for perch and bass, while nearby wedding parties take photos. Early in the morning during much of the year, actual blue herons can be spotted from the bridge, perched on rocks in the river, doing some fishing of their own."[37] On a recent evening, a planner for the MDC approached a group of teenagers who were lingering on the bridge and asked them what they thought of the structure. One of them said, "This place rules."

Continuing Conflicts

As the first segment of the greenway neared completion in 1997, a spate of newspaper articles described the project and cited some of the conflicting opinions about maintenance and access. Would the new path be maintained?

The president of a business that had agreed to pay thirty-four hundred dollars in damages observed that the MDC has "a half-dozen parks that have deteriorated to the point they are unusable. Is that what is going to happen to this one?"[38]

Arguments were made that the greenway would be too accessible—or not accessible enough. Neighbors in Newton reportedly felt that the greenway might "be a threat to the area's security because it would increase access from Boston to the surrounding suburbs."[39] A Watertown police sergeant acknowledged that his department was aware that some of the locations along the path were isolated, but he said that officers on foot or on bicycles would patrol the riverfront.[40] The city-owned Mount Feake Cemetery in Waltham determined not to grant formal access to any part of the cemetery. The cemetery commissioner said, "It's a cemetery. It's not a riverwalk. I don't believe a cemetery and a riverwalk go together." But the superintendent suggested that joggers and bicyclists were usually left alone if they were "behaving properly."[41] Theresa Boudrot, whose husband had objected five years before in one of the first public meetings, said that she was "against it right from the very start and still am, but now that it's there, I use it every day—even in the rain."[42] She later asked the MDC to cut a "window" in the dense vegetation she had initially insisted on, so she could visit with all the people she had met along the riverwalk. And in a recent public meeting, Boudrot stated that every so often "you have to admit you were wrong."

Given the up-and-down state of the commonwealth's budget during the years of planning and construction, it is surprising that there were no written objections to the cost of the project, or the possibility that the state should not be in the business of developing local greenways. Letters to the agency have been universally favorable. One nearby resident wrote to the MDC commissioner and offered to donate money and to help raise money for plantings along the path. "For once," he added, "I feel that my tax dollars have been wisely spent."[43]

Over the next several years, as subsequent phases of the greenway were completed, supporters acknowledged the conflicts that had been raised but argued that as the work continued, the evidence favored the riverwalk. A columnist in an essay on "Favorite Places" in the newsletter of the Massachusetts chapter of the Sierra Club wrote that "there is more vegetation alongside the trail, and the overlooks provide unobstructed views of the river and its denizens. . . . It doesn't feel like a city park."[44] A Waltham city councillor agreed that maintenance would be a big problem, but the agency has to "work with what they're given by the Legislature." His primary concern about the project was "just that it gets done."[45]

John Rosenthal, who donated land to the MDC at Bridge Street in Newton to fill a gap in the path, acknowledged to a Watertown reporter that he was "a huge supporter of public access to the river because maintaining this pivotal resource is extremely important to me as a person. Right now, I am sitting in my office, looking at a great blue heron standing in the middle of the river, fishing and getting dive-bombed by two house finches. What could be a more wonderful thing to look at while at work?"[46] The connection of the riverwalk with the environmental history of the region was explicitly addressed in a letter written to the agency soon after the first segment in Watertown was completed. After stumbling upon the first segment, the writer said, "I was so impressed that I returned that same morning to show my wife. . . . It has often been felt that the Charles River's history is not sufficiently emphasized, its beauty not fully appreciated, and its presence underutilized. This new riverwalk helps resolve that."[47]

A letter to the editor of the *Boston Globe* at the end of December 2004 lamented the increased human traffic along the river, the noise, and the heightened threat to wildlife. In response, Brookline resident Louise Bowditch described a walk along the Charles taken by three generations of her family on New Year's Day, 2005: "Previously no access had existed for those wanting to enjoy the great natural resource of the Upper Charles; much of the area was covered in industrial pollution and asphalt. . . . Now we can all enjoy the urban wilds of the Charles in ways never possible before. Blight has been eliminated, and wildlife can continue to delight as well in the meandering jewel in our midst."[48]

THE CLIMATE AND THE WEATHER

11

Boston's Weather and Climate Histories

William B. Meyer

THERE is no part of their environment that Bostonians talk about more than the weather and climate, and for good reason. For one thing they have plenty to talk about. The city's location in the path of a set of major storm tracks gives it an unusually wide and fast-changing repertoire of weather—atmospheric events and conditions with lifespans ranging from hours to days.[1] Not only in time, but also in space at any given moment, Boston's weather can exhibit some of the "sumptuous variety" that Mark Twain ascribed to that of New England in general. On hot summer days the famous sea breeze can cool the immediate coastline deliciously without reaching the inland suburbs or even the city's western neighborhoods.[2] In winter rain in the city center and along the coast often falls as snow farther west.

These spatial patterns also appear in statistics of climate, which average the weather over a number of years into profiles of typical conditions. Differences in elevation, in proximity to the ocean and to the city core, and in local land cover produce somewhat different climates in different parts of the urban area. The Blue Hill Meteorological Observatory, only ten miles from the national weather station at Logan Airport, but 635 feet above sea level and located within one of metropolitan Boston's largest parks, gets half again as much snow in an average winter as does downtown, which in turn gets half again as much as ocean-girdled Cape Cod. Variety appears also in some climatic features common to the whole urban region, such as four strongly contrasting seasons. ("Who lives one year in Boston," Emerson wrote, "ranges through all the climates of the globe."[3]) Still, there is enough unity in diversity to give the city's climate overall a definite character, and one that falls short of most people's ideal today. Therein lies the second reason Bostonians talk so much about the weather: they have plenty to complain about. The "Athens of America," like another "second Athens," Edinburgh, is not a sunny Mediterranean city but distinctly a northern one. The pictures of it that come most readily to mind, cheery postcards for tourists notwithstanding, are chilly and gray, swept by ocean gusts and northeast storms. In some ways reality outdoes the images; Boston is wetter than Seattle, windier than Chicago, foggier than San Francisco, hotter in July than San Diego, and colder in January than on the coast of Iceland.

Boston, then, has plenty of weather and climate. In what sense does it have a weather and climate history? In several senses, as it happens, such that one would do better to speak in the plural of Boston's weather and climate *histories.* The city has at least three sets of them, interacting in many ways but none reducible to any of the others. First, Boston weather has one history that lies largely within the domain of the natural sciences. They reconstruct and analyze major weather events; identify the climatic changes that those events, in the aggregate, display; trace those changes to natural forces and/or to local- and global-level human-induced impacts; and investigate the effects that those changes have had on other natural systems. Meteorological historians, such as the late David Ludlum, have documented the origins, spatial extent, and physical dimensions of such events as severe storms, heat waves, and cold spells.[4] Climate historians, working at longer time scales, have shown that the region was distinctly colder in the seventeenth and eighteenth centuries than it has been since, and it is currently in its warmest phase of the whole era.

Within the colder period, as the historian William Baron and colleagues have documented, some decades were distinctly warmer or colder, wetter or drier than others. Baron and his collaborators went beyond early instrumental records of temperature and precipitation, which are scanty and uneven in quality, to glean insights from qualitative records (for example, comments in manuscript diaries and newspapers on such important phenomena as early and late harvests, frosts, storms, and droughts).[5] Abraham J. Miller-Rushing and Richard B. Primack, in their chapter in this book, likewise use nontraditional data (in their case, herbarium specimens, photographs, and personal journals) to document an advance in times of plant flowering and the arrival of migratory birds as the Boston area has warmed during the past century and a half.

When questions are raised about what weather events or climatic trends meant for people and why, we enter a second kind of weather history.[6] Unlike the first, it unavoidably involves the social and human as well as the natural sciences. How any atmospheric event, any climatic feature, or any shift in climate affects a society, whether for better or for worse, and for whom, will depend on that society's ways of life. Changes in technology, in livelihood, in population size and settlement patterns, in affluence, in gender roles, and in innumerable other variables can transform the role played by the weather and climate even while they remain stable and will likewise determine what any change in the climate will mean. The longer growing season Miller-Rushing and Primack indicate brings less economic gain now than it would have done in the past because of the precipitous twentieth-century decline of farming in the metropolitan area.

Each of the weather disasters so frequent in New England lore was catastrophic not only because of the weather event involved, but also because of a particular set of activities that made people vulnerable to its effects. The utopian writer and reformer Edward Bellamy (1856–1898) suggested many ways in which the replacement of a competitive by a cooperative organization of society might lessen Bostonians' problems with the weather without changing the weather itself. Such variables as cultural norms governing acceptable clothing will affect the way people experience heat or cold, a point the Harvard researchers C. P. Yaglou and Anne Mesmer neatly illustrated in 1941. The men and women in their studies consistently differed in the indoor temperatures they judged comfortable: not, they discovered, because of any inherent differences between the sexes, but rather because of the different way each was expected to dress.[7]

Finally, the ways in which people interpret or understand the weather form the subject matter of a third kind of history, one with closer affinities to the humanities than to the natural or social sciences. Certainly it overlaps a great deal with the other two. The research that gives us the first kind, natural-science inquiry into climatic change, is itself one way in which people think about the weather, with an interesting history of its own. Changes in people's attitudes toward the weather can often be traced simply enough to history of the second kind, changes in activities of theirs on which the weather impinges. When an eighteenth-century Boston diarist recorded "a splendid Rain" and "A most Delightful Rainy morning," he was not expressing an aesthetic judgment, but rather the normal reaction to the breaking of a dry spell in a premodern peninsular seaport that depended on frequent downpours to replenish its cisterns and wash its streets. When another observed that "the snow which has fallen, will be very useful in the roads: it fell very even, and has filled up the bare spots," he was speaking quite rationally for an age when horse-drawn vehicles and unpaved roads made snow cover an asset for easy and rapid movement.[8]

But the ways in which people view the weather are not strictly imposed either by the weather and its fluctuations or by the practical challenges of living with it. History even records discontinuities in what people have classified as "meteorological" and "climatic" events, as Lauri Bauer Coleman shows in her chapter in this book. She documents and explains eighteenth-century interpretations of such events and their meaning whose prevalence in their day, evolution over time, and eventual disappearance owed much more to changes in the climate of opinion than they did to any determinants associated with the climate in the literal sense of the word or any change in the effects the climate had on people. New England's writers have made much use of the weather and seasons, but in ways that have differed significantly from one cultural era and writer to another.[9] Present-day concern over global warming stems from changes in beliefs and values regarding the environment as well as from concerns about immediate practical consequences of climate change.

As the companion contributions ably illustrate the first and third of these forms of history, this overview devotes more space to the second. It follows Brian Donahue's chapter in this volume in using as its main organizing framework the three eras—eotechnic, paleotechnic, and neotechnic—which Patrick Geddes and Lewis Mumford identified and which Mumford presented most fully in his 1934 book *Technics and Civilization.*[10] The periodization rests on a sequence of key energy sources and the complexes of machines and social institutions through which they have been exploited. Energy in the eotechnic era came from wind, wood, running water, and the muscles of draft animals. The adoption of the steam engine and of coal for energy marked the shift to the paleotechnic era. Neotechnic society substituted electricity, petroleum, and natural gas for coal in many of its activities and powered them through the electric motor and the internal combustion engine. Given the key role of technology in mediating relations between people and their physical environment, it is as plausible a framework as any that could be used.

The roles played by the weather in Bostonians' lives and livelihoods have changed in ways that Mumford's periodization helps us to understand. So too have Bostonians' contributions to weather and climate change, and so too, in a much less determinate way, have their perceptions of the weather and its meanings. Of course the framework is not all-sufficient, as no single framework could be. It necessarily emphasizes technologically driven changes over others. Nor do its categories mark off separate and self-contained periods of history. The eras overlapped, as Mumford himself took pains to point out.

Eotechnic Boston

The English settlers who colonized the shores of Massachusetts Bay in the early seventeenth century brought the eotechnic civilization of their homeland with them. It could be said, indeed, that it brought them there as well, in the form of the compass-guided sailing vessels that carried the Puritan emigrants to the New World. To their Amerindian predecessors, who lacked the technology of sail power, the wind had been no asset at all; to the new settlers it was an essential resource that could also be a terrifying hazard. According to the wind's whims, the passage from seventeenth-century England to New England could last as little as five weeks or several times as long; it could be a placid crossing or a tempestuous one.[11] Shipping, the mainstay of the new port town's economy, was entirely dependent on the wind, and powerful storms after 1630 were a threat to life and property in ways they had not been in Amerindian times. Shelter from such storms had something to do with the choice of the Shawmut Peninsula as a site for settlement; the enclosed harbor behind its capes and islands offered ships a better haven in heavy weather than other, less protected possible locales did.[12] Such, more generally, was the relation of eotechnic-era Bostonians to the weather. Its importance for them as a resource made them, in the same measure, concerned with its extremes and its vagaries.

These extremes and vagaries began with the climate that they encountered in the vicinity of present-day Boston. It was different, and on the whole markedly colder, than today's. In what climate historians have dubbed the

"Little Ice Age," beginning in the fourteenth century and lasting well into the nineteenth, mean temperatures were lower by several degrees Celsius than they are currently, even in the parts of the metropolitan area least affected by modern urban warming. The difference translated into a number of implications for the colonists' activities. The growing season for farmers was shorter, and the risk of damaging frosts at the beginning or the end was greater. The cold of winter was more severe. Boston Harbor froze over, locking shipping in and out, more frequently than it would do in the nineteenth and twentieth centuries.[13] William Baron has proposed a number of research questions, which still invite study, about how New Englanders dealt with the severer climate of the Little Ice Age. What effects did it have, and what (if any) responses did it evoke in such areas as architecture, clothing, settlement patterns, or agricultural methods?[14]

The early colonists were, of course, coming from an England itself in the grip of the Little Ice Age, lessening the severity of the change that they experienced in relocating. It was not so great as to prevent them from largely reproducing their Old World ways of life in the New. They had expected some differences that they did not find, and found some that they had not expected. As the historian Karen Ordahl Kupperman has shown, the European settlers of the North American Atlantic Coast had taken it for granted that conditions would resemble those in Europe at the same parallels of latitude.[15] Massachusetts Bay lay well to England's south, and Boston, they thought, at the latitude of Marseille, should have had its seasons. It turned out instead that though summers were indeed hotter than England's, winters were not warmer but rather much colder. The weather, moreover, was more variable and its extremes—of drought, of storm, and of changeability—were greater. Moving to New England exposed Pilgrims and Puritans to a change of climate that was an abrupt and, on the whole, unwelcome surprise.

For a time the colonists trusted in divine intervention to change the features they most disliked. The hope grew out of the same providentialist religious beliefs that led them to regard daily weather events as signs of supernatural approval or chastisement (see Coleman's chapter in this book). The extremes of the seasons (by comparison to the English standard) troubled them the most, and they particularly looked forward to a moderation in the severity of winter: "Tame Thou the Rigour of our Clime," prayed the Boston merchant Samuel Sewall at the end of the seventeenth century.[16] Yet they did not do for themselves what they could easily have done to keep warmer. The same seasons were not equally severe for the colonists and the

Amerindians. The early settlers noticed that the open, airy summer dwellings of the Indians were cooler than their own frame houses and that the tightly insulated wigwams were warmer in the cold months.[17] Nonetheless, they held to their own ways, even though the open fireplaces and chimneys on which, as good Englishmen, they relied for heating were hopelessly outmatched against a Massachusetts winter.

Open fires furnished warmth mostly through direct radiation from the burning wood; powerful drafts through the chimney carried away most of the heat that they generated and pulled in cold air from outside. Even quite well-to-do Bostonians such as the Revolutionary-era merchant John Rowe could not escape the indoor temperatures on winter days so cold that, as he noted a number of times in his journal, "the ink freezes as I write." Settlers in North America from France, Germany, and Sweden, as they had done at home, heated their dwellings with closed stoves, which wasted much less fuel in heating the outdoors and kept the interiors as warm as desired. Yet for a full two centuries after the founding of Boston, its Anglo-American residents clung stubbornly to the open fireplace, even as the fuelwood that it consumed so profligately was becoming more and more expensive to bring to town.[18]

The extremes of heat, if less uncomfortable than those of cold, were at least as dangerous to health.[19] New Englanders of that era could be grateful for a climate that, if it did not support the rich staple crops of the English colonists in Virginia, the Carolinas, and the Caribbean, granted them longer lives. Eotechnic-era Boston was less healthy than the surrounding towns, and its peak season for deaths was in late summer, when warmer weather combined with crowding and poor sanitation to promote infectious diseases.[20] Human influence on the climate, in the form of the urban heat island (higher temperatures in town than in the surrounding countryside), also played a role. Bostonians began to note the phenomenon in summertime as early as the end of the eighteenth century, tracing it to the blocking of breezes by the crowded buildings and "the sun's heat . . . greatly increased by reflection and refraction from pavements, buildings, &c," and to discuss it mainly as a factor adversely affecting the town's health.[21]

The combination of the climate and the colonists' way of life divided much of Boston's economy into distinct seasons. Involuntary idleness and poverty rose in the winter, worsened by the higher cost of living. Employment and pay in such important trades as construction and public works became scarcer as the days grew cold and dark and the ground froze. The port was busiest, and shipping-related jobs most numerous, in the warmer months. When the seventeenth-century colony experimented with wage regulation, it permitted lower pay in the winter than in the summer. Eighteenth-century Boston's poorhouse averaged twice as many inmates in the winter as in the summer. Nonetheless, seasonal rhythms of work and life in town overall were more subdued than in the countryside. Marriages in rural colonial Massachusetts were few during the busiest months of agricultural labor; they were much more evenly distributed in Boston.[22]

For trade with its backcountry, eotechnic Boston relied on the modest power of animal muscles, a fact that, coupled with the poor state of the roads at most seasons, made nature's low-friction surfaces—waterways in the warmer seasons, snow and ice in the colder ones—exceptionally valuable for transportation of heavy loads or for rapid movement of people. Boston's very poor water routes inland were only modestly improved by the completion in 1803 of the Middlesex Canal. Commerce with the outlying towns was easiest in winter; snow did much to keep New England's metropolis cheaply and conveniently supplied with fuel and provisions. "In the winter-time, when the ground was hard with frost or covered with snow, clumsy carts and sleds, drawn mainly by oxen, were kept busy bringing loads of cordwood from the woodlots, or carrying corn, potatoes and other farm produce to market at Boston," wrote a historian of eighteenth-century Quincy. During the siege of 1775–76, only the fortunate chance of adequate snow permitted Henry Knox of the American army to haul the heavy cannon captured at Fort Ticonderoga eastward to Boston, where their emplacement on Dorchester Heights forced the British to evacuate. Bostonians were urged in 1784 to plant trees along the Neck to prevent the "useful Snow" from being blown off the road that linked them to the mainland.[23]

A legislative report in 1827 asked the state to promote the new technology of steam rail transportation because the snow cover that trade depended on was proving less and less reliable (indeed, as William Baron's climate reconstructions show, the preceding decade had been an unusually warm one). A series of milder winters that today would mean easier conditions for transportation meant alarmingly the opposite in the 1820s, jeopardizing both the provisioning of the city and the livelihood of the surrounding countryside: "For many years previous to the past season, the snow of our winters in Massachusetts has been gradually diminished. Instead of sleighing for four months in a winter, it is now rare for so many weeks; and in the neighborhood of Boston, during the twelve years previous to the present, the sleighing has not upon an average been more than sufficient for two weeks. Our winters have often been wet; our roads have been rendered soft, and nearly impassable for heavy loaded teams; and this has occurred at that season of the year, when the great agricultural interest has most regard to the means of intercourse with the market."[24]

Paleotechnic Boston

The use of fossil coal for fuel and of the steam engine for power marked the transition from the eotechnic regime to the paleotechnic. Both changes greatly expanded the energy available for human activities, and both brought many changes in weather-society relations, not all for the better. One of the first forms that the paleotechnic phase took in the Boston area was the steam railroad, and one of its first effects on weather-society relations was not so much to reduce the weather's importance as to transform it: to turn snow from a resource into a nuisance.

The tracks on which both steam railroads and horse-drawn streetcars ran in Boston had to be kept clear of packed snow and ice, but these vehicles long had to coexist with others that, for their part, found snow as useful as ever. "Steam engines," in historian Clay McShane's words, "the solution to long-distance travel and to powering factories, could not meet the need for enhanced movement within cities," for they were never successfully adapted to small, short-range vehicles, yet "burgeoning railroad networks generated increasing demand for local deliveries of freight and passengers."[25] The hired hacks, delivery wagons, and private carriages that met that demand continued to swap wheels for runners when winter came, for to them, as the historian Samuel Eliot Morison put it, "snow was a blessing rather than a curse."[26]

"The omnibuses are sleighs," marveled the English actor Henry Irving; "the grocer's cart is a sleigh—the express-wagons are sleighs—it is a city of sleighs!" A snowstorm that would have meant meager attendance at the theaters in London, he noted, had no such effect in Boston. Snow was even more prized for pleasure driving. Two famous engravings of sleighing in Boston's streets and suburbs by the young Winslow Homer in the 1850s convey some of the thrill of speed that could be experienced only in winter and on a snowy surface; so does an episode among the "swiftly dashing sleighs" in

William Dean Howells's *The Rise of Silas Lapham.* Public opinion strongly resisted the use of salt by the streetcar companies to melt snow from their tracks because it ruined the sleighing for the streets' other users.[27]

The application of steam power to navigation liberated ships from the whims of the wind, but as on land, the eotechnic and the paleotechnic coexisted well into the twentieth century. Sailing vessels remained a major part of the harbor's traffic, particularly in the coastal trade in such heavy, nonperishable items as coal, stone, and timber. For such commerce, the wind's cheapness as an energy source long outweighed its unreliability, even as the steamships quickly took over the passenger business and others in which speed and punctuality were more important.[28]

Much of the anthracite coal brought through the harbor was for home heating. Beginning in the 1820s, the paleotechnic complex of coal and iron brought about the most drastic and rapid climatic change (albeit confined to the indoors) the Boston area has seen to date. The adoption of closed cast-iron stoves and of central heating by furnaces, both much more effective than open fires, raised the winter temperature of most dwellings far higher than it had been before. Well-to-do Bostonians, for the most part, added furnaces while retaining their fireplaces, shunning closed stoves because they did not provide adequate ventilation. Poorer Bostonians adopted closed stoves once mass production lowered their cost. These powerful and fuel-efficient devices came into general use not when they might have seemed most necessary, but rather as the Little Ice Age gave way around midcentury to a worldwide warming that has lasted with minor interruptions to the present, intensified in the city by the urban heat island effect.[29]

The new and artificial indoor climate that stoves and furnaces created left their users uneasy on several counts. They worried about stale air and gave much thought to means of ventilation; they worried about excessive and perhaps debilitating warmth; they worried about the extreme dryness of heated winter air.[30] They learned to worry as well about interruptions in the supply of heating fuel from distant sources on which they had now become dependent. What the severest winters were on a scale of human experience cannot be told from a record of outdoor temperatures. They were the ones in which coal became scarce and costly, due to strikes in the anthracite fields in 1902 and several times in the 1920s, and the disruption of trade in the wartime winter of 1917–18, when brutal cold made the fuel shortage particularly inopportune and floating ice in the harbor added to the difficulties of transportation.[31]

Ice gathered on land became a valuable resource for coping with hot weather. Just as "coal famines" could make a hard winter more severe, mild winters could mean "ice famines" in summer, with higher prices or the faster spoilage of food. The ice trade was another business in which a long transition from the eotechnic to the paleotechnic gradually insulated Bostonians from the effects of weather fluctuations. The trade began on an entirely eotechnic basis in the early decades of the nineteenth century with the cutting near winter's end of ice from local ponds (beginning with Fresh Pond in Cambridge), its storage in insulated warehouses on their banks, and its shipment both to the cities of the Northeast and to ports in warmer climates such as the Caribbean and India. The inroads of the paleotechnic began with the use of railroads to move ice and ended with the use of steam engines to make it. Yet as late as 1900, although the shift from natural to manufactured ice was in full spate in most of the United States, it had not yet reached Boston, which lay in the northern zone, where the artificial product still could not compete with the natural in price. Only during the next few decades did improvements in ice manufacturing finally abolish winter cold's value as a natural resource for an important Boston-area trade, while solidifying residents' defenses against the challenges of summer heat.[32]

Steam, which made ice more reliably but less cheaply than the weather did, was also used for a time to raise crops on a more generous schedule than the natural seasons allowed. The commercial production of vegetables and flowers in artificially heated interiors, under glass and out of season, though it went back to the early nineteenth century in and around Boston, expanded in the postbellum decades and in the early twentieth century, the result of rising urban demand coupled with transportation still too slow to bring fresh produce from warmer regions. The controlled climate that boilers and pipes maintained allowed suburban Massachusetts farmers to distance the South at a bound and enjoy a growing season of 365 days a year. Though far costlier than conventional outdoor growing of the same items, it paid so well under the conditions of the period that greenhouses occupied a considerable acreage in Essex, Middlesex, Norfolk, and Suffolk counties by the early twentieth century.[33] While it lasted, it offered an example of a highly profitable form of agriculture liberated from what we usually assume to be the despotism of the seasons over farm livelihoods.

For farmers who did not work under glass, of course, the round of the seasons still dominated, and something of the same pattern existed too even for city workers. Passengers on the city's late-nineteenth-century street railways, a good index to business activity, were only half as numerous in January as they were in June. "In the labor force as a whole," the historian Alexander Keyssar has written, "there was always more idleness in winter than in the spring, summer, or autumn," although, he adds, different trades had different schedules of busy and slack seasons. Boston had less seasonal unemployment than smaller cities and towns, but jobs for its many construction workers, longshoremen, and day laborers dwindled in the cold months, coinciding with higher expenses for food, fuel, and clothing. An 1856 obituary for Nathaniel Wyeth, one of the pioneers of the ice trade, praised him in particular for having created paying jobs in ice-cutting at a time of the year when work was scarce. Extreme urban poverty continued to wax and wane with the seasons. The Boston almshouse's male population routinely doubled in the winter compared with the summer, and the overseers observed in 1899 that admissions closely followed the curve of average daily temperatures through the year, rising as it fell and falling as it rose.[34]

If the winter remained the most difficult time economically for most Bostonians in the nineteenth century, summer remained the deadliest season, chiefly for the youngest of them. Over time the mortality peak shifted from late to midsummer, and increasingly it reflected the results of the gastrointestinal scourge known as "cholera infantum" or "the summer complaint." In the late nineteenth century, adult death rates in Boston varied only modestly by season, but there was an enormous July–August peak in deaths below the age of five and an even higher peak for those in the first year of life. As in other American cities, this resulted from the interaction between the rapid growth of pathogens in the heat; the increased warmth of the crowded urban environment; the absence or inadequacy of refrigeration and sanitary protection for milk, water, and food; and the weak resistance of the very young to bacterial infection. A popular charity, the Boston Floating Hospital, launched in 1894, sought to help vulnerable children by exposing them to cooler air: "The hospital boat makes daily trips down the harbor during the months of July, August, and September, thus covering those summer months which are so baneful to children, especially those under five years of age, and so dreaded by mothers forced of necessity to live in the tenement districts of the hot cities."[35]

A summer exodus to the seashore or the mountains—cool, pleasant, and salubrious refuges from both the city's heat and odors and the sickness that was generally blamed on them—became a habit among upper-class Bostonians. It inverted the traditional seasonal calendar of work and idleness still in force among Boston's poor and foreshadowed the one prevalent today. Not only did urban heat every summer in this period take the lives of hundreds of young children, but it also could pose a threat to the adult poor who stayed and sweltered in the city. The usual images of severe weather in New England are ones of cold, snow, and high wind. Yet the deadliest weather event in its twentieth-century history, although it has never taken a secure place in the region's lore alongside the famous hard winters, hurricanes, and nor'easters, was the heat wave of July 1911. Within Boston, death certificates cited the weather as a cause of 346 deaths in that month and as the principal cause of 234 of them. The mortality rate was much lower in nonurban areas, where air movement was freer and the heat was less intense.[36]

Thus the challenges of cold, on the whole, had been met better than those of summer heat, which were of the most concern to Bostonians when they worried about how they might be changing their climate. At hearings held in 1869 and 1894 on the future of the Charles River flats, one of the chief fears expressed was of even sultrier summers to come if part of the water surface and its cooling effect were lost.[37] Elite Bostonians had once looked to divine favor to soften the severity of winter. Over the course of the nineteenth century and into the twentieth, however, they came to see matters more in naturalistic terms: chiefly in ones that emphasized the shaping of people and of races by their environments and the salutary sternness of the cold. "Not a luxurious climate," Emerson summed up Boston, "but wisdom is not found with those who dwell at their ease."

The Harvard professor Robert DeCourcy Ward (1867–1931) was both America's leading academic climatologist and one of its leading advocates of curbs on immigration from non-"Nordic" lands.[38] Yet new standards of thermal comfort fostered by furnace and stove heat increasingly belied New Englanders' self-image as a hardy northern race. They were prone now to complain bitterly about the indoor chill in England, where the open fireplace still held sway, while British visitors to Boston, for their part, complained about the overheated buildings.[39] Already, one could not, without considerable qualification, speak of most Bostonians as living in a cold climate.

Many elements of the weather by the end of the nineteenth century played different roles in Bostonians' lives than they had done a hundred years earlier, and extreme weather events were no exception. A number of geographers in the early 1980s proposed what they called the "lessening" and "worsening" hypotheses as two guides to assessing the changing effects of recurrent extreme natural events over time. Do societies become less or more vulnerable to such occurrences?[40] The latter thesis, at least, was adumbrated by some newspaper comment in the late nineteenth century. "That there have been storms of larger extent and more accumulation there is no doubt," observed Lynn's *Daily Evening Item* after a major blizzard in February 1899, "but the conditions are now so different than those of yore that the inconvenience is proportionately increased." Looking from a distance at the Massachusetts Bay gale of November 1898 (the "*Portland* Storm," named for a paddle-wheel passenger steamer that disintegrated in the winds off Cape Cod), the *New York Tribune* suggested that "with the growing complexity of civilized life the greater is the dependence of man on appliances which are at the mercy of the elements."[41]

Though perhaps colored by nostalgia, these assessments were not without some foundation. Schedules had tightened and travel, including regular commuting to work, had increased; people relied more than before on frequent purchases of provisions, whose supply storms could disrupt; and they relied more, too, on utility networks (telegraph, telephone, and electricity) that wind, heavy snow, sleet, and ice could easily snap. But the shift to steam at sea had reduced vulnerability to violent weather (the *Portland*'s fate notwithstanding), as had the provision and rapid dissemination of storm warnings and weather forecasts. The *Portland* disaster illustrated several conflicting tendencies at once.[42] Severe-storm signals were flying at the docks when it departed, and its captain had been urged to stay in port: testimony to the value of forecasts. The fact that the ship left nonetheless had much to do with new pressures for punctuality that had grown out of the unprecedented reliability of rail and steam travel, pressures that would only intensify in the coming era.

Neotechnic Boston

The neotechnic era is most of all characterized by the adoption of electricity and the internal combustion engine. It has witnessed an unprecedented rise of mobility and of spatial dispersion of Boston's population—greatly expanding the urbanized area's extent—which in turn have changed the weather's role in human life. Neotechnic-era Bostonians are much freer than their predecessors to choose their place of residence within the region. In 1954 the geographer Edward Ullman predicted that environmental amenities would play a growing role in settlement patterns and regional prosperity in an ever more mobile society. He specifically forecast that Cape Cod, with its mild maritime climate, would grow faster than the rest of the Boston area, and this prediction has been fully borne out.[43]

A pleasant climate to live and play in now weighs more than a healthful one (the colonial-era gap in lifespan and health between North and South has disappeared) or a supposedly bracing and stimulating one (so has belief in the stimulating powers of the cold). The change works to Boston's disadvantage, though balanced by the nonclimatic amenities of heritage, culture, and coastal location. Changes in work have also made people more mobile within the calendar of the seasons, freer to schedule their activities as they wish. New patterns have emerged, mostly reflecting a preference for time off during the summer, which is now (except in resort areas) the period of maximum idleness. In a different way greater mobility helped to undercut the urban-fringe zone of greenhouse agriculture, both because fresh vegetables could be brought quickly from afar by truck or airplane and because pressures of suburbanization made residential development a more lucrative use of the land. Some changes in what extreme weather means for Bostonians emerge from a comparison of the 1898 "*Portland* Storm" with the severest winter storm of the twentieth century, the blizzard of 1978.[44] The 1978 storm combined high winds, high seas, and heavy snowfall (twenty-seven inches) on top of an already large accumulation. The 1898 storm was similar in all respects other than the amount of snow (it brought twelve inches to Boston in the first major fall of the season). Wind speeds were similar and so were minimum barometric pressures; high tides accompanied both storms (higher in 1978), with severe coastal flooding and erosion.

The simplest indicator of impact, loss of life as a proportion of the exposed population, clearly supports the lessening hypothesis. The 1898 storm claimed between four hundred and five hundred lives in New England (all but two dozen at sea); the severer 1978 storm, about a tenth of that number from a larger population. Of the many shipwrecks along the Massachusetts coast in the earlier storm, the *Portland*'s alone accounted for close to two hundred deaths; most of the rest were on small fishing and cargo schooners. Only the five crew members of a small government rescue vessel died at sea in the more powerful 1978 blizzard. The difference in impact may in part reflect improved weather forecasting and communications. It certainly has much to do with the twentieth-century decline in coastal shipping of passengers and freight, which has been replaced by less weather-sensitive land-based transportation by road and rail, and with the switch from sail to engine power on the boats that remain. Fishing, one of the few sea-based livelihoods still present in the Boston area, is still one of the most exposed activities, as the book and film *The Perfect Storm* remind us, but it too is notably less vulnerable than in the past.[45] There were some notable and heroic rescues on the shores of Boston Bay in November of 1898, but nothing comparable to what modern technology makes possible.

In other respects, however, vulnerability does not so clearly seem to have lessened over time. The disappearance of horse-drawn vehicles and the advent of paved roads has made snow a hazard for everyone and put an end to the old controversy over its removal. The low-density suburbanization the automobile has fostered in some ways intensified the impact of winter storms. The majority of the deaths blamed on the 1978 blizzard were directly related to automobile use (mostly from exposure or carbon monoxide poisoning in stranded cars and heart attacks while shoveling driveways).[46] The region enjoys a greater capacity for snow removal today but also a much more extensive need for it. Dispersed, automobile-based suburbanization gives snow more potential to disrupt routines than concentrated, public transit–oriented settlement. How much faster and more reliable than cars Bostonians know the MBTA trains to be in winter, no one needs to be told who has seen how quickly they fill when there is heavy snow on the ground or in the forecast. Cars, paved roads, and speed have likewise made fog and rain into hazards that they were not before, just as the advent of steamships had made fog more dangerous in the harbor.

Neotechnic Boston has succeeded much better than paleotechnic in meeting the challenges of summer heat. Higher temperatures are not nearly so catastrophic in their consequences for health than they were in the nineteenth and early twentieth centuries. The overall mortality peak, a far lower one, now appears in winter. Boston summers have not grown notably cooler since the early twentieth century—on the contrary, they have shared, though to a lesser degree than winters, in the warming both of the globe and of the metropolitan region—yet high summer mortality among infants and young children has vanished altogether, the result of cleaner food, milk, and water and of better refrigeration. The spread of air-conditioning, the dispersal of population, and the availability of emergency response services have further lessened the impacts of extreme heat. Such electric appliances as household refrigerators and freezers have also done away with the manufactured ice trade just as it did with the natural product. But again, not all changes have been for the better. Air-conditioning in particular has transformed a onetime winter peak in electricity demand into a summer peak that now sets new records and places new stresses on power generation on hot days. The defining technology of neotechnic society, the automobile, is responsible for a new seasonal hazard: high concentrations of ground-level ozone, generated by car exhaust and summer conditions, which are especially prevalent in such inner-city areas as Roxbury.

All of this is of particular importance inasmuch as climate forecasts call for warmer days to come. Rising temperatures, all else equal, may more than offset lowered vulnerability to their effects. The already unfolding manifestations of global climate change will also include changes in other dimensions of weather and climate. What these changes will mean for future Bostonians will have much to do with how they choose to live. The greater volume of movement within the neotechnic region, if it continues, makes flooding of low-lying roads by sea-level rise and more frequent heavy rains a potentially disruptive element of projected climatic changes in the coming century.[47] Another trend, the rise in amenity-based coastal development, already exposes more people and property to high winds, storm surge, and erosion at the water's edge, as the 1978 blizzard and some later hurricanes and nor'easters have shown. Two projected consequences of global climate change, sea-level rise and an increase in the number and strength of coastal storms, would only worsen matters.

What a warmer climate will mean for Boston's water system remains unclear. On the supply side, some models have projected a drop in precipitation coupled with increased loss from evaporation in the northeastern United States, others a higher overall yield of water to municipal systems.[48] On the demand side, such measures as pricing water, fixing leaks, and promoting the adoption of water-efficient appliances allowed Boston to cope with droughts and population growth since the mid-twentieth century without seeking new sources of supply. Less encouraging is the story documented by the geographer Colin Polsky of the rising frequency of "suburban drought" in some parts of the metropolitan area even though dry weather has not on average become more frequent or more severe to date. The same water supply that was once adequate now experiences stress during dry spells because of the explosion in large-lot suburban settlement, with a rise of withdrawals for lawn watering and a reduction in groundwater recharge.[49]

In any case weather now plays a new role in Bostonians' mental lives, as something that they may be helping to unsettle globally as well as locally. Not divine agency as in the seventeenth and eighteenth centuries, but physical processes set in motion by human actions are today scrutinized as possible causes for any unusual season or event. As in earlier times, such awareness now prompts a good deal of self-examination. Most Bostonians, like most Americans, have contributed far more than their share to global climate change. Almost every resident of neotechnic Boston produces much higher carbon dioxide emissions than Bostonians of the past through auto travel, air travel, heating, and electricity generation for air-conditioning, lighting, and other appliances. Although Boston's public transportation system is better than that in most other cities, eastern Massachusetts ranks surprisingly high on a comparative sprawl index (a measure, albeit imperfect, of automobile dependence and, through it, of carbon dioxide emissions).[50] However, the city and its legislators have been prominent in efforts to put emissions reductions on the local, regional, and national political agenda.

ONE celebrated Bostonian was fond of saying that "when one damns the weather, he only signifies that he is ill adjusted to his environment, not that the environment is bad."[51] Few people, perhaps, could have agreed entirely with him.[52] No one, probably, could ever design a way of living perfectly adjusted to any place's weather, let alone weather as varied as Boston's. But on an important point, the remark is correct, as history demonstrates. Good and bad weather are human creations, less because people have altered the weather (although they have to some degree, and they have entirely transformed the weather indoors where they pass most of their lives) than because it is people's changing wants and abilities that have repeatedly redefined certain conditions as helpful or harmful. An old myth holds that it is their weather that has made Bostonians what they are: cold, bleak skies producing cold, bleak people. It would be much closer to the truth, in several senses, to say that Bostonians have made their weather what it is.

12

"Rain Down Righteousness"

INTERPRETATIONS OF NATURAL EVENTS IN MID-EIGHTEENTH-CENTURY BOSTON

Lauri Bauer Coleman

> About an Hour after [three in the morning,] I was awakened, or rather alarmed, by the shaking of my Bed, and of the House; the Cause, wherefore, I immediately concluded, could be nothing but an EARTHQUAKE.
>
> *Boston Gazette,* 1755

ON November 18, 1755, an earthquake abruptly roused the residents of Boston. The tremor was centered just off the coast of Massachusetts, but it shook cities, towns, and hamlets all the way from western New York to Nova Scotia to Annapolis, Maryland. Despite the intensity of the quake, which had a moment magnitude of 5.8, no Bostonians were killed or even hurt.[1] The early hour meant that most people were still in bed and not the unwary targets of crumbling chimneys and flying roofing tiles. Even without deaths to add to its toll, people in and around Boston were eager to record their impressions of this relatively rare and alarming natural disaster. Newspapers provided a number of accounts like the above description of a learned gentleman, which the *Boston Gazette* printed several days after the event.

Well-known Congregational ministers in Boston, such as Thomas Prince and Charles Chauncy, published sermons in response to the quake, while across the countryside many of their colleagues recorded their thoughts and sermon topics in diary entries.[2] According to Isaac Backus, the Baptist minister of Titicut, Massachusetts, the earthquake "put me in the mind of the dissolution of all things."[3] The farmwife Experience Wight Richardson of Sudbury, Massachusetts, wrote in her diary that she hoped for a moral reformation on account of the quake.[4] And John Winthrop IV, Hollis Professor of Mathematics and Natural Philosophy at Harvard College, composed a lecture on the natural causes and consequences of this surprising phenomenon.[5] The plethora of print the earthquake and other alarming natural events generated in the mid-eighteenth century provides a useful window into the ways in which the English colonists made sense of the world around them.

Early Americans as diverse as Professor Winthrop and Goodwife Richardson considered natural events such as the earthquake of 1755 to be signs in need of interpretation. In other words, a disaster had some larger significance beyond an immediate threat to life and limb. What did it mean when the earth trembled? How could the deeper implications of a comet or threatening weather be determined? Who had the authority to do the interpreting? To answer these questions, New Englanders both learned and common drew on a set of beliefs that historians have since labeled "supernatural rationalism."[6] This cumbersome term seems at first to combine contrary ideas. But in the eighteenth century a rational approach to the natural world went hand in hand with the recognition of God's supernatural powers.

For the learned no less than the uneducated, science was, first and foremost, a religious pursuit. In the words of John Winthrop, "the consideration of a DEITY is not peculiar to *Divinity*, but belongs also to *natural Philosophy*. And indeed the main business of natural Philosophy is, to trace the chain of natural causes from one link to another, till we come to the FIRST CAUSE [God]; who, in Philosophy, is considered as presiding over, and continually actuating, this whole chain and every link of it."[7] Eighteenth-century intellectuals such as Winthrop considered the book of nature to be a manifestation of God's will. Furthermore, they believed the words of God in the Bible and the actions of God in nature to be in accord.[8] The learned study of scripture and empirical inquiry into nature were distinct but complementary ways to comprehend God. Each represented a rational approach to the supernatural.

This form of Christian apologetics was first conceived in late-seventeenth-century England in response to the philosophical speculation of Sir Isaac Newton and John Locke and in the social context of constant religious strife. Newton's mathematical proofs showed the regularity and order evident in the natural world; these demonstrations also spoke to God's continual and active participation in the system of nature. For who else, Newton concluded, could create and sustain such perfection?[9] At the same time that Newton sought to identify the fundamental principles of the God-driven universe, Locke suggested that human reason and sensory input, rather than innate ideas, provided the most accurate form of all knowledge, including religious belief. These two complementary philosophies offered a new learned approach to understanding God that harmonized with Christian revelation, one that raised human reason and empirical method over inner grace and divine illumination. In the aftermath of the religious instability that had plagued seventeenth-century England, in which enthusiasts claimed special knowledge and authority derived from God, church and state leaders were quick to adopt this new, more reasonable form of natural religion in an effort to buttress their own authority.[10]

New England scholars gained exposure to these ideas and an aversion to enthusiasm through required reading at Harvard as well as through such scientific publications as the *Philosophical Transactions of the Royal Society.*[11] The Royal Society, based in London, fostered a transatlantic communications network that offered the latest empirical findings and observations on the natural world.[12] Educated colonists read the *Transactions* regularly, and many would-be natural philosophers submitted their observations for publication. The Reverend Thomas Prince, for one, dedicated his sermon on the drought of 1749 to the Royal Society.[13] This sermon, as well as those he wrote on the Boston earthquakes in 1727 and 1755, embodied the goals of the Society and of the new Christian natural philosophy: scientific inquiry for the cause of God. The introduction to his 1755 earthquake sermon declares the piece: "a Discourse on that Subject Wherein . . . is offer'd a Brief Account of the *natural, instrumental, or secondary* Causes of these Operations of the Hand of God. After which, Our Thoughts are led up to HIM, as having the Highest and principle *Agency* in this stupendous WORK."[14]

The less educated population of Boston and its environs was audience to many such sermons reflecting the principles of supernatural rationalism. Fast Day and Thanksgiving Day sermons, such as Prince's offering on the drought, were particularly resonant. Unlike regular Sunday sermons, which tended to focus on such topics as personal salvation and grace, pastors used these occasional sermons to elucidate the meanings behind God's messages in the more tangible natural and political worlds.[15] Beyond what their pastors might say about the relationship between God and nature, the common folks' own well-read Bibles clearly articulated the importance of the book of nature as a source of God's judgments. Psalms 18:7 identified earthquakes as messages from an angry God: "The Earth Shook and Trembled, and the Foundations of the Hills Moved and were Shaken, because He was Wroth."[16] Yearly almanacs, which circulated widely across New England, also discussed such unusual natural events as comets and threatening weather. Almanac author John Jerman included the following poem, which praised all of nature as a reflection of God's glory: "O great Creator, King . . . Let Thunder-Clouds that float from Pole to Pole . . . salute thee as they roll. . . . Let Hail and Rain, let Meteors form'd of Fire, . . . in this blest Work conspire."[17]

The system of beliefs scholars now call supernatural rationalism was nearly ubiquitous. From almanac writers to farmers to Harvard-educated pastors and professors, almost everyone believed that the natural world reflected God's will and that to study nature was to study God. As such, they assumed unusual and threatening natural events to be messages from God that required interpretation and empirical analysis. However, to agree that these events should be interpreted did not mean that all members of society discerned the same significance in an earthquake or a drought. Supernatural rational teachings provided a template for the relationship between religious belief and the natural world, but determining the meaning of God's words and deeds was much more complicated. As the religious historian Conrad Wright has explained, "supernatural rationalism never appeared by itself, but always in association with other doctrines."[18] These other teachings, which coexisted within the larger framework of supernatural rationalism, point to the complexity of eighteenth-century attitudes toward the natural world. When they interpreted alarming events, early New Englanders revealed different styles of argument and divergent beliefs in the nature of God.

For instance, an argument based on biblical history was unlike one drawn from empirical evidence gathered from the natural world even if scripture and nature were assumed to be in concert. Redemptive scriptural history suggested that all events in the natural world were acts in the play of sacred time. This type of history was teleological, revealing God's purpose in nature's design, as well as deductive.[19] Students of the Bible knew the final act in the play: revelation and redemption. Alarming events were signposts and markers along the way to the end times and, as such, their meaning was predetermined. In contrast, those who studied surprising phenomena from an empirical perspective collected sensory evidence that would ultimately reveal the natural laws by which God governed the universe. Natural philosophers based their method on Sir Francis Bacon's inductive research agenda of comparative empiricism.[20] They conducted their studies from the bottom up, using specific instances to formulate general rules rather than assuming a preordained purpose behind each event.

Those who argued from the standpoint of biblical history were also more likely to claim that God intervened directly in the natural world, altering the order of nature to suit his purposes and send portentous messages. This argument for God's providential authority was much different than an argument from design. In the latter, natural philosophers discerned God's will in the regular course of nature rather than in aberrations. These interpreters considered God to be the "first cause" in the perfect system that he designed and operated. Natural laws, as his "second causes," were the instruments he used to carry out his moral agenda.[21]

These distinctions between the arguments from authority versus those from design were closely related to different understandings of the nature of God himself. The God of the Old Testament was very clearly an authoritarian figure, prone to fits of wrath, who ruled by terror as often as by love. Many New Englanders understood their relationship with this wrathful God in terms of a continual cycle of sin, anger, and forgiveness. In this conception natural disasters gave voice to God's anger and his moral authority. They required repentance on the part of a collective group of sinners to earn forgiveness. Specifically, God reordered that natural world through threatening events to punish the transgressions of sinful "nations"—which could alternately be defined as cities, states, or regions—just as he had in biblical times. God as wrathful corrector of corporate sin stood in contrast to God as benevolent and orderly father figure whose goal was the ultimate well-being of his people and the universe writ large. In the latter, God demonstrated his will through ordinary rather than extraordinary nature and related to the residents of Boston and its environs on an individual rather than a corporate level. Natural disasters may have conveyed the will of God, but he intended his message to lead to personal reflection on sin not corporate repentance. To those who subscribed to this view of God, an earthquake or a comet inspired awe rather than terror and reflection on God's wisdom in creating and sustaining the universe rather than on his punishing wrath.[22]

Differing views on the nature of God and the nature of argument concerning disastrous events were not necessarily oppositional. Instead, they represented the variety of supernatural and rational constructs that Bostonians drew on in turn to make sense of the natural world in the mid-eighteenth century. The popular almanac author Nathaniel Ames was not unusual in alternating his depictions of God. He sometimes praised the designing wisdom of God the Creator: "who can tell what wonderful Changes have happen'd by the mighty Operations of Nature, such as Deluges, Vulcanoes, Earthquakes, &c.!" And he also warned fearfully of God's impending wrath in the form of a comet "Soon to blaze along the Skie, Portending (sure) some dire Event is nigh."[23] Ames was neither intentionally nor inadvertently inconsistent. A wrathful deity made more sense in the threatening circumstances of war and an impending comet in 1758. At less portentous moments Ames was free to contemplate the more "wonderful" effects of God's actions in the natural world.

As Ames's example demonstrates, the interpretation of natural events relied on more than a cluster of supernatural rational beliefs; the meaning of unusual nature was also firmly grounded in the social and cultural context of eighteenth-century Boston. Such current events as the French and Indian War and a series of transatlantic earthquakes as well as such spiritual upheavals as the Great Awakening significantly influenced responses to disaster. Although the Great Awakening ended in the mid-1740s, religious historians contend that it had lasting interpretive effects. The historian Frank Lambert has argued that the Awakening was not only a series of revivals but also a struggle over meaning—in his term, a "cultural war."[24] For those who participated in the Awakening, mass meetings and impassioned audiences spoke to the presence of God's grace. To detractors, these occurrences appeared to be signs of dangerous social disorder and unchecked "enthusiasm," like that which had run rampant during the religious upheavals in England the century before. Enthusiasm, in these cases, was defined as asserting and acting on personal illumination or visions from God. The question then became, who had the authority to interpret God's will? Those who claimed direct communications with God, or those applying human reason to the scriptures and the book of nature? Those who possessed recognized positions of authority, such as church pastors and Harvard professors, or self-proclaimed instruments of God, such as itinerant preachers and lay exhorters?[25]

Born out of the Awakening and seventeenth-century English religious strife, these same questions of cultural authority and social responsibility applied to the interpretation of threatening natural events in the mid-eighteenth century. Many public authority figures worried that enthusiastic responses to unusual events would incite the passions of the common people and spell a return to the societal instability that characterized the years of the Awakening. Others, both learned and unlearned, believed that droughts, earthquakes, and comets were a legitimate cause of fear and spiritual trembling and that to treat them as anything less would be offensive to God. Enthusiasm, therefore, became the linchpin of contention in debates over the meaning of disasters. Although belief in the complementary relationship between God and nature—and in the complex of doctrines now called supernatural rationalism—was widely shared, the issue of enthusiasm and the competition for authority that it engendered drove a wedge in the interpretive climate.

How did Bostonians apply supernatural rational apologetics, their beliefs about the nature of God, and their opinions about interpretive authority to occurrences in the natural world? What were their goals in interpreting nature for themselves or for society at large? Did they want personally to prepare for God's judgment? Impart popular knowledge? Provide entertainment? Improve the morality of society? Instill fear? Increase collective knowledge about the laws of nature? Encourage a religious revival? The interpretation of alarming and unusual phenomena could be used for all of these ends, and the natural world in the mid-eighteenth century provided a number of events worthy of examination. Colonists in and around Boston struggled to make sense of one of the worst droughts in early American history, a series of transatlantic earthquakes, and the first predicted return of Halley's comet.

The midcentury interpretive environment was not one of consensus, nor one in which scientific or "rational" ideas outweighed religious or "supernatural" explanations.[26] There was instead near-unanimous agreement that God was the first cause and ultimate mover in the universe, and that the book of nature bespoke his will as clearly as the Bible. However, the particular meaning of these God-driven events—his purpose in sending and directing calamity—was less clear and therefore subject to debate. Examining these eighteenth-century natural events reveals a complex climate of interpretation, characterized by religious conviction, empirical aspirations, and the struggle to establish authority and cultural meaning.

The Drought of 1749

Colonists became anxious about unusually dry conditions in the fall of 1748, when agricultural goods were already scarce because of the military requisitions of King George's War.[27] By the following June, Bostonians were complaining of wells and streams drying up and of a "disagreeable smell" near the rivers caused by dead fish that had apparently succumbed to record heat and dry conditions.[28] Goodwife Richardson confided in her diary in early July that "the extreme drought . . . has cut off a great many things. The Lord seemed to smook against his people['s] prayers."[29] And by August the Reverend Ebenezer Parkman wrote: "Last Year there was Drought; This Year a much greater—such an one as the oldest doth not remember; perhaps the Next may be a Proper Famine."[30] The severity of the dry weather, which lasted from 1748 well into the summer of 1749, may have been unusual, but the absence of rain was not. Drought marked about a quarter of the years between the mid-seventeenth century and the mid-eighteenth.[31] Between its frequency and the significance accorded rainless spells in a drier climate in Old Testament stories, the drought of 1749 triggered a built-in communal response, a societal supplication before God.

On June 2, in response to a petition from the General Court, Governor William Shirley of Massachusetts declared June 15 a day of fasting and humiliation "taking into Consideration the frowns of divine Providence we are under." By this "public Acknowledgement of our Sins," the leaders of Massachusetts sought to reform God's people and appease his anger, so that he might (play on words intended) "Rain down Righteousness upon us."[32] The connection between fast days and drought was long-standing.[33] New Englanders consistently interpreted drought as a message from an angry God to his People; fast days, a meaningful communal rite, reaffirmed the corporate nature of sin and sent a return message to God promising reformation.

When June 15 arrived, the Reverend Aaron Smith of Marlborough, Massachusetts, delivered two discourses, later published, in which he interpreted God's purpose in sending the drought.[34] Other pastors in the Boston area, including Ebenezer Parkman, Smith's counterpart in Westborough, also preached about this natural sign of divine anger. Though on different sides in the recent Awakening, Smith and Parkman both drew on supernatural rationalist concepts when it came to interpreting the drought. Mirroring Governor Shirley in their reliance on Old Testament passages and covenant language, they emphasized the themes of corporate sin and an angry God. Smith specifically compared his congregants to those who were in covenant with the Lord in biblical times: "In like Manner has God proceeded with us, our Conduct resembling theirs. . . . He rebukes our rebellious Spirit by Corrections and Frowns, but we revolt yet more the more."[35] And Parkman remarked of his sermon in his diary: "I undertook to Shew [them] that tho God had inflicted on us many Judgments yet we did not turn to Him as he justly might expect we Should."[36]

According to Parkman and Smith, the drought was a clear sign in nature that God was passing judgment on his people. Smith's address quoted a number of relevant biblical passages, mostly from Psalms, that showed how God expressed his will through nature. God could turn "Rivers into Wilderness, and the Water Springs into dry Ground . . . a fruitful Land into Barrenness, for the Wickedness of them that dwell therein." The Lord could even make it rain in one city while preventing it in another.[37] In their sources and style of argument, Parkman and Smith demonstrated two elements of their common worldview: teleological reasoning based on biblical precedent and an angry God's use of the natural world to send messages that required a communal, rather than individual, response.

Two months later, in a sermon commemorating the end of the drought, the Reverend Thomas Prince reaffirmed these first two elements and also addressed the empirical study of nature in the cause of God, a third aspect of supernatural rationalism.[38] Prince, the pastor of the Old South Church in Boston, kept up with the latest philosophical theories from London and took seriously his public role as philosopher-divine. As a learned authority figure, he considered it his duty to share all of his knowledge concerning God's will—both philosophical and scriptural—with his congregation and the larger community through his many publications. His sermon was thus filled not only with biblical passages but also with scientific explanations of the weather as well as with passages from the Royal Society's *Philosophical Transactions*.[39]

Prince's interest in contemporary philosophical discussion did not, however, prevent him from explicating God's providential authority as expressed in nature. "I desire you not to think, I am going to give you a Lecture of mere Philosophy," he instructed. "No! I am going to treat on a noble Subject of Divinity; viz. on the wise, mighty and constant *Operations* of God."[40] In carrying out his moral government of punishments and rewards, God could work through second causes, but he could also alter the natural world supernaturally as he had done in biblical times. God molded the weather to fit his needs: "Such like reserved Cases seem to be Earthquakes, Hurricanes, & Tempests; Storms of Wind, Rain, Thunder, Lightning, Snow & Hail, with the raging of the Seas; & indeed all Kinds of Weather, hot & dry, cold & wet; the rising of noxious Vapours out of the Earth. . . . Making all Kinds of Substances indeed, his fitting Instruments and using them; but whenever he sees Occasion, *giving additional Degrees of Powers and different Directions, or abating their Degrees of Power* in other Ways than in his common Course of Nature, which is only his usual Way of Operation in them."[41]

For Prince and his colleagues, God's power was both immediate and outside the laws of nature. To restrict God to acting through second causes, or the "*mere Operation of created Instruments,*" would be a form of "Atheism," wrote Prince.[42] The separatist minister Isaac Backus captured the belief in God's immediate intervention most clearly: "the Earth was exceedingly dried, so That it seemed as if there would be nothing Left Either for Man or Beast. Then the Lord appeared and gave refreshing showers and things were remarkably revived."[43] Through his mere appearance, God could as easily bestow rain as withhold it. Those, like Backus, who recorded their thoughts on the dry weather of 1749, understood that the drought was a meaningful sign from God, expressed through the natural world. They employed both religious and scientific means to explain that message and used their interpretations to encourage both a reformation in behavior and further inquiry into the God-driven order of nature.

Some also used God's supernatural message for more secular purposes, as a way to hold down the rising price of agricultural goods. One Boston man wrote to the *Boston Evening-Post* complaining about the "dishonest" practices of country merchants who inflated prices, "selling Butter and Milk, and other things for the support of Life, without any Regard to *Mercy.*" He feared that the lack of "Christian Morality" in these transactions would prevent God from listening to the pleas of others calling for mercy from the drought.[44] The newspaper editors themselves commented that many "covetous" persons had made the drought "which was a sore Judgment of Heaven . . . an *Excuse* and *Cloak,* to cover their abominable Extortion."[45] Country farmers, for their part, defended their pricing decisions with an appeal to supply and demand and a criticism of town residents for wasting scarce hay supplies to feed carriage horses used only in leisure pursuits.[46] Farmers and town folk, editors and ministers, may all have agreed that the drought was a message from God to reform their communities, but how to deal with economic ramifications of this weather calamity was a point of contention.

The Earthquake of 1755

As both an economic threat and a judgment from God, the great drought of 1749 was an inauspicious entrance into the decade of the 1750s, and the public mood was about to grow even worse. In 1754 the French and the English embarked on a war that would have destructive consequences for the colonies. For the English the bleakest year of the French and Indian War was 1755. There were staggering colonial losses and the complete breakdown of government in the backcountry.[47] It was in this context that Boston's most severe earthquake struck on November 18, 1755, killing none but shaking the collective conscience of New England.

According to New England's ministers, the earthquake and the war were not unrelated. Reverend Eliphalet Williams, the fiery pastor of the Congregational church in East Hartford, Connecticut, sought to convince his flock that God "Frowned upon us" through the "awful Event" of the earthquake and by "letting loose . . . a Barbarous, Blood-thirsty, Heathen Enemy."[48] In less colorful language, Charles Chauncy also suggested that God was using the tools at hand to punish the transgressions of his people.[49] The Reverend Jonathan Mayhew of Boston's West Church concluded, in a sermon so popular that the *Boston Gazette* reprinted it: "such is the present critical Situation of our [political] affairs, such the Aspects of Providence towards us, and so numerous our Sins against Heaven, that all who value their Lives, Liberties or Estates, not to say their Souls, had need to fear God, [and] secure his favor and protection."[50]

The meaning of the earthquake could not have been clearer. Taken together with the war, it was a statement of the Lord's anger toward a sinful people and a warning of worse to come if New Englanders did not immediately humble themselves before God. In the vivid words of Reverend Thomas Foxcroft, New Englanders had "had a loud Lecture on the Existence, Providence, and Perfection of GOD . . . a practical Sermon, preached to us in very awful Accents, upon the absolute Necessity of Repentance."[51] In expounding this message, the ministers of Boston, so recently and bitterly divided over the Great Awakening, presented a surprisingly united front in which biblical history and teleology loomed large. Boston's clerical establishment argued that the recent earthquake and the tremors recorded in the Bible had identical meanings. Nearly every minister cited Psalms 18:7 as proof that earthquakes expressed God's wrath.[52] One zealous preacher went so far as to list every anger-induced quake in the Bible; for Mather Byles Jr., at least, history was merely repeating itself in New England in 1755.

The Bible also provided ministers with a second and larger meaning for earthquakes, one that suggested the new tremor might be more than a repeat warning. It might also foretell impending disaster or even the end times. The Reverend Timothy Harrington of Lancaster, Massachusetts, retold the story of Israel's punishments to persuade his parishioners to repent immediately and not give God cause again to send "Successive Calamities" giving notice of "approaching Ruin."[53] Mather Byles Jr. cited John's revelations to remind his listeners that earthquakes would precede the end times. According to Revelations, God would send "Earthquakes in divers places . . . [and] shake the Heavens, and the Earth, and the Sea, and the dry Land."[54] Byles's father, Mather Byles Sr., also a minister and a noted literary figure, had composed a poem in 1727 entitled *The God of Tempest and Earthquake*; it was reprinted in response to the 1755 earthquake. The senior Byles contended that signs in the earth, notably earthquakes as God's "marching Foot-Steps," were a sure prediction of his Second Coming.[55] Other ministers were not convinced that the end times were at hand, but just in case they urged their listeners to repent. Eliphalet Williams reasoned, "what Calamities this may presage, is not for me to say. But suppose it Forebodes no future Event, and yet under the apprehension that it does, we . . . are put upon using the proper Means to Obtain the reconcil'd Favour of God. . . . What Hurt accrues from our Mistake?"[56]

Whether they argued that the 1755 quake was a timely warning to abjure sin or an event leading up to the end times, New England's ministers emphasized a teleological message. Biblical precedent or prophecy revealed the meaning of the earthquake. A wrathful God enraged by the corporate sins of his people in both contemporary and biblical times demanded a reformation of morality and religious sentiment within the church. As Mather Byles Jr. proclaimed, "how happy would it be, if this awful Providence should be blessed by God, to the producing a universal Reformation of Manners, and a Revival of Religion among us!"[57] The use of earthquakes in the cause of religious revival was not unprecedented. As recently as the 1727 quake, ministers had capitalized on the fear of a shocking natural event to convert sinners in large numbers.[58] The Reverends Charles Chauncy and Thomas Prentice, in their 1755 sermons, expressed the hope that this earthquake might effect a more lasting moral reform, since it was much greater and occurred within living memory of the last tremor.[59]

Ministers such as Charles Chauncy considered the earthquake to be more than a spur to moral reform. Drawing on the beliefs of the day, they also treated it as a natural phenomenon worthy of philosophical study. The Reverend Thomas Foxcroft compared the "Flame of devouring of Fire" described in the Bible with the "Eruption of subterraneous Fire" that naturally created the recent earthquake.[60] Mather Byles Jr. also treated earthquakes as natural events with prophetic meaning, since God intentionally "formed our Globe of such a Constitution as to be liable to Earthquakes. And he produces them at such Times, in such Places, and in such Degrees of Violence & Severity as are . . . best calculated to accomplish the Designs of his Providence, in the government of the natural and the moral World."[61] For Byles, Foxcroft, and Chauncy there was a hierarchical relationship between God's authority and the operations of the natural world. As Chauncy explained, "tho' earthquakes are *ordinary* produced, in a natural way, by a concurrence of suitable and sufficient second causes; yet it is God who produces them: He is more properly their *Cause,* than any *secondary* one. . . . For they are all his; he made them; he upholds them."[62]

For these ministers recent philosophical speculation about the natural or second causes of earthquakes in no way hindered God's ability to demonstrate his will in the natural world. The Lord simply used the regular operations of nature—taking advantage of the ready-made combustible makeup of the globe—to carry out his wishes. God, according to Chauncy, was the "supreme regent of nature"; second causes were his tools.[63] By incorporating natural explanations into their interpretations of the quake, New England's ministers were not becoming less supernatural or more secularized.[64] Instead, they were applying a rational approach to the supernatural by using contemporary knowledge to explicate God's actions and intentions. Mather Byles Jr. in his early twenties and only recently removed from the halls of Harvard, clearly expressed how religion and reason complemented one another: "If we should neglect to . . . make a suitable [spiritual] Improvement of an Event of this Kind, we should undoubtedly act against the Light of Nature, the first Principles, and the most evident Dictates of Reason."[65]

Many of Byles's senior colleagues were less willing to speculate in their sermons on the natural causes of the earthquake. Neither Eliphalet Williams nor Thomas Prentice discussed second causes at all, choosing instead to stress that the earthquake was an assertion of the Lord's immediate and unaided power in nature. The recent earthquake was a "special instance of Judgment" in which God chose to "go out of the settled Order and Succession of Operations, in his Government of the World, and perform . . . some sudden, and unforeseen and wonderful Works" in order to shake sinners out of their spiritual lethargy.[66] Thomas Foxcroft reasoned that the recent quake must have been a "special Interposition" because such convulsions were inimical to the perfection and regularity of God's earth, which "Tend[ed] to the Good of the Whole."[67] Instead, the sins of man must have triggered the "Finger of God," to use Thomas Prentice's phrase.[68]

Those ministers who emphasized the immediate nature of God's role in sending the earthquake did not necessarily dismiss second causes out of hand. Some, such as Williams, feared that dwelling too much on the nature of the earthquake would distract his listeners from its real meaning: God's anger at human sin.[69] Others, such as Foxcroft, considered God's message to be more important and more properly the subject of a sermon than an extended discussion of the operations of nature: "Theology points us to GOD, as the first Cause, Whatever be the secondary Causes."[70] This was a common opinion, held even by those more inclined to discuss natural means. Charles Chauncy urged his audience to "seriously carry your tho'ts beyond all *second Causes* and fix your eye on *God.*"[71] And Mather Byles Jr., who allotted some space to natural philosophy, discussed these matters on only four pages of a thirty-page sermon in which he spent almost twice as long listing biblical earthquakes.

The exception to this rule was, once again, the Reverend Thomas Prince, who devoted more time and effort to the consideration of natural causes than any other clergyman. As in his sermon on the drought in 1749, Prince demonstrated his belief in the interconnection of theology and natural philosophy in what he called "divine Philosophy." On the one hand, he agreed with his contemporaries that the recent earthquake was a reflection of God's righteous anger. His sermons ably depicted the sins and suffering of those whom the Lord had visited with earthquakes in biblical times along with the terror of his wrath in the end times. Nor did Prince doubt that the Lord could immediately intervene and interrupt the regular order of nature to express his wrath: God "can shake the Earth out of her Place in a moment without any Instruments; for he needs them not, nor is confined to them." On the other hand, Prince also asserted that the Creator often worked through second causes because, in "his perfect Wisdom," he had designed the earth and all of its functions to carry out his moral government.[72]

In 1727, Prince suggested that God used the earth's cavernous quality to cause the recent quake. In 1755 he modernized this interpretation by appending a discussion of electricity and its likely role in shaking the earth to his reprinted 1727 sermon, *Earthquakes the Works of God.* He concluded his appendix with the suggestion that Benjamin Franklin's iron rods, which had been erected all over Boston to prevent lightning strikes, could have electrified the ground so much that Boston was shaken more severely than neighboring areas. But, lest his readers think that Bostonians could have somehow avoided the earthquake, he warned: "O! there is no getting out of the mighty Hand of God! If we seek to avoid it in the *Air,* we cannot in the *Earth.*" Repentance, he reminded his audience, was the only sure way to appease God's wrath.[73]

While Prince's fellow New England preachers may not have been willing to delve as deeply into "divine Philosophy," his scientific speculations were akin to those of religious scholars in the transatlantic philosophical community. The English ministers Stephen Hales and William Stuckely published scientific papers on the 1750 London earthquakes in the Royal Society's *Philosophical Transactions.*[74] Stuckely judged: "nor need we lose Sight of the theological Purpose of these amazing Alarms, whilst we endeavor to find out the Philosophy of them." For who but God, he asked, "*guides the* Thunder, *and directs the* Storm?"[75] Cosmopolitan ministers such as Prince and Stuckely, immersed as they were in the learned supernatural rationalism of the day, authoritatively discussed both the natural causes and the moral ramifications of earthquakes.

Though not a clergyman, Professor John Winthrop was no less well read. As a scientist, he approached the study of earthquakes differently than the clergy. His *Lecture* is empirical, rather than biblical, containing his own experiments, critical analysis of previous theories, and quotations from such contemporary philosophers as Robert Boyle and Isaac Newton, who had recently studied comets. Winthrop used these sources to pursue scientific knowledge rather than moral reformation, although he believed that God's moral and natural purposes were in accord. As he explained: "though I make no doubt, that the laws of nature were established, and that the operations of nature are conducted, with a view, *ultimately* to *moral* purposes . . . yet it would be improper for me to enter into these disquisitions at this time, since my province limits me to consider this subject, only in the relation which it bears to *natural philosophy.*"[76]

Despite their different professions and divergent purposes, Winthrop and Prince had much in common. Both sought to examine empirical evidence and participate in the larger transatlantic discussion of the causes and consequences of natural phenomena. Prince had dedicated his drought sermon to the Royal Society, and Winthrop published an account of the Boston quake in the *Philosophical Transactions* in 1757.[77] Both men also recognized the prophetic significance of the earthquake, sharing the belief that God was the "First Cause" of the earthquake, while secondary causes were merely instruments, which God designed and directed to carry out his moral agenda. According to Winthrop, all "true philosophy [ascribes] what we call *natural* effects, to the agency of GOD."[78]

These two intellectuals each employed the argument from design in their speculations as well. Although Prince believed that God could escape the natural order to send his moral messages, he agreed with Winthrop that the

Lord's will was equally evident in the regular order of nature. For Prince, God had designed the earth in his "perfect Wisdom," so that he knew "every Effect" that his natural laws would have, even before he created the earth and set it in motion.[79] While Prince's theological emphasis on God's wrath prevented him from recognizing any positive effects from earthquakes, Winthrop was willing to extend the argument from design and contended that God could use tremors both as religious punishments and natural blessings. These natural events, said Winthrop, "have indeed occasioned most terrible desolations, and in this light may justly be regarded as the tokens of an incensed DEITY; yet it can by no means be concluded from hence, that they are not of real and standing advantage to the globe *in general.*"[80]

He suggested that earthquakes served such an advantageous purpose by performing a kind of underground tillage. The earth, "opening its pores," facilitated the growth of minerals underground and improved the quality of soil on the surface. Thinking empirically, Winthrop pointed as evidence to the well-known fertility of volcano-prone areas such as Mount Vesuvius and further contended that only a lack of adequate knowledge prevented philosophers from grasping the larger, beneficial functions of earthquakes.[81] But whatever advantages God as benevolent creator had built into the system, Winthrop was quick to establish that they "do not in the least detract from the majesty, or from the justice, of GOD."[82] Ultimately, for Winthrop as well as for Prince, speculating about the natural causes and consequences of the quake did not make it any less terrible a warning from an angry God. Their views about God's power by design versus his arbitrary authority may have differed, but they were not necessarily oppositional.

Despite their shared reliance on supernatural rationalist constructs, Prince and Winthrop became embroiled in a well-publicized debate that had more to do with social responsibility than their respective interpretations of the earthquake. Their argument did stem, in part, from a disagreement over the specific natural cause of the tremor. Both men were interested in empirical speculation, but Winthrop accused Prince of being swayed by electricity, a "fashionable" cause, rather than continuing to support what he considered the true scientific reason for the quake: the violent release of compressed gases from underground caverns.[83] The greater problem with Prince's philosophical speculation, at least from Winthrop's perspective, also had to do with electricity. Winthrop worried that Prince's allegations against Franklin's lightning rods "would fill with unnecessary terrors the minds of many persons . . . and that it would discourage the use of the *iron-points* . . .

which, by the blessing of GOD, might be a means of preventing many of those mischievous and sorrowful accidents, which we have so often seen to follow upon thunder storms."[84]

Winthrop was concerned that Prince had misused his substantial influence as a well-known public figure to prejudice his readers against lightning rods. Even more, Winthrop worried that Prince's attack on the rods would spread panic. In further correspondence with Prince, in which he described the properties of electricity in great detail, Winthrop expressed the hope that he had "rescued this worthy Divine . . . [and] a great number of others, especially of the more *timorous* sex (so extensive is his influence!) who have been thrown into unreasonable terrors, by means of a too slender acquaintance with the laws of electricity."[85] Clearly, Winthrop's comment was meant to insult Prince by comparing his fear to female panic, but there is also a significant social meaning behind Winthrop's ridicule. He worried that the popular Prince would stir up the passions of his audiences to the point that "unreasonable terror" could result in societal instability.[86]

Winthrop was not the only one to voice this concern. Public leaders in both England and the colonies struggled against enthusiastic visions and prophecies, based on alleged direct communications from God during periods of religious upheaval. In the midst of the Great Awakening, Charles Chauncy had gone so far as to publish a sermon on the threat of "imaginary" religion and enthusiasm, which he considered to be "one of the most dangerous enemies to the church of GOD, and has done a world of mischief."[87] Eliphalet Williams gave an explicit warning against passionate and unreasonable reactions to the 1755 earthquake. He counseled his listeners against excessive fear or anger at God, which would only lead to "Despair and Distrust of the ALMIGHTY; And 'tis therefore pernicious, and baneful, and contrary to the Design of GOD."[88] While such ministers as Chauncy and Williams certainly made use of the terror engendered by the quake to reform the morals of their congregants, they felt, like Winthrop, that overwhelming fear and panic was excessive and even dangerous, providing fertile ground for enthusiastic demagogues.

There were also transatlantic reasons for anxiety in the American colonies. In London in 1750 two quakes occurred exactly one month apart. The unusual timing of these relatively minor tremors led one very frightened London man to prophesy publicly that God had warned him that a third quake would destroy the capital one month later, on April 4. Despite the assurances of London's clerical establishment that this prophecy was a result of enthusiastic visions rather than the will of God, panic reigned. On the eve of the prophesied date, thousands fled London. Inns in neighboring towns made a fortune, and many refugees camped in fields, believing that their community, which the clergy had consistently told them was guilty of many sins, would receive the brunt of God's wrath.[89]

The panicked response of the English to this potential disaster was nothing compared with the aftermath of the earthquake in Lisbon, Portugal. Lisbon, unlike London, was destroyed when a devastating tremor hit on November 1, 1755. Incapacitated by fear for their lives, many of the survivors of the dreadful quake and accompanying fire listened to the words of religious visionaries, who claimed that this event marked the beginning of the end of the world. More concerned about the state of their souls than their material well-being, survivors refused to build shelters or help the injured. Instead, they could be found praying and confessing night and day.[90] One English merchant wrote of the Lisbon quake that "as you pass'd along the Streets, you saw . . . some alive, crying out from under the Ruins; others half buried; others with broken Limbs, in vain begging for Help: And they were passed by Crowds [seeking to escape], without the least Notice or Sense of Humanity."[91] In both London and Lisbon those in positions of authority lost control over responses to the earthquakes. Reforming fear, which community leaders sought to use to improve public morals, turned into panic and social irresponsibility. In this atmosphere those who relied on enthusiastic visions for their authority gained popular influence, much to the frustration of ministers and secular leaders whose own cultural authority suffered. In these situations of natural crisis, popular leaders and established authorities had much different purposes. The former, more enthusiastic, sought to prepare humankind for the end of time; the latter, more conservative, tried to reform society from within.

In the context of turmoil both at home and abroad, New England's clerical and secular elite sought to remain firmly in control of the meaning of their own earthquake. By sharing their views in the pulpit and in print—in all, some fifteen sermons were printed on the earthquake—Boston's clergy worked to make their own words heard over popular interpretations. The press published sermons alongside such popular pamphlets as reprints of "very circumstantial accounts" of the Lisbon quake, a letter enumerating in graphic detail all of England's many sins, and a poem on the "great earthquake," which allegedly encompassed all of North and South America as well as Lisbon.[92] In all of these publications, doomsday prophesies and visions of sin and suffering now and at the end times fed the fears of already frightened Bostonians.

In the competition between learned authority figures and the popular press over the meaning of the earthquake, neither theological nor philosophical doctrines were in question. Bostonians of all social levels thought earthquakes were signs requiring interpretation. Most also believed that the recent quakes were a warning to repent, sent by an angry God, and that earthquakes could be signs of the end times. Contention instead revolved around the issues of social authority and appropriate responses to the fear born out of these natural events. Such public authority figures as Professor Winthrop and Reverend Prince might disagree on the effects of electricity, but they concurred that people should direct their fears into socially acceptable channels, whether further philosophical inquiry into the nature of earthquakes or participation in public fast days and communal prayer. In contrast, the learned community considered anger, disregard for the suffering of others, and the spreading of "unreasonable terror" born of enthusiastic visions—or misguided philosophical speculation—to be socially irresponsible reactions to disaster.

The Comet of 1759

Underlying this critique was the learned community's belief that reasoned application of scripture and philosophy was more valid than interpretations based on alleged divine illumination or fear-induced prophesy. As Professor Winthrop explained, "astrological or enthusiastical predictions of these or any other unusual phaenomena deserve no regard, but are to be treated as the dreams of visionaries. And yet, trifling and contemptible as they are, such have been made both in former and later times."[93] Winthrop could have written these words about responses to the earthquake, but instead he was referring to the predictions that had historically surrounded comets. With the return of Halley's comet in April of 1759, Winthrop had sufficient cause to promote the scientific method over enthusiasm; for the first time a comet's reappearance matched an empirical prediction.

In his published lecture, Winthrop emphasized the importance of empirical data on observable cometary paths for formulating predictions, and he shared his own calculations in the *Boston Evening-Post.*[94] His lecture also summarized historical comet theories and the more recent speculations of

Isaac Newton and Edmund Halley concerning comets' composition and function. Winthrop was as fascinated with the possible benefits that God had built into comets as he had been with earthquakes. He admitted that he knew little about comets' "*natural* purposes," but he noted that Newton had proposed that "the tails of the Comets may be intended to recruit the perpetual waste of fluids in the Planets by vegetation and putrefaction."[95] As a man of his times, Winthrop believed that earthquakes could be both heaven-sent warnings and an ingenious system of underground tillage just as comets could be both "ministers of divine justice" and interplanetary garbage collectors.[96]

For all his optimism about the purpose of comets and his frustration with those "who see with unphilosophic eyes . . . [and] turn every unusual phaenomenon into a presage of some dire event," Winthrop could not ignore the historical correlation of comet sightings and natural disasters.[97] Most significantly, he and other natural philosophers, including Benjamin Franklin and Yale president Thomas Clap, were persuaded by a contemporary theory that a comet had caused the biblical flood.[98] Based on proven natural laws, Winthrop also acknowledged that if a comet were to become trapped in the earth's gravitational pull, it could cause extensive damage.[99] However, he became annoyed when popular pamphleteers, such as the anonymous author of *Blazing Stars Messengers of God's Wrath,* "instead of searching into the nature of things, indulge their own imaginations."[100] This particular pamphlet suggested that comets not only "Prognosticated evil Things" but were proven "Messengers of Death" and the cause of multiple military defeats, earthquakes, droughts, and floods.[101] One newspaper also reported that the comet sighting along with an eerie repetition of loud explosions in the air "have given much Uneasiness to timorous People, especially *Women,* who will needs have it that they portend some dreadful Judgments to this our Land."[102] Winthrop found such extravagant claims both philosophically unfounded and liable to provoke social instability—the very same criticisms he had leveled at Prince concerning lightning rods.

In his conclusion Winthrop admitted, on the one hand, that to dismiss the possibility that a comet was a divine warning of catastrophe would be arrogant and irreligious, but on the other hand, he warned that to succumb to disorderly panic would "betray a weakness equally unbecoming a reasonable being."[103] Without denying the devastating potential of comets or God's ability to control them to his own ends, Winthrop and fellow natural philosopher Thomas Clap chose instead to emphasize God's role as the "omnipotent creator" of all moving bodies in the universe who had, in his "consummate wisdom," prevented any comets from striking the earth.[104]

Clap's essay on meteors, written in 1755, when he was the president of Yale College, shows that clergy members applied the principles of natural philosophy to comets, but there were no fast days or sermons published on account of the 1759 appearance.[105] Though many, including Winthrop and Clap, considered comets to be religiously significant, they did not trigger the same response as earthquakes and droughts, which had greater biblical implications. Instead, aside from a few brief newspaper reports and the lectures of Clap and Winthrop, the vast majority of speculation concerning the comet appeared in the pages of yearly almanacs.[106]

Almanacs were a particularly good way to reach a population concerned about the approaching comet because of the annuals' broad readership, which was much greater than for newspapers or printed sermons. In fact, during the century the circulation of almanacs exceeded that of all other printed materials combined.[107] In this sense almanacs were "popular" literature because they appealed to readers across all social strata, not only to the less educated. In the 1750s the authors of almanacs themselves revealed the diverse interest in this printed medium. Benjamin Franklin, alias Poor Richard, was a laboring printer as well as an important name in philosophical circles as a result of his experiments with electricity; David Sewall, who wrote a very learned New England almanac, was a graduate of Harvard; and Dr. Nathaniel Ames was far and away the most popular almanac writer in the eighteenth century, although he had no personal claim to educational or philosophical fame. As the historian Elizabeth Carroll Reilly has suggested in her study of eighteenth-century print culture, the very appeal of almanacs lay in their eclectic nature; "they succeeded in holding a kind of tension between competing tastes and ideologies . . . in which the ingredients never dissolved into a simple solution."[108] In regard to the comet of 1759, almanacs provide insight into the varying beliefs that made up the supernatural rationalism of popular culture in mid-eighteenth-century New England.

Like other public interpretations of unusual events, almanacs demonstrated the widespread conviction that natural events were religiously meaningful and also worthy of philosophical examination. The Boston writer George Wheten, in an almanac article of unusual length, explained the first (or divine) and secondary (or physical) causes of most natural phenomena, including hail, earthquakes, and the rainbow. The last he defined as both a sign of faith to Christians and an uncommon interaction of sunlight and clouds.[109] Almanac coverage of comets echoed this dual causation; writers typically provided a philosophical explanation of natural events followed by a discussion of God's moral intentions as expressed through the natural world. In their mechanical descriptions of the movement and composition of comets, almanac authors relied heavily on their readings of the theories of Edmund Halley and Isaac Newton, particularly Newton's laws of motion and gravitation. As information middlemen, they often simplified and rewrote the ideas published by Winthrop and other learned philosophers for public consumption. Sounding much like Winthrop, David Sewall praised Newton for "at once solv[ing] all the perplexing Phaenomena which has been observed of these surprizing Bodies" by demonstrating that they were subject to the same natural laws as planets.[110]

God, in his role as benevolent designer rather than wrathful judge, was as important as Newton in almanac writers' philosophical explications of comets and the larger order of nature. In his description of the "grand Machine of the Universe," the almanac writer James Davis praised the "Almighty Architect" whose "Miracles of Mechanics" abounded throughout the planetary system.[111] Using remarkably similar language, Sewall quoted from a poem that referred to God as a "Great Artist! Thou, whose Finger set aright this exquisite Machine, with all its Wheels."[112] These writers considered comets to be one of God's awe-inspiring marvels, which one could account for in philosophical terms. As conceived in eighteenth-century almanacs, the universe and all of its parts were testament to God's power and wisdom as "Omniscent Architect": "Let Hail and Rain, let Meteors form'd of Fire, And lambient Flames, in this blest Work conspire."[113]

For the extremely popular Nathaniel Ames, like his fellow almanac authors, comets were much more than testaments; they were also instruments of God's moral government. In the regular workings of nature, Ames identified comets as a "sort or kind of *Planets* because they revolve about the same *Sun* the *Planets* do . . . and are subject to the same Law of Gravitation which the *Planets* are," though their orbits are "Excentrical" rather than regular. Yet, he argued, God in his wrath could "change the Course of Nature, and reduce [a planet] to the unhappy Circumstances of *Blazing Stars*—A doleful Inheritance, reserve'd perhaps for the Punishment of their ancient guilty Inhabitants!"[114] Other writers echoed this view of God's arbitrary power, although some did suggest that he could just as easily use "second Means to Work his Ends."[115] According to Benjamin Franklin, God might even bring about the conflagration at the end times by directing a comet to collide with the earth.[116] Comets, as these writers described them, could therefore serve multiple purposes, as displays of God's designing power and omniscience, "Instruments of Divine Vengeance" in such instances as the biblical flood, and possible precursors or even causes of the end times.[117]

Some almanac authors also argued that the series of unusual and threatening events in mid-eighteenth-century New England lent an even greater significance to the approaching comet. According to John Tobler's calculations, the comet was one of the "Signs of the Times," along with the earthquake and the faltering war effort that "loudly calls for Repentance and Amendment of Life."[118] Ames also emphasized the contemporary pattern of woe in a poem adorning the front page of his almanac for 1758: "MINORCA's gone! OSWEGO too is lost! These sad Events have silenced my Muse, The Prince of Day eclipses at the News, And Comet soon will blaze along the Skie, Portending (sure) some dire Event is nigh."[119]

As those who spread "superstitious fancies" about comets as harbingers of "dire events," Ames and Tobler were the sort of popular authors whom Winthrop accused of instigating enthusiastic visions and disorderly panic in response to unusual natural phenomena.[120] Almanac writers, however, did not share Winthrop's concerns, because their purposes in addressing the comet could not have been more different. Winthrop, on the one hand, was attempting to increase scientific knowledge about comets through empirical method and reasoned analysis. By publishing his *Lectures,* he also sought to buttress his own reputation as a learned philosopher. Enthusiastic and fearful reactions to the comet could negate both of Winthrop's goals by deflecting interest from philosophy and conveying social authority to less-educated but more popular spokesmen.

Almanac authors, in contrast, stood to benefit from popular interest and fearful anticipation of the comet, which would only increase the sales of their yearly publications. As purveyors of information, they did not seek to formulate new knowledge but instead claimed only to provide easily accessible science for the masses. As entertainers, the more dramatic prophesy and attention-grabbing stories of former catastrophic comets contained in their almanacs, the better. Religious conviction and concern over the ominous "signs of the times" on the part of many almanac authors also led them to emphasize the moral implications of comets even as they discussed God's designing omnipotence. Winthrop, in the interests of natural philosophy, chose instead to focus on God's benevolent and creative purposes, although he did not deny the role of comets as tools of divine judgment.

Ultimately, for Winthrop natural events such as the comet or the earthquake demonstrated a "mix'd state; in which there are a variety of purposes, *natural* as well as *moral*, in prosecution at the same time."[121] The interpretations of almanac writers were not dissimilar. According to the almanac writer Roger Sherman, comets "are supposed to be portentous of great Revolutions and Overtures, both in the natural and moral Worlds."[122] Both the learned philosopher and the popular author recognized the inherent religious and philosophical purposes of comets, of their moral and natural significance. However, their divergent intentions in writing about these natural events led them to characterize comets very differently: Winthrop's benevolent air-purifier and Sherman's fearful portent of revolution were worlds apart.

Common Ground and Conflict of Purpose

The perspectives of both Winthrop and the almanac writers in the 1750s reflected the common understanding of the relationship between God and nature in midcentury New England. This discourse provided Bostonians of all social levels with a way to combine their sincere religious convictions with philosophical speculation into the inner workings of the natural world. Groups as diverse as popular purveyors of information, learned philosophers, clerical scholars, and lay exhorters marshaled both the scriptures and the latest scientific theories in support of their beliefs about the supernatural government of God. Rational study and supernatural power, natural occurrences and moral implications, were all interrelated in the minds of eighteenth-century Bostonians. In effect, the doctrines scholars have called supernatural rationalism formed a shared starting place for the interpretation of the natural world, a flexible discourse that loosely held together a variety of argument styles and portrayals of God: from belief in the corporate nature of sin and the teleological use of the Bible to the inductive approach and empirical method of analyzing nature, from a wrathful God capable of immediate intervention to an awe-inspiring and benevolent designer who worked through the regular order of nature.

These varied approaches to interpreting the natural world were not necessarily oppositional. They could be cumulative. Charles Chauncy described earthquakes as the "most terrible" judgment of a wrathful God who could intervene directly in the natural world but "more commonly makes use of second causes."[123] When Benjamin Franklin wrote about the predicted comet in his almanac, he cited the empirical calculations of Newton and Halley, but he also suggested that God used a celestial messenger as "the Cause of the Deluge."[124] Thomas Prince's "divine philosophy" led him both to praise the designing wisdom of God the Creator and to interpret the earthquake and the drought as messages from an angry and arbitrary moral governor. As John Winthrop wisely recognized, God himself was not bound to any one purpose in directing natural events. He concluded, in the very last line of his earthquake lecture, "how 'excellent in working' is that BEING, who can bring good out of the greatest evils; and can answer intentions, the most widely differing, by one and the same dispensation of His providence!"[125]

Despite the inherent flexibility of the complex of ideas now labeled supernatural rationalism and the complementary intentions of God in nature, opposing viewpoints did surface when interpreters attempted to assign cultural meaning to natural events in the mid-eighteenth century. Explanations of the drought, the earthquake, and the comet in the 1750s were far from objective and, against a social context that included war and earthquakes in important commercial cities including Boston itself, as well as London and Lisbon, the meaning of natural events was of great cultural significance. Ultimately, opposition and conflict stemmed from differences in purpose and attempts to establish interpretive authority. Such pastors as Thomas Foxcroft in Boston and Aaron Smith in Marlborough attempted to use natural events to reform the morals of their congregations. John Winthrop, professor of natural philosophy at Harvard, recognized the moral implications of the comet and the earthquake, but his primary purpose was to promote empirical inquiry. As societal leaders, pastors and professors sought to maintain their authority and control the meaning of natural events, harnessing heightened interest in the causes and consequences of earthquakes to their own ends, whether communal reform or science, for the cause of God.

In contrast, less-educated Bostonians, who expressed their perspectives in places such as almanacs, pamphlets, and newspapers, sought to achieve different goals in their interpretations of natural events. Farmers and merchants struggling to maintain their livelihood in the midst of dry conditions and war used the drought of 1749 to control prices. Authors of almanacs capitalized on popular interest and fear about the impending return of Halley's comet in 1759 to increase sales of their publications, in which they provided nearly equal amounts of philosophical and religious speculation designed to appeal to a wide readership. And finally, doomsday prophesiers and pamphlet writers interpreted earthquakes that rocked the Atlantic world between 1750 and 1755 as sure signs of the end times, inspiring panic among those who gave credence to these ideas. Taking advantage of the destabilizing effects of unusual natural events, these more popular spokesmen grasped at the social control they did not normally possess and threatened those who held established positions of authority. Professor John Winthrop's frustration with popular explanations of the earthquake and the comet from those who "indulge their own imaginations" suggests that contention did exist between learned and popular interpretations of events; his public argument with the Reverend Thomas Prince indicates that even the learned had their interpretive disagreements, especially if social stability and the dangers of "unreasonable terrors" and enthusiasm were at issue.[126]

In the mid-eighteenth century a relatively minor earthquake went beyond a momentary fright, a drought was not just a threat to crops, and a comet was more than an unusual visual spectacle. Instead, Bostonians from all levels of society believed that such phenomena were communications from God that required exegetical, moral, and philosophical examination. Speaking through the shared language of supernatural rationalism, interpreters nevertheless diverged in their efforts to control the meaning and determine the purpose of Boston's natural events.

13

Biological Responses to Climate Change in Boston

Abraham J. Miller-Rushing
and Richard B. Primack

THE behaviors of plants and animals are changing in response to warming temperatures. In recent years biologists have observed birds wintering farther north, tropical frog populations declining, and insects relocating to higher altitudes on mountain slopes because of changes in climate.[1] Yet the most convincing evidence that living organisms are responding to global warming comes from the timing of spring events. Analyses of these records show conclusively that wild and cultivated plants are flowering earlier, birds are migrating earlier, and ponds are ice-free earlier than in the past because of warmer spring temperatures.[2] Historical records from the Boston area provide some of the best evidence for these changes in the United States.

Observations of changes in the timing of spring events are particularly compelling when examined over very long time frames—the longer the better. Variability in temperatures can be quite large, even during the course of a decade, so research into temperature-related events such as spring flowering or migrations is stronger when supported by several decades' worth of observations. Much of the data currently available, however, has come from the notebooks of professional biologists who have made observations for the purpose of documenting basic ecology. Despite the value of these data for understanding how certain species in certain places have already responded to global climate change, too few long-term data sets exist to make them useful in predicting future impacts on many species in many locations.[3] To build a more taxonomically and geographically complete picture, scientists need more information.

The additional information that scientists need can often be found in sources unusual for biological studies but typical in historical research. For example, dated historical photographs, collections of botanical specimens (also known as herbarium specimens), and journals and records of amateur naturalists can document when birds migrate or when plants flower. This chapter presents three case studies from our work in the Boston area. The first describes our use of herbarium specimens to track changes in flowering times.[4] The second shows the value of personal journals—in this case the journals of the naturalist Kathleen Anderson of Middleborough, Massachusetts.[5] The third describes changes as shown in photographs from Concord, Massachusetts, and the Arnold Arboretum in Jamaica Plain.[6] These case studies highlight the utility of herbarium and archival collections in climate change research and set the stage to find even more.

Boston's Naturalist History

Even without taking into account its storied past as a center of U.S. history, Boston is an especially good place to look for naturalist observations and records. From the perspective of climate change studies, the strength of this tradition lies in the length and continuity of naturalist activities in the Boston area. For example, the botanist Edward Jarvis documented the composition of the flora of Concord, Massachusetts, in 1834, a feat that has been repeated four times since. The famous philosopher and naturalist Henry David Thoreau observed flowering times and ice-out at Walden Pond. Later botanists, Alfred Hosmer, Richard Eaton, and ourselves have since followed in Jarvis's and Thoreau's footsteps.[7]

The Boston area also has outstanding weather records to match the naturalist tradition. The Blue Hill Meteorological Observatory is the longest continuously running meteorological station in North America.[8] It maintains weather records that date back to 1831 and ice-out records for

Houghton's and Ponkapoag ponds that date back to 1886.[9] Today the Observatory maintains a written record of the temperature, precipitation, cloud, wind, barometric pressure, and humidity observations that have been made, often on an hourly basis, since the station's founding. In addition to Blue Hill, fifty-nine other weather stations are located within forty kilometers of Boston. Many of these sites have records dating back to 1950 or earlier and are still in operation, including stations in Boston (at Logan International Airport), Bedford (at Hanscom Field), Norwood, and South Weymouth. Other archives contain weather records dating back to 1786.[10] Yet even with its widely known history of climatologic and naturalist data collection, we have discovered that the wealth of naturalist records in Boston includes several sources of data that have not previously been used to document species' responses to climate change. These sources of data include herbarium specimens, personal journals, and photographs.

Herbarium Specimens

For examining the utility of herbarium specimens, we turned to one of the largest and best-known herbaria in the region, the Arnold Arboretum, established in 1872. Although the Arnold Arboretum did not keep systematic records of the phenological behavior of plants in its collection until our work in 2002, its herbarium provides a source of phenological data extending well over a hundred years. The herbarium contains eighty thousand dried plant specimens, most of which were taken from the woody plants of the living collections as part of the arboretum's standard documentation process. The record that accompanies each dried and pressed specimen includes the name of the species, the identification number of the plant, and the date of collection. One can also readily see the phenological state of the plant on that date, such as flowering, past flowering, or in fruit. A large portion of the herbarium specimens was collected when the plants were in flower. Many of the plants from which the specimens were taken are still among the fifteen thousand plants growing on the grounds of the arboretum (figure 13.1). Together with the eighty thousand herbarium specimens, the arboretum's living collection provided a potential sample size large enough to compensate for any species-specific phenological changes unrelated to temperature change. Thus we could use the herbarium collection to detect general patterns of flowering response to climate change common to most of the species.

Another advantage of using the plants at the arboretum for climate research derived from the controlled environment found there. Individual

Figure 13.1. A conceptual demonstration of the utility of herbarium specimens in climate change research. The authors hold a herbarium specimen collected in 1938 from the exact same azalea (*Rhododendron vaseyi*) that is flowering in the background. Note that the plant is flowering eight days earlier in 2004 than it did in 1938. Comparing herbarium specimens like this one with current flowering times can demonstrate dramatic changes in flowering times as a result of climate change. Dated photographs might be used in the same manner.

plants were well spaced and grown under conditions considered ideal for their species, with the grounds being carefully mulched, weeded, fertilized, and kept free of pests. This level of care probably reduced the possibility of nonrepresentative flowering times that might result elsewhere from crowding, scarcity of nutrients or light, or other suboptimal conditions. Based on all of these factors, we regarded the arboretum herbarium as a rich potential source of phenological data.

Personal Journals

A second source of climatic data came from the personal records of naturalists, who often make observations more for their own amusement and curiosity than for use in scientific study. Such records, sometimes kept as personal journals, can be another valuable source of data on spring events. These journals frequently hold information on when birds arrive, fish spawn, or salamanders are active in the spring. Naturalists' journals are relatively common and may be found in private homes, museums, and libraries throughout the world.

To determine just how useful data from personal journals might be, we worked with a well-known naturalist, Kathleen S. Anderson, who owns a farm in Middleborough, Massachusetts. Anderson has been involved for many decades in conservation issues at the local, regional, and national levels. From 1970 to 2003 she kept informal records of the birds, flowering plants, butterflies, and amphibian choruses that she heard or saw on her forty-hectare farm. She made observations while on walks through her farm and while sitting on her back porch. Anderson never intended for her diaries to be used in a scientific study, and she was surprised that we would consider analyzing them in this way. She did not collect the data based on a consistent schedule or a regular manner, which may have resulted in unintended variation in the data sets and may have obscured biologically significant trends. Nevertheless, her observations were consistent enough to gain a strong sense of when bird migrations, plant flowering, and other springtime events took place over time.

Photographs

Perhaps the most abundant source of data we examined were dated photographs. Much like herbarium specimens, they provided estimates of when particular events occurred in the past. Photographs have the additional benefit that they are much more common than herbarium specimens. Vast collections of photos exist at museums, libraries, research institutes, and in private homes. Thus photographs could greatly increase the amount of data available for climate change research.

We gathered photographic data from two collections. The first consisted of images of cultivated woody plants in flower at the Arnold Arboretum. Staff photographers, other staff members, and amateur photographers had taken the photographs between 1904 and 2004. We found many of these images, particularly those taken by staff photographers, at a central location, the Arnold Arboretum library. We found many additional photographs scattered among personal collections and at other institutions, however. All told, we used 251 dated images of forty-eight species in our study. The second collection we examined featured wild plants in flower in Concord, Massachusetts. These photographs spanned the years 1900 to 1921 and were taken by the landscape photographer Herbert Wendell Gleason. Gleason had focused on plants and places mentioned in the journals of Henry David Thoreau. Gleason's collection was smaller than the Arnold Arboretum's—thirty-four dated photographs of seventeen species—but it was big enough for our purposes.

Climate Changes

When examining changes in phenological events over time, it is important to have a strong understanding of long-term climatic changes in the region of interest. In the case of Boston, there is a long history of temperature and other climatic records. These records detail a 1.5-degree Celsius (2.7-degree Fahrenheit) increase in its average annual temperatures over the past hundred years—more than in less urbanized parts of the world (figure 13.2). Less than half of Boston's increase corresponds to the increase in the average global temperature.[11] The other portion is probably related to urbanization—more paved surfaces to absorb sunlight and radiate it as heat, less plant cover to reflect light and to remove heat through transpiration, and more sources of greenhouse gas emissions, such as buildings and cars—phenomena collectively labeled as the "urban heat island effect"[12]). As a result, the increase in Boston's average temperatures has already reached the magnitude expected for the entire planet later in this century and shows no signs of slowing down. Thus Boston provides a preview of the warming that will occur elsewhere in the world.

Herbarium Specimens

Using temperature data as a touchstone for comparison with other data, we set out to determine whether our resources accurately reflected the demonstrable changes in temperatures in the Boston area. In 2003 we examined changes in the flowering times of 229 living plants encompassing thirty-five different genera.[13] Major genera (represented by ten or more individuals) included *Amelanchier* (shadbush), *Cornus* (dogwood), *Corylopsis, Enkianthus, Halesia* (silverbells), *Magnolia, Malus* (apple), *Prunus* (cherry),

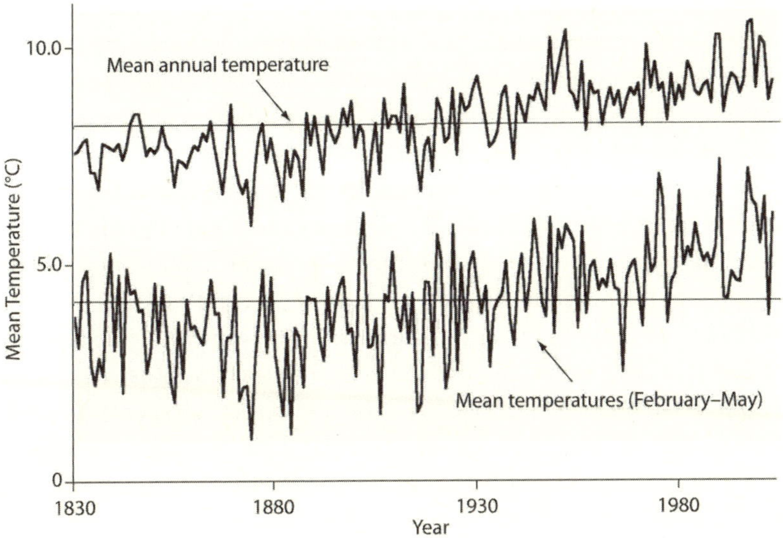

Figure 13.2. Boston temperatures, 1831–2004, as reported by the Blue Hill Meteorological Observatory. The top series represents mean annual temperatures. The bottom series represents mean temperatures in February, March, April, and May. The two horizontal lines represent the long-term mean temperatures for each series (annual 8.2 degrees Celsius; February–May 4.1 degrees Celsius). Note that in the early years, temperatures were generally colder than the long-term mean. In later years, temperatures were warmer than the long-term mean. Almost all of the warmest years have occurred in the past thirty years.

Rhododendron (including azalea), and *Syringa* (lilac). The Arnold Arboretum staff had often collected multiple herbarium specimens from the same plant in different years, so we found a total of 372 herbarium specimens taken from these 229 plants. We had a record of when plants flowered on the grounds of the Arnold Arboretum from 1881 to 2002.

During the spring and summer of 2003, we observed the individually numbered plants weekly. We determined a single date of full flower for each plant, although this date could have missed the true flowering peak by several days because of sampling only once a week. Once we determined the date of full flowering in 2003 for each plant, we subtracted it from that of the corresponding herbarium specimen to estimate a change in flowering time. We then investigated how changes in spring flowering times correlated with the temperature differences between individual years. To our surprise, the spring of 2003 (February through May) was colder than in any previous year since 1967, with temperatures typical of the early twentieth century, whereas in four of the five previous years—1998, 1999, 2000, and 2002—we had experienced extraordinarily warm springs. As a result, the 229 plants examined in our study flowered at about the same time in 2003 as they had between 1900 and 1920. In contrast, typical plant flowering times between 1980 and 2002 were about eight days earlier than in 2003, and eight days earlier than between 1900 and 1920, clearly showing a significant trend toward earlier flowering over the past hundred years (figures 13.3A and 13.3B).

Although we found a significant correlation between changes in temperature and flowering times, the use of herbarium specimens raised several questions. First, we could not be certain that the herbarium specimens were all collected on the exact day of peak flowering. Specimens may have been collected when the plant first started to flower or when it had nearly finished flowering. In the case of short-flowering plants, on the one hand, the error would be small: if the collector took a sample from a plant with a one-week flowering period, for example, the collection date would be only three or four days away from the actual date of peak flowering. In the case of a plant with a twenty-day flowering period, on the other hand, the amount of error would be about ten days. However, our statistical tests found no evidence of this type of bias; data were no more variable over time for long-flowering plants than for short-flowering plants.

A second concern was that outlying data points might have obscured biologically important trends. Examples of outlying data points include a dogwood (*Cornus mas*) that flowered twenty-seven days later in 1965 than in 2003 and a cherry tree (*Prunus apetela*) that flowered twenty-four days later in 1987 than in 2003. These apparent anomalies may have been caused by someone collecting the specimen at the very end or very beginning of an unusually long flowering season in those years. But again, our large sample size prevented the few outlying data points from having a statistically significant effect on the results. Finally, we were concerned about the uneven collection of herbarium samples: in different years different numbers of herbarium specimens were collected, creating gaps in the period between 1940 and 1960 that could give disproportionate weight to certain years in our analysis. We resolved this problem by dividing the data into two subsets, one on either side of the 1940–60 gap, and separately analyzing each subset using the same methods that we employed for the entire group. In each subset we found the same significant trend toward earlier flowering times, indicating that the irregularity in specimen collection (that is, the gap in collection) did not affect the outcome.

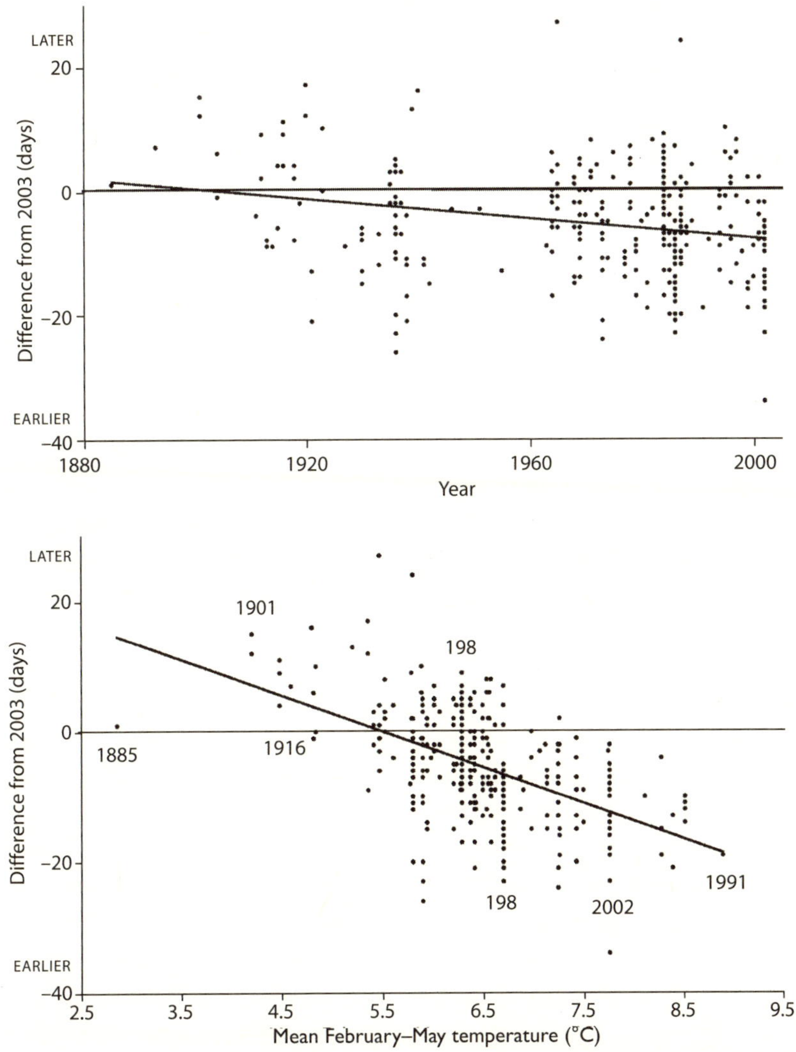

Figure 13.3A and Figure 13.3B. Changes in flowering times of plants at the Arnold Arboretum of Harvard University in Boston (figure 13.3A) over time and (figure 13.3B) as temperatures increase. Each dot shows the difference between when a plant flowered in the past (based on herbarium specimens) and when it flowered in the benchmark year of 2003. Negative values indicate that a plant flowered on an earlier date than it did in 2003. The lines are the best-fit lines for the series as determined by regression analysis. Figure 13.3A shows that plants are flowering progressively earlier over the past hundred years. Figure 13.3B shows that plants flower earlier in warmer years. Years are indicated for certain years with many specimens and years with extreme temperatures. *Source:* D. Primack, C. Imbres, R. B. Primack, A. J. Miller-Rushing, and P. Del Tredici, "Herbarium Specimens Demonstrate Earlier Flowering Time in Response to Warming in Boston," *American Journal of Botany* 91 (2004): 1260–64. Figures courtesy of the *American Journal of Botany.*

Given these concerns about the quality of data from herbarium specimens, our results are quite striking and show clearly that plant flowering times are highly responsive to changes in average temperatures in the four months before and during flowering. For the spring-flowering species we studied, we used the mean temperature for February, March, April, and May to calculate changes. In general, flowering times advanced about four days per 1-degree Celsius (1.8-degree Fahrenheit) increase in mean spring temperature, as calculated using multiple regression, a statistical technique that considers the flowering time of plants in warm and cold years. This rate falls within the range of findings of other published studies from the United States and Europe, which record flowering times occurring from two to ten days earlier for each 1 degree Celsius of increase in temperature.[14] Because Boston's temperatures in February through May have warmed approximately 1.5 degrees Celsius over the past hundred years, the recorded temperature increase can account for nearly six of the eight days that flowering time has advanced over the past hundred years. Factors other than the temperature increase recorded at the Boston weather station must account for the other three days of increase.

These other factors might include temperatures during other months of the year or other climatic variables such as rainfall and humidity. Local conditions within and around the arboretum may also be affecting flowering times. Construction of roads and buildings on adjacent land, for example, may have led to a very localized increase in temperature that was not registered at the Boston weather station. Finally, it is possible that plants may have flowered over a longer period as they increased in size and age and may have consistently been collected at the beginning of their flowering periods. If both of these circumstances occurred, plants may have appeared to flower earlier over time even without a temperature effect. Further investigations are needed to determine the relative importance of these factors.

Personal Journals

With respect to Anderson's journal entries, we knew that her records could contain serious biases because she did not make her observations in a consistent manner. Nevertheless, her observations contained valid information that was potentially useful in climate research. From each of her narrative-style journals, we extracted the date she first observed particular plants flowering, birds arriving, butterflies emerging, and frogs and toads calling for twenty-four common species.[15] We compared Anderson's records from 1970 to 2003 with temperatures for the region to determine whether these spring events were occurring earlier over time and whether earlier events were tied to the 2.3-degree Celsius (4.1-degree Fahrenheit) increase in February–May temperatures at the nearest weather station.

Of the sixteen bird species that Anderson observed in most years, five—chipping sparrows (*Spizella passerine*), house wrens (*Troglodytes aedon*), ovenbirds (*Seiurus aurocapillus*), ruby-throated hummingbirds (*Archilochus colubris*), and wood ducks (*Aix sponsa*)—are arriving significantly earlier than they did in the past (table 13.1). For example, wood ducks are now arriving thirty days earlier than they did thirty years ago because the ponds that they occupy are melting earlier each spring (figure 13.4). Overall, these spring events are occurring about eight days earlier now than they did in 1970, a trend that closely correlates to the increase in spring temperatures over that period. A total of twenty-two of the twenty-four species are active earlier in warmer springs.

Although there was no rigorous scientific methodology to Anderson's observations, her journal entries nevertheless convincingly showed that birds are migrating, plants are flowering, amphibians are calling, and butterflies are emerging much earlier than in the past. Moreover, her observations show that these earlier spring events are in response to higher temperatures. Her data agree with other scientific analyses of spring events in Europe and North America.[16] We believe other journals that contain naturalist observations will tell similar stories. We are continuing to analyze other diaries of flowering-time records from Concord and elsewhere in Massachusetts.

Photographs

Our study of the photographs from the Arnold Arboretum indicates that plants are flowering about eleven days earlier on average than they were a century ago. The rate of change was about four days for each increase of 1 degree Celsius in mean February–May temperatures. In other words, plants are flowering earlier because the temperatures in the months before flowering were getting warmer over time. On average, mean February–May temperatures at Blue Hill Observatory warmed 2.1 degrees Celsius (3.8 degrees Fahrenheit) from 1904 to 2004. In the particularly cold springs of 1916, 1923, 1924, and 1926 (mean February–May temperatures less than 3 degrees Celsius or 5.4 degrees Fahrenheit), plants flowered nine days later than

TABLE 13.1

Number of days earlier each species is active in spring per decade and per 1-degree Celsius increase in spring temperatures.

Species	*Decade*	*1 degree Celsius*
BIRDS		
Black-and-white warbler (*Mniotilta varia*)	–0.76	0.20
Baltimore oriole (*Icterus galbula*)	1.23	0.81
Barn swallow (*Hirundo rustica*)	0.57	–0.81
Gray catbird (*Dumetella carolinesis*)	0.89	0.31
Chipping sparrow (*Spizella passerina*)	**7.81**	0.54
Common grackle (*Quiscalus quiscula*)	–3.37	2.75
Hermit thrush (*Catharus guttatus*)	4.35	0.87
House wren (*Troglodytes aedon*)	**2.59**	0.57
Ruby-throated hummingbird (*Archilochus colubris*)	**5.75**	–0.55
Ovenbird (*Seiurus aurocapillus*)	**1.48**	0.31
Pine warbler (*Dendroica pinus*)	5.19	0.65
Red-winged blackbird (*Agelaius phoeniceus*)	2.14	**2.54**
Tree swallow (*Tachycineta bicolor*)	2.73	1.04
Wood duck (*Aix sponsa*)	**10.05**	**6.21**
Wood thrush (*Hylocichla mustelina*)	1.22	0.14
Yellow warbler (*Dendroica petechia*)	5.2	0.08
PLANTS		
Goldthread (*Coptis trifolia*)	1.93	**2.99**
Spicebrush (*Lindera benzoin*)	0.83	0.45
Wood-anemone (*Anemone quinquefolia*)	0.71	1.13
AMPHIBIANS		
American toad (*Bufo americanus*)	2.86	0.44
Spring peeper (*Pseudacris crucifer*)	5.41	**3.75**
Wood frog (*Rana sylvatica*)	1.23	2.86
INSECTS		
Mourning cloak (*Nymphalis antiopa*)	0.36	0.13
Spring azure (*Celastrina ladon*)	1.43	1.40

For example, the wood duck has arrived ten days earlier per decade, and arrives six days earlier for each 1-degree Celsius increase in spring temperature. Negative values indicate that a species arrived later over time or as temperatures warmed. Changes that are statistically significant are indicated in bold.
Source: Data from the personal journals of Kathleen Anderson as described in A. Ledneva, A. J. Miller-Rushing, R. B. Primack, and C. Imbres, "Climate Change as Reflected in a Naturalist's Diary, Middleborough, Massachusetts," *Wilson Bulletin* 116 (2004): 224–31.

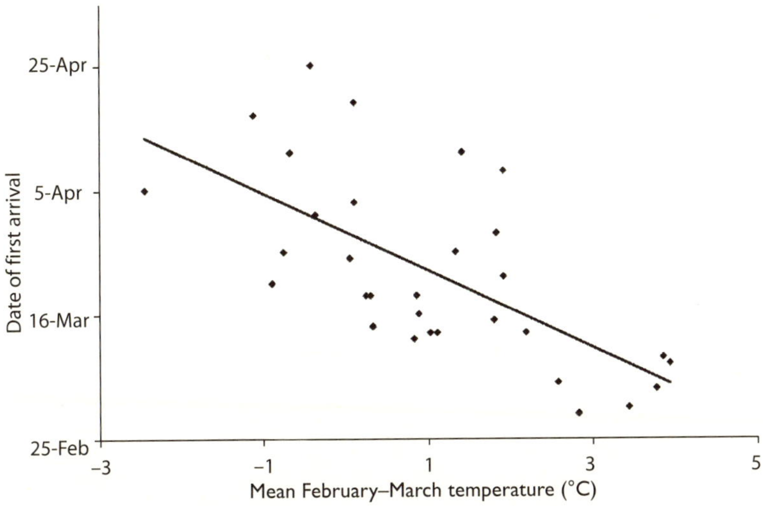

Figure 13.4. Wood duck (*Aix sponsa*) arrival dates in Middleborough, Massachusetts, 1970–2002, in relation to mean springtime (February–March) temperatures. Line represents best-fit using regression analysis. The slope of the line shows that wood ducks arrive much earlier in warmer years. *Source:* A. Ledneva, A. J. Miller-Rushing, R. B. Primack, and C. Imbres, "Climate Change As Reflected in a Naturalist's Diary, Middleborough, Massachusetts," *Wilson Bulletin* 116 (2004): 224–31. Figure courtesy of the *Wilson Bulletin.*

average. In the particularly warm springs of 1976, 1977, 1981, 1991, 2002, and 2004 (mean February–May temperatures greater than 6.0 degrees Celsius or 10.8 degrees Fahrenheit), they flowered two days earlier. This rate closely matched the response to temperature change that we had found in our previous study of herbarium specimens at the Arnold Arboretum (figure 13.5).

The study of wild species in Concord confirmed these findings. Flowering times as recorded in the Concord photographs did not change during the period 1900 to 1921; this was to be expected since temperatures at Blue Hill Observatory did not increase between those years. However, during these years the photographic record showed plants flowering about five days earlier for each 1-degree Celsius increase in spring temperatures. In warm years, such as 1903, plants flowered earlier than in cool years, such as 1916. In the particularly cold springs of 1901, 1916, 1917, and 1920 (mean February–May temperatures less than 3.0 degrees Celsius), plants flowered eight days later

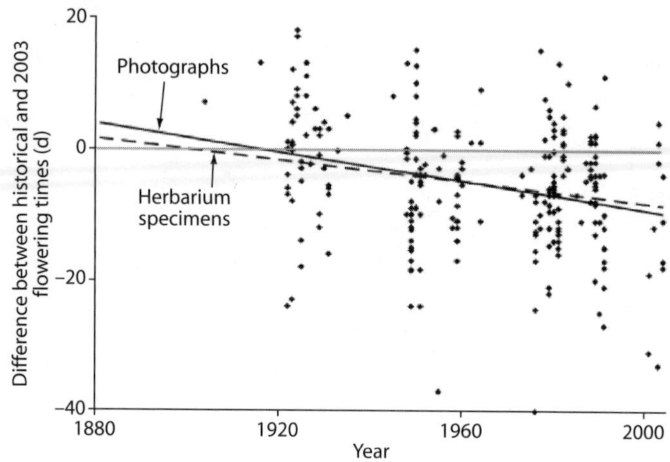

Figure 13.5. Changes in flowering times of woody plants at the Arnold Arboretum of Harvard University in Boston, 1904–2004. Each point represents the difference between the date a historical photograph showed a specimen in flower and the date that the same species was in flower in 2003 (historical date–2003 date). Negative values indicate historical flowering times that were earlier than flowering times in 2003. The solid line represents the best fit via ordinary least squares regression (slope = –0.11 d/yr, $R^2 = 0.08$, $P < 0.001$). For comparison, the dashed line represents regression using independent data from herbarium specimens and field observations in 2003 (slope = –0.08 d/yr, $R^2 = 0.06$, $P < 0.001$). The individual data points from the herbarium specimen data are not shown. *Source:* A. J. Miller-Rushing, R. B. Primack, D. Primack, and S. Mukunda, "Photographs and Herbarium Specimens as Tools to Document Phenological Changes in Response to Global Warming," *American Journal of Botany* 93 (2006): 1667–74. Figure courtesy of the *American Journal of Botany.*

than average. In the particularly warm spring of 1903 (mean February–May temperature more than 6.0 degrees Celsius), they flowered eight days earlier.

We verified these findings in Concord by comparing them to a set of unpublished observations of flowering times in Concord made by Alfred Hosmer from 1888 to 1902.[17] Hosmer's observations indicated that the same species flowered about five days earlier for each degree Centigrade warming in spring temperature. The results from the two sets of photographs and from Hosmer's observations are statistically indistinguishable (figure 13.6). The results also reflect the disparity in dates: Hosmer recorded first flowering dates, whereas Gleason photographed plants on their peak flowering

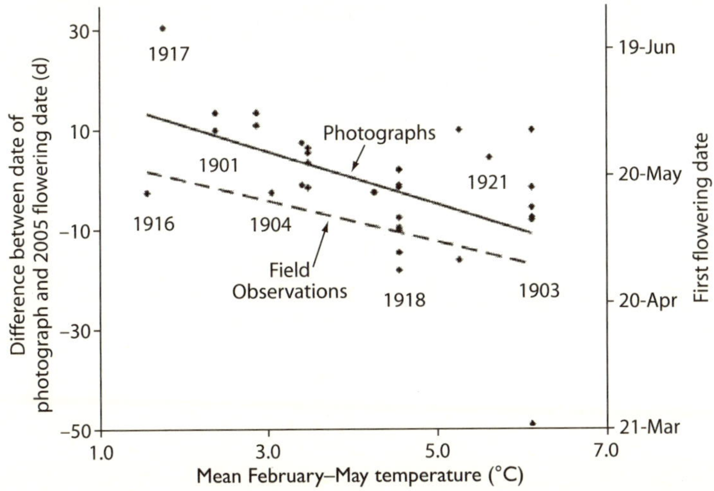

Figure 13.6. Changes in flowering times in response to changes in mean spring (February–May) temperatures for wild plants in Concord, Massachusetts, 1900–1921. Each point represents the difference between the date a historical photograph showed a specimen in flower and the date that same species was in flower in 2005 (historical date–2005 date). Negative values indicate historical flowering times that were earlier than flowering times in 2005. The solid line represents the best fit via ordinary least squares regression (slope = –5.3 d/°C, R^2 = 0.28, P = 0.001). The dashed line represents regression using independent data from field observations of first flowering dates collected by A. W. Hosmer between 1887 and 1903 (slope = –4.8 d/°C, R^2 = 0.08, P < 0.001). The individual data points for Hosmer's observations are not shown. If the point in the bottom right is removed, the trend remains significantly different from zero (P = 0.003) and not significantly different from the trend determined from the field observations (P = 0.69). Although the *y*-axes are different, the slopes for the two regression lines are in the same units, change in flowering date (days) per change in mean spring temperatures (degrees Celsius). *Source:* Miller-Rushing et al., "Photographs and Herbarium Specimens," 1667–74. Figure courtesy of the *American Journal of Botany.*

dates. Hosmer's observations are therefore dated several days earlier than Gleason's photographs.

We also noted an example of how photographs can be used to document changes in the timing of leaf-out as well as flowering (figure 13.7). The striking photograph was taken in the Lowell Cemetery in Lowell, Massachusetts, on Memorial Day, May 30, 1868. In the photo the trees have not yet leafed out, despite the late date and people are wearing heavy clothing. The

A
20 June 1926
7 May 2003
B
22 June 1917
17 May 2005
C
30 May 1868
30 May 2005

Figure 13.7. Representative comparisons of historical and recent photographs of phenological events. (A) Fringe trees (*Chionanthus virginicus* L.) flowered later in 1926 (mean February–May temperature = 2.5 degrees Celsius) than in 2003 (3.8 degrees Celsius) at the Arnold Arboretum, Harvard University, in Boston. (B) Wild specimens of pink lady's slipper (*Cypripedium acaule* Ait.) flowered later in 1917 (1.8 degrees Celsius) than in 2005 (4.7 degrees Celsius) in Concord, Massachusetts. (C) Leaf-out at Lowell Cemetery in Lowell, Massachusetts; the trees lack leaves in the 1868 (1.9 degrees Celsius) photograph, but the same trees were fully leafed in 2005 (4.7 degrees Celsius). Left to right, photographs were taken by (A) E. H. Wilson, R. Mayer, (B) H.W. Gleason, A. J. Miller-Rushing, (C) unknown, and R. B. Primack. *Source:* Miller-Rushing et al., "Photographs and Herbarium Specimens," 1667–74. Figure courtesy of the *American Journal of Botany.*

photograph taken on the same date in 2005, at the same location, shows the trees fully leafed out. At least two of the large, leafless trees in the 1868 photo were still alive and had fully leafed out in 2005. An exceptionally cold spring probably caused the delayed leaf-out in 1868; the mean temperature from February to May of that year was 2.2 degrees Celsius (4.0 degrees Fahrenheit) lower than the average over the past 150 years and 2.7 degrees Celsius (4.9 degrees Fahrenheit) colder than February to May 2005.

Are the Results Useful?

Although the different data showed some variation in historical timing of flowering, bird migrations, and other natural events, each of the methods clearly shows that spring events are occurring earlier than they did in the past. Moreover, each method shows that earlier natural events are strongly correlated with warmer years, providing strong evidence that warming in Boston has caused spring events to occur earlier over time. Herbarium specimens, photographs, and traditionally collected scientific field observations, strikingly, show nearly identical changes in flowering dates—on average, plants flower four or five days earlier for each 1-degree Celsius increase in spring temperatures. The naturalist's diary that we used showed a different rate of change, a little more than one day earlier for each 1-degree Celsius increase in spring temperatures, but the diary included observations of bird migrations, butterfly emergence, and amphibian calling in addition to flowering. We know that different taxonomic groups respond differently to climate change, so it is not surprising that we see a different change when we include several types of animals along with plants.[18]

These changes in the timing of natural spring events may not seem substantial, but they constitute a major development. For plants that flower for three weeks, a change of eight days, as we observed at the Arnold Arboretum, represents more than one third of the entire flowering season. Changes of this magnitude can significantly affect relationships between plants and the animals that pollinate their flowers, eat their leaves and seeds, and disperse their fruits. In many cases these relationships rely on a synchrony between the phenology of the plant and that of the animal—for instance, between the flowering of a plant and the activity of a pollinator—which may be disrupted if the timing of events shifts too quickly. Evidence from elsewhere in the world shows that some of these relationships are already changing.[19]

Many data sets from scientific observations, photographs, herbarium specimens, personal journals, and local publications could be assembled from around the world, covering the past 100 to 150 years. These records could provide invaluable insight into how the natural world is responding to climate change. A more detailed record of biological changes will allow us to determine which species and regions of the world will be most sensitive to climate change. We will be able to make more accurate predictions of how climate change will affect forests, grasslands, and other ecosystems. We hope that our work will contribute to the ongoing discussion of global climate change and encourage others to take advantage of these novel methods of documenting biological response to climate change.

NOTES

1. Introduction: Boston from Peninsula to Metropolis

1. William A. Newman and Wilfred E. Holton, *Boston's Back Bay: The Story of America's Greatest Nineteenth-Century Landfill Project* (Boston: Northeastern University Press, 2006), 20.

2. Ibid.

3. Ibid., 189.

4. Ibid.

5. On industrial pollution in Pittsburgh, see Joel A. Tarr, ed., *Devastation and Renewal: Environmental History of Pittsburgh and Its Region* (Pittsburgh: University of Pittsburgh Press, 2003).

6. See, for example, John B. Blake, *Public Health in the Town of Boston, 1630–1822* (Cambridge: Harvard University Press, 1959); Eric Jay Dolin, *Political Waters* (Amherst: University of Massachusetts Press, 2004); Sarah S. Elkind, *Bay Cities and Water Politics: The Battle for Resources in Boston and Oakland* (Lawrence: University of Kansas Press, 1998); and Fern F. Nesson, *Great Waters: A History of Boston's Water Supply* (Hanover, N.H.: University Press of New England, 1983).

7. Peter S. Rosen and Duncan M. FitzGerald, "The Drowning of Boston Harbor and the Development of the Shoreline," in this volume.

8. Michael J. Rawson, "What Lies Beneath: Science, Nature, and the Making of Boston Harbor," in this volume.

9. William Cronon, *Nature's Metropolis: Chicago and the Great West* (New York: Norton, 1991).

2. The Drowning of Boston Harbor and the Development of the Shoreline

1. C. A. Kaye and E. S. Barghoorn, "Late Quaternary Sea-Level Change and Crustal Rise at Boston, Massachusetts, with Notes on the Autocompaction of Peat," *Bulletin of the Geological Society of America* 75 (1964): 63–80.

2. H. Knebel, R. Rendigs, R. Oldale, and M. Bothner, "Sedimentary Framework of Boston Harbor, Massachusetts," in C. Fletcher and T. Wehmiller, eds., *Quaternary Coasts of the United States: Marine and Lacustrine Systems,* Project No. 274 Quaternary Coastal Evolution, Spec. Pub. No. 48, SEPM (Tulsa, Okla.: Society for Sedimentary Geology, 1922), 35–44.

3. C. A. Kaye, *The Geology and Early History of the Boston Area of Massachusetts: A Bicentennial Approach,* Geological Survey Bulletin 1476, (Washington, D.C.: Government Printing Office, 1976).

4. Lawrence LaForge, *Geology of the Boston Area, Massachusetts,* Bulletin 839 (Washington, D.C.: U.S. Geological Survey, 1932).

5. D. M. FitzGerald, P. S. Rosen, and S. Van Heteren, "New England Barriers," in R. A. Davis, ed., *Geology of Holocene Barrier Island Systems* (New York: Springer Verlag, 1994), 305–94.

6. P. M. Colgan and P. S. Rosen, "Quaternary Environments and History of Boston Harbor, Massachusetts," in D. P. West and R. H. Bailey, eds., *Guidebook for Geological Field Trips in New England,* 2001 Annual Meeting of the Geological Society of America (Boston: Geological Society of America, 2001), I1–I20.

7. P. S. Rosen, B. Brenninkmeyer, and L. Maybury, "Holocene Evolution of Boston Inner Harbor, Massachusetts," *Journal of Coastal Research* 9 (1993): 271–311.

8. F. Johnson, "The Boylston Street Fishweir," *Robert S. Peabody Foundation for Archaeology* 2 (1942): 212; and F. Johnson "The Boylston Street Fishweir II," *Robert S. Peabody Foundation for Archaeology* 4 (1949): 133.

9. Rosen, Brenninkmeyer, and Maybury, "Holocene Evolution of Boston Inner Harbor" 271–311.

10. E. A. Himmelstoss, D. M. FitzGerald, P. S. Rosen, and J. R. Allen, "Bluff Evolution of Coastal Drumlins: Boston Harbor Islands, Massachusetts," *Journal of Coastal Research* 22, no. 5 (2006): 1230–40.

11. P. Rosen, D. FitzGerald, E. Himmelstoss, and J. Allen, "Morphology and Evolution of Boston Harbor Island Shorelines, Massachusetts, USA," Fifth International Symposium on Coastal Engineering and Science of Coastal Sediment Processes, Coastal Zone 03, American Society of Civil Engineers, *Proceedings* (Clearwater Beach, Fla.: American Society of Civil Engineers, 2003).

12. F. F. Escoffier, "Traveling Forelands and the Shore Line Processes Associated with Them," *Bulletin of the Beach Erosion Board* 8, no. 4 (1954): 11–14.

13. Himmelstoss et al., "Bluff Evolution of Coastal Drumlins," 1230–40.

14. William A. Newman and Wilfred E. Holton, *Boston's Back Bay: The Story of America's Greatest Nineteenth-Century Landfill Project* (Boston: Northeastern University Press, 2006).

15. A. Heintzelman-Muego, "Construction Material and Design of Nineteenth Century and Earlier Wharves: An Urban Ecological Concern," paper presented at the Society for Historical Archaeology and Council for Underwater Archaeology, Denver, Colorado, 1983.

16. O. Bray, "Restoring Historic Wharf at Salem, Massachusetts," *Civil Engineering* 10, no. 2 (1940): 105–7.

17. P. S. Rosen and D. B. Vine, "Evolution of Seawall Construction Methods in Boston Harbor, Massachusetts," *Proceedings, Institute of Civil Engineers* 110 (1995): 239–49.

3. What Lies Beneath: Science, Nature, and the Making of Boston Harbor

This essay, originally presented at the conference "Remaking Boston: The City and Environmental Change Over the Centuries," appears by arrangement with the *Journal of Urban History,* which published it in volume 35, no. 5 (July 2009).

1. John S. Sleeper, *Address on the Encroachments in the Harbor of Boston* (Boston: Rockwell & Churchill, 1872), 12. Sleeper spent much of his career in the merchant service sailing ships out of Boston. He retired in 1830 at the age of thirty-six and subsequently worked as a newspaper editor and author of sea tales under the pseudonym "Hawser Martingale."

2. Two works show some interest in the natural environments of urban harbors: Donald Squires and Kevin Bone, "The Beautiful Lake: The Promise of the Natural Systems," in Bone, ed., *The New York Waterfront: Evolution and Building Culture of the*

Port and Harbor (New York: Monacelli Press, 1997), 16–35; and Ann L. Buttenwieser, *Manhattan Water-Bound: Planning and Developing Manhattan's Waterfront from the Seventeenth Century to the Present* (New York: New York University Press, 1987). In general, however, there is almost no critical literature on the history of harbor engineering in America. A good place to begin exploring the topic would be the U.S. Army Corps of Engineers's *Occasional Papers* and *Port Series Reports,* as well as its factual histories of various American harbors and engineering works. There is far more literature on Great Britain, whose poor natural harbors often required massive engineering works to make them navigable. See Gordon Jackson's classic study, *The History and Archaeology of Ports* (Tadworth, Great Britain: World's Work Limited, 1983); and Adrian Jarvis, ed., *Port and Harbour Engineering,* Studies in the History of Civil Engineering, vol. 6 (Aldershot, Great Britain: Ashgate Publishing, 1998). For a more international perspective, see Josef W. Konvitz, *Cities and the Sea: Port City Planning in Early Modern Europe* (Baltimore, Md.: Johns Hopkins University Press, 1978); and Nicholas C. Kraus, ed., *History and Heritage of Coastal Engineering: A Collection of Papers on the History of Coastal Engineering in Countries Hosting the International Coastal Engineering Conference, 1950–1996* (New York: American Society of Civil Engineers, 1996).

3. A recent call for environmental historians to historicize the oceans can be found in W. Jeffrey Bolster, "Opportunities in Marine Environmental History," *Environmental History* 11 (2006): 567–97. Harbors do receive some attention in Todd Shallat, *Structures in the Stream: Water, Science, and the Rise of the U.S. Army Corps of Engineers* (Austin: University of Texas Press, 1994), but most environmental historians interested in navigable waterways have focused on the Mississippi River. See, for example, George S. Pabis, "Subduing Nature through Engineering: Caleb G. Forshey and the Levees-only Policy, 1851–1881," in Craig E. Colten, ed., *Transforming New Orleans and Its Environment: Centuries of Change* (Pittsburgh: University of Pittsburgh Press, 2000), 64–83; George S. Pabis, "Delaying the Deluge: The Engineering Debate over Flood Control on the Lower Mississippi River, 1846–1861," *Journal of Southern History* 64 (1998): 421–54; and Martin A. Reuss, "Andrew A. Humphreys and the Development of Hydraulic Engineering: Politics and Technology in the Army Corps of Engineers, 1850–1950," *Technology and Culture* 26 (1985): 1–33. Also see Martin A. Reuss, "The Art of Scientific Precision: River Research in the United States Army Corps of Engineers to 1945," *Technology and Culture* 40 (1999): 293–323. An important work of environmental history that pays close attention to the relationship between a city and its waterfront is Ari Kelman, *A River and Its City: The Nature of Landscape in New Orleans* (Berkeley: University of California Press, 2003), although it does not attend to issues of harbor engineering.

4. For general histories of Boston Harbor, see Work Projects Administration, *Boston Looks Seaward: The Story of the Port, 1630–1940* (Boston: Bruce Humphries, 1941); Samuel Eliot Morison, *The Maritime History of Massachusetts, 1783–1860* (1921; reprint, Boston: Houghton Mifflin, 1941); and W. H. Bunting, *Portrait of a Port: Boston, 1852–1914* (Cambridge, Mass.: Belknap Press, 1971). For recent histories of pollution in Boston Harbor, see Eric Jay Dolin, *Political Waters: The Long, Dirty, Contentious, Incredibly Expensive but Eventually Triumphant History of Boston Harbor—A Unique Environmental Success Story* (Amherst: University of Massachusetts Press, 2004); and Charles M. Haar, *Mastering Boston Harbor: Courts, Dolphins, and Imperiled Waters* (Cambridge: Harvard University Press, 2005).

5. For the concept of "scientific revolutions," see Thomas S. Kuhn, *The Structure of Scientific Revolutions* (Chicago: University of Chicago Press, 1962). For "ecological

revolutions," see Carolyn Merchant, "The Theoretical Structure of Ecological Revolutions," in Char Miller and Hal Rothman, eds., *Out of the Woods: Essays in Environmental History* (Pittsburgh: University of Pittsburgh Press, 1997); and Carolyn Merchant, *Ecological Revolutions: Nature, Gender, and Science in New England* (Chapel Hill: University of North Carolina Press, 1989).

6. See, for example, Nancy Langston, *Forest Dreams, Forest Nightmares: The Paradox of Old Growth in the Inland West* (Seattle: University of Washington Press, 1995); and Joseph E. Taylor III, *Making Salmon: An Environmental History of the Northwest Fisheries Crisis* (Seattle: University of Washington Press, 1999).

7. Descriptions of journeys through Boston Harbor in the nineteenth century appear in Nathaniel B. Shurtleff, *A Topographical and Historical Description of Boston,* second edition, revised (Boston: Holmes and Company, 1872), 582–87; and M. F. Sweetser, *King's Handbook of Boston Harbor,* third edition (Boston: Moses King Corporation, 1889), 23–28. The most detailed guide to the harbor islands remains Edward Rowe Snow, *The Islands of Boston Harbor: Their History and Romance, 1626–1935* (Andover, Mass.: Andover Press, 1935).

8. For bragging about the size of the harbor see, for example, Elias H. Derby, "Commercial Cities and Towns of the United States: City of Boston," *Hunt's Merchant's Magazine* (November 1850): 483.

9. Commonwealth of Massachusetts, *Fifth Annual Report of the Board of Harbor Commissioners, January, 1871,* House No. 53 (1871), 78–81.

10. For the removal of stone and gravel for ballast, see City of Boston, *Report on Memorial of the Boston Marine Society,* reprinted in Commonwealth of Massachusetts, Senate No. 4 (1847), 4–5; Benjamin A. Gould, *Boston Harbor: A Series of Communications to the Boston Daily Advertiser* (Boston: J. E. Farwell & Company, 1863), 36–37; Commonwealth of Massachusetts, *Fifth Annual Report of the Harbor Commissioners,* House No. 53 (1871), 82; and Commonwealth of Massachusetts, Committee on the Judiciary, Senate No. 39 (1862). The estimate of the amount of ballast removed is from William A. Baker, *A History of the Boston Marine Society, 1742–1981,* second edition (Boston: Boston Marine Society, 1982), 172.

11. Gould, *Boston Harbor,* 32–37; and Commonwealth of Massachusetts, *Report of the Commissioners in Relation to the Flats in Boston Harbor, January 1850,* Senate No. 3 (1850), 66.

12. Commonwealth of Massachusetts, *Report of the Commissioners,* Senate No. 3 (1850), 22, 58, 64.

13. William Whiting, *The Destruction of Boston Harbor: Argument of William Whiting, Esq., Before the Committee of the Legislature, April 17, 1851, Against an Application for Leave to Fill Up Flats in Mystic River* (Boston: J. M. Hewes & Co., 1852). The same pamphlet appeared the prior year under the title *Application of John C. Tucker and Others for a Charter for the Mystic River Rail Road* (Boston: J. M. Hewes & Co., 1851).

14. For an overview of Boston's shipping and railroad networks at this time, see Bunting, *Portrait of a Port,* 6–17.

15. Commonwealth of Massachusetts, *Report of the Commissioners for the Survey of Boston Harbor,* Senate No. 47 (1837), 21. One of the more detailed discussions of the theory of tidal scour is Whiting, *Destruction of Boston Harbor.* For more recent analyses of water circulation in Boston Harbor, see G. B. Gardner, R. F. Chen, and S. Rudnick, "Estuarine Circulation in Boston Inner Harbor," *Eos Trans.* AGU 79 (17), spring meet. suppl. (1998), S187; and Richard P. Signell, "Tide- and Wind-Driven Flushing of

Boston Harbor, Massachusetts," *Estuarine and Coastal Modeling: Proceedings of the Second International Conference Sponsored by the Waterway, Port, Coastal, and Ocean Division of the American Society of Civil Engineers* (New York: American Society of Civil Engineers, 1992), 594–606.

16. Thomas Stevenson, *The Design and Construction of Harbours: A Treatise on Maritime Engineering,* second edition (Edinburgh: Adam and Charles Black, 1874), 223. For the history of hydraulics, see Enzo Levi, *The Science of Water: The Foundation of Modern Hydraulics,* translated by Daniel E. Medina (New York: American Society of Civil Engineers, 1995); Hunter Rouse and Simon Ince, *History of Hydraulics,* revised edition (New York: Dover Publications, 1963); and Hunter Rouse, *Hydraulics in the United States, 1776–1976* (Iowa City, Iowa: Institute of Hydraulic Research, 1976). For the challenges faced by engineers trying to understand flowing water, see Reuss, "Art of Scientific Precision."

17. Sleeper, *Address on the Encroachments in the Harbor,* 11.

18. Commonwealth of Massachusetts, *Report of the Commissioners,* Senate No. 3 (1850), 68.

19. The definitive study of land making in Boston is Nancy S. Seasholes, *Gaining Ground: A History of Landmaking in Boston* (Cambridge: MIT Press, 2003).

20. For the kinds of legal problems that the ordinance had created by the nineteenth century, see Nathaniel I. Bowditch, *Wharf Property; or, The Law of Flats; Being Remarks Before the Judiciary Committee of the Senate of Massachusetts, April 14, 1852* (Boston: J. Wilson & Son, 1852).

21. Seasholes, *Gaining Ground,* 420. For land making in New York City, see Anne-Marie Cantwell and Diana diZerega Wall, *Unearthing Gotham: The Archaeology of New York City* (New Haven, Conn.: Yale University Press, 2001), chapter 13. For Seattle, see Matthew Klingle, "Changing Spaces: Nature, Property, and Power in Seattle, 1880–1945," *Journal of Urban History* 32 (2006): 197–230.

22. Commonwealth of Massachusetts, *Report of the Commissioners,* Senate No. 3 (1850), 27–28; and Commonwealth of Massachusetts, Joint Committee on Mercantile Affairs and Insurance, Senate No. 119 (1850), 9–10.

23. *Letters to His Excellency, George S. Boutwell, on His Veto of the Bill to Create the Eastern Avenue Corporation, in the Session of 1851* (Boston: Office of the Daily Courier, 1851), 17.

24. Aubrey Parkman, *Army Engineers in New England: The Military and Civil Work of the Corps of Engineers in New England* (Waltham, Mass.: U.S. Army Corps of Engineers, New England Division, 1978), 44–45.

25. Seasholes discusses the South Bay project in *Gaining Ground,* 261–74. For criticism of the city, see Commonwealth of Massachusetts, *Report of the Commissioners,* Senate No. 3 (1850), 45. The Joint Committee on Mercantile Affairs and Insurance responded to the criticism in Commonwealth of Massachusetts, Senate No. 119 (1850), 4.

26. Commonwealth of Massachusetts, Committee on Mercantile Affairs and Insurance, Senate No. 117 (1848), 4.

27. Commonwealth of Massachusetts, *Report of the Commissioners,* Senate No. 47 (1837). For a sample of subsequent efforts to draw new harbor lines and amend existing ones, see Commonwealth of Massachusetts, *An Act to Preserve the Harbor of Boston and to Prevent Encroachments Therein,* House No. 63 (1837); Commonwealth of Massachusetts, *Report of the Commissioners,* Senate No. 8 (1840); Commonwealth of Massachusetts, Senate No. 25 (1847); and Commonwealth of Massachusetts, Senate No. 53 (1849). The courts upheld the legality of harbor lines as part of the state's police power in *Com-*

monwealth v. Alger, 7 Cush. 53 (1851). For a brief history of the "public trust doctrine" in Massachusetts, see Jack H. Archer et al., *The Public Trust Doctrine and the Management of America's Coasts* (Amherst: University of Massachusetts Press, 1994), chapter 5.

28. Commonwealth of Massachusetts, Committee on Mercantile Affairs and Insurance, Senate No. 134 (1848) and Senate No. 119 (1849), 2.

29. Seasholes, *Gaining Ground,* 292–95 and 397–401.

30. The most detailed history of the Back Bay is William A. Newman and Wilfred E. Holton, *Boston's Back Bay: The Story of America's Greatest Nineteenth-century Landfill Project* (Boston: Northeastern University Press, 2006). Also see Walter Muir Whitehill and Lawrence W. Kennedy, *Boston: A Topographical History,* third edition (Cambridge: Harvard University Press, 2000), chapter 7; Seasholes, *Gaining Ground,* chapter 7; and Bainbridge Bunting, *Houses of Boston's Back Bay: An Architectural History, 1840–1917* (Cambridge: Harvard University Press, 1967).

31. *Letters to His Excellency, George S. Boutwell.*

32. City of Boston, *The Mayor's Communication in Relation to a Scientific Survey of Boston Harbor,* City Doc. No. 64 (1859). General Richard Delafield replaced Totten after his death in 1864.

33. Gould, *Boston Harbor,* 12; and Parkman, *Army Engineers in New England,* 87.

34. *Tenth Report of the United States Commissioners on Boston Harbor,* City Doc. No. 50 (1866), 84.

35. Merchant, "Theoretical Structure of Ecological Revolutions," 20.

36. Quoted in Commonwealth of Massachusetts, *Fifth Annual Report of the Harbor Commissioners,* House No. 53 (1871), 91.

37. The recommendations of the U.S. Commissioners are scattered throughout their ten reports. *See Preliminary Report of the Commissioners on Boston Harbor,* City Doc. No. 37 (1860); *Second Report of United States Commissioners on the Condition of Boston Harbor,* City Doc. No. 97 (1860); *Special Report of United States Commissioners on Boston Harbor, on the Relation of Mystic Pond and River to Boston Harbor,* City Doc. No. 12 (1861); *Fourth Report of the United States Commissioners on Boston Harbor,* City Doc. No. 62, (1861); *Fifth Report of United States Commissioners on Boston Harbor,* City Doc. No. 35 (1863); *Sixth Report of United States Commissioners on Boston Harbor,* City Doc. No. 53 (1863); *Seventh Report of the United States Commissioners on Boston Harbor,* City Doc. No. 33 (1864); *Eighth Report of the United States Commissioners on Boston Harbor,* City Doc. No. 34 (1864); *Ninth Report of the U.S. Commissioners on Boston Harbor on the Subject of Compensation,* City Doc. No. 28 (1865); and *Tenth Report of the United States Commissioners on Boston Harbor,* City Doc. No. 50 (1866).

38. Ibid. The kinds of harbor works suggested by the U.S. Commissioners were more common in Great Britain and its colonies. For a taste of the political, economic, and engineering problems posed by such works, see Adrian Jarvis, "Beyond the River Wall: The Attack on the Mersey Bar, 1890–1923," and L. Twyman, "The First Harbour Works at Port Natal: The Role of John Milne from 1849–1857," reprinted as chapters 14 and 15 in Jarvis, *Port and Harbour Engineering.*

39. *Tenth Report of the United States Commissioners,* City Doc. No. 50 (1866), 43–44. The U.S. Commissioners devoted their entire ninth report to the issue of compensation; see *Ninth Report of the United States Commissioners,* City Doc. No. 28 (1865). The report also appears as an attachment to Commonwealth of Massachusetts, *Third Report of the Commissioners on the Harbors and Flats in the Commonwealth, February, 1865,* Senate No. 58 (1865).

40. Gould, *Boston Harbor;* and Charles Francis Adams Jr., "Boston," *The North American Review* 106, no. 218 (January 1868): 16.

41. Commonwealth of Massachusetts, Acts of 1866, chapter 149.

42. *Report of the United States Advisory Council to the Board of Harbor Commissioners upon Recent Changes in the Bed of Boston Harbor, to Which Is Added an Opinion upon Compensation,* in Commonwealth of Massachusetts, *Second Annual Report of the Board of Harbor Commissioners, January, 1868,* House No. 10 (1868), 32–38.

43. Ibid.

44. Commonwealth of Massachusetts, *Report of the Committee Appointed Under Chapter 93 of the Resolves of 1867, in Relation to the Commonwealth Flats near South Boston,* House No. 76 (1868), 56.

45. Ibid., 64.

46. Ibid., 75.

47. Ibid., 83.

48. Commonwealth of Massachusetts, *Report Relative to the Flats and Water Areas of the Commonwealth, by the Committee Appointed Under the Order of the House of Representatives, June 17, 1869,* House No. 240 (1870).

49. Senate No. 25 (1847); and *Fourth Report of United States Commissioners,* City Doc. No. 62 (1861), 10.

50. Commonwealth of Massachusetts, *Seventh Annual Report of the Board of Harbor Commissioners, January, 1873,* House No. 65 (1873), 31–38.

51. Commonwealth of Massachusetts, *Fifth Annual Report of the Harbor Commissioners,* House No. 53 (1871), 16–17.

52. Commonwealth of Massachusetts, *Annual Report of the Board of Harbor Commissioners for the Year 1878,* House No. 33 (1879), 16.

53. Commonwealth of Massachusetts, *Fourth Annual Report of the Board of Harbor Commissioners, January, 1870,* House No. 55 (1870), 106–9; and Commonwealth of Massachusetts, *Fifth Annual Report of the Harbor Commissioners,* House No. 53 (1871), 12–15.

54. Commonwealth of Massachusetts, *Seventh Annual Report of the Harbor Commissioners,* House No. 65 (1873), 42–48.

55. For a full account of the Charles River Dam controversy, see Karl Haglund, *Inventing the Charles River* (Cambridge: MIT Press, 2003), chapter 5.

56. Commonwealth of Massachusetts, *Annual Report of the Harbor and Land Commissioners for the Year 1894,* Public Doc. 11 (1895), 16; and Commonwealth of Massachusetts, *Report of the Joint Board Consisting of the Metropolitan Park Commission and the State Board of Health,* House No. 775 (1894), 18–22.

57. Commonwealth of Massachusetts, *Annual Report of the Harbor and Land Commissioners for the Year 1894,* Public Doc. 11 (1895), 21, 18.

58. Freeman as quoted in Commonwealth of Massachusetts, *Report of the Committee on Charles River Dam* (Boston: Wright & Potter, 1903), 20. The John Ripley Freeman Papers are held in the Massachusetts Institute of Technology's Institute Archives and Special Collections.

59. Ibid., 22–24, 100–102. Freeman's findings came at the tail end of a period when civilian engineers were questioning the competence of military engineers. See Reuss, "Andrew A. Humphreys and the Development of Hydraulic Engineering," 23–26.

60. Freeman as quoted in Commonwealth of Massachusetts, *Report of the Committee on Charles River Dam,* 22. I am using the freighted term "paradigm shift" purposefully, because the replacement of one scientific model by another in the case of Boston

Harbor closely parallels the path described in Kuhn, *Structure of Scientific Revolutions.* The existing scientific model, in this case the theory of tidal scour, was growing progressively incapable of explaining the observable facts. Yet, as Kuhn predicts, it was judged invalid only after someone suggested a new model that better explained the facts.

61. Bunting, *Portrait of a Port,* 48.

4. Remaking Boston Harbor: Cleaning Up After Ourselves

The author thanks Caroline Howland of the University of Massachusetts–Boston for her assistance.

1. Eric Jay Dolin, *Political Waters: The Long, Dirty, Contentious, Incredibly Expensive but Eventually Triumphant History of Boston Harbor: A Unique Environmental Success Story* (Amherst: University of Massachusetts Press, 2004), 143.

2. Ellen Swallow Richards, "Ellen Swallow Richards Defines Human Ecology," in Carolyn Merchant, ed., *Major Problems in American Environmental History* (Boston: Houghton Mifflin Company, 2005), 428–30.

3. Dolin, *Political Waters,* 6–7.

4. Mary Gaymon, "A Glimpse into London's Early Sewers," reprinted from *Cleaner Magazine,* n.d. Also available online at http://swopnet.com/engr/londonsewers/londontext1.html (accessed on March 22, 2007).

5. Martin V. Melosi, *The Sanitary City: Urban Infrastructure in America from Colonial Times to the Present* (Baltimore, Md.: Johns Hopkins University Press, 2000), 19.

6. Dolin, *Political Waters,* 6, 11.

7. *Report of the Record Commissioners of the City of Boston* (Boston: City of Boston, 1869–1909), vol. 13.

8. Nancy S. Seasholes, *Gaining Ground: A History of Landmaking in Boston* (Cambridge: MIT Press, 2003), 38, 50.

9. *Report of the Record Commissioners,* 125–26, 161, 136.

10. John B. Blake, *Public Health in the Town of Boston, 1630–1822* (Cambridge: Harvard University Press, 1959), 104.

11. Ibid., 168

12. James Kirby Martin et al., *America and Its Peoples: A Mosaic in the Making* (New York: Addison Wesley Longman, 2001), 522.

13. Clay McShane and Joel A. Tarr, *The Horse in the City: Living Machines in the Nineteenth Century* (Baltimore, Md.: Johns Hopkins University Press, 2007), 16, 26.

14. Josiah Quincy, *A Municipal History of the Town and City of Boston During Two Centuries* (Boston: Charles C. Little and James Brown, 1852), 65

15. Eliot C. Clarke, *Main Drainage Works of the City of Boston,* third edition (Boston: City of Boston, 1888), 16

16. Charles Rosenberg, *The Cholera Years: The United States in 1832, 1849, and 1866* (Chicago: University of Chicago Press, 1962), 16, 91.

17. Dolin, *Political Waters,* 17.

18. Seasholes, *Gaining Ground,* 172, 8.

19. Massachusetts Sanitary Commission, *Report of a General Plan for the Promotion of Public and Personal Health: Devised, Prepared, and Recommended by the Commissioners Appointed Under a Resolve of the Legislature of Massachusetts. Relating to a Sanitary Survey of the State* (Boston: Dutton & Wentworth, 1861; reprint, New York: Arno Press, 1972.)

20. Edwin Chadwick, *Report on the Sanitary Condition of the Labouring Population of Gt. Britain* (Edinburgh: University Press of Edinburgh, 1842).

21. Dolin, *Political Waters,* 29.

22. Michael Rawson, "What Lies Beneath: Science, Nature, and the Making of Boston Harbor," in this book.

23. Ibid.

24. "Robert Koch" in *Encyclopedia Britannica* (2008). Available online at http://www.britannica.com/EBchecked/topic/320834/Robert-Koch (accessed June 5, 2008).

25. Massachusetts Medical Commission, *The Sanitary Condition of Boston: The Report of a Medical Commission to the Boston Board of Health,* quoted in Sarah S. Elkind, *Bay Cities and Water Politics: The Battle for Resources in Boston and Oakland* (Lawrence: University of Kansas Press, 1998), 53.

26. Board of Health of the City of Boston, *First Annual Report,* City Doc. 84 (1873), 22–23.

27. Boston Board of Aldermen, Committee on Sewers, "Report on the Present System of Sewage in the City of Boston," City Doc. 94 (1873), 7, 8.

28. Board of Health of the City of Boston, *Second Annual Report,* City Doc. 63 (Boston: City of Boston, 1874), 47.

29. Ibid., 51.

30. Board of Health of the City of Boston, *Report on the Necessity of Improved Sewerage in the Roxbury Canal,* City Doc. 112 (1874), 6. The report does not indicate why the "salt" aspect of the water was focused on and if it was merely a physical observation or folk knowledge. Today we know that estuarine processes are beneficial both chemically and biologically, but it is unlikely that the science was known in 1874.

31. Board of Health of the City of Boston, *Third Annual Report,* City Doc. 85 (Boston: City of Boston, 1875), 10.

32. Elkind, *Bay Cities and Water Politics,* 54.

33. *Boston Daily Globe,* September 10, 1875, p. 4.

34. Elkind, *Bay Cities and Water Politics,* 56.

35. Clarke, *Main Drainage Works,* 26.

36. Ibid.

37. Massachusetts General Court, *Resolves of the General Court* (1900), chapter 65.

38. Massachusetts State Board of Health, *Report upon the Discharge of Sewage into Boston Harbor* (Boston: Commonwealth of Massachusetts, 1900), 5.

39. Massachusetts General Court, *Report of the Special Legislative Commission on the Unsanitary Conditions in Savin Hill Bay* (Boston: Commonwealth of Massachusetts, 1928).

40. Massachusetts General Court, *House Report Number 1600* (Boston: Commonwealth of Massachusetts, 1937), 23.

41. *Report of the Special Commission Investigating Systems of Sewerage and Sewage Disposal in the North and South Metropolitan Sewerage Districts and the City of Boston* (Boston: Commonwealth of Massachusetts, June 15, 1939), 16.

42. Ibid., 72–80.

43. U.S. Environmental Protection Agency, *Health Effects Criteria for Fresh Recreational Waters* (Research Triangle Park, N.C.: Environmental Protection Agency, 1984), 7–9.

44. Elkind, *Bay Cities and Water Politics,* 167.

45. Massachusetts Water Resources Authority (MWRA), *Combined Sewer Overflows,* April 14, 2008. Available online at http://www.mwra.state.ma.us/03sewer/html/sewcso.htm (accessed June 4, 2008). When the Main Drain was designed, it was recognized that storm events would lead to overfilling of the intercepting sewers so provisions were made at the original outfalls to allow overflow directly to the rivers and the harbors.

46. David Doneski, "Cleaning up Boston Harbor: Fact or Fiction," *Boston College Environmental Affairs Law Review* 559 (1985): 6–7.

47. Dolin, *Political Waters,* 61, 70–71.

48. U.S. Environmental Protection Agency Region 1, *State of the Harbor: Boston Harbor Cleanup History.* Available online at http://www.epa.gov/region01/ra/bharbor/state.html (accessed June 5, 2008).

49. Doneski, "Cleaning up Boston Harbor," 3–4.

50. Dolin, *Political Waters,* 107–8.

51. Andrew T. Savage, "Boston Harbor: The Anatomy of a Court-Run Cleanup," *Boston College Environmental Law Review* 22 (1995): 365–412.

52. David G. Aubrey and Michael S. Connor, "Boston Harbor: Fallout over the Outfall," *Oceanus* 36 (1993): 61–71.

53. MWRA, *A History of the Sewer System,* available online at http://www.mwra.com/03sewer/html/sewhist.htm (accessed June 4, 2008).

54. The sludge, or settled materials from the primary- and secondary-settling tanks, is digested by bacteria in the most visible entity of the Deer Island Treatment Plant, the eggs, and pumped to the treatment plant in Quincy, where it is dewatered, dried, and formed into pellets.

55. C. Zago, A. E. Giblin, and A. Bergamesco, "Changes in the Metal Content of Surficial Sediments of Boston Harbor Since the Cessation of Sludge Discharge," *Marine Environmental Research* 51 (2001): 389–415.

56. E. D. Gallagher and K. E. Kray, "Organism Sediment Contaminant Interactions in Boston Harbor," in K. D. Stolzenbach and E. E. Adams, eds., *Contaminated Sediments in Boston Harbor* (Cambridge: Massachusetts Institute of Technology Sea Grant Publications, 1998), 89–132.

57. N. J. Maciolek, R. J. Diaz, D. T. Dahlen, and I. P. Williams, *2005 Boston Benthic Monitoring Report* (Boston: Massachusetts Water Resources Authority, 2006).

58. Save the Harbor, Save the Bay Advisory Committee (SHSB), "Why the Beaches Close," 2004. Available online at http://www.savetheharbor.org/downloads/beachreport.pdf (accessed June 4, 2008).

59. Ibid., 3

60. Steven Rudnick, "In-Situ Fluorescence Detection of Polycyclic Aromatic Hydrocarbons in the Marine Environment," Ph.D. dissertation, University of Massachusetts–Boston, 1998.

61. M. F. Sweetser, *The Kings Handbook of Boston Harbor* (Cambridge, Mass.: Moses King 1882), 21.

5. *In Search of the Shawmut Peninsula: Using Modern Cartographic Analysis to Discover the "Original" Boston Shoreline*

The views and conclusions expressed in this chapter are the author's and in no way reflect the official position or policy of the Massachusetts Office of Coastal Zone Management (CZM) or other state agencies. *Epigraphs:* Herman Melville, *Moby Dick, or The Whale* (London: Penguin Books, 1986), 150; C. Beaglehole, ed., *Journals of Captain James Cook on His Voyage of Discovery with Addenda and Corrigenda to Volume I and Volume II* (Cambridge: Published for the Hakluyt Society at the University Press, 1955–74), volume 2, 509fn5, quoted in Beaglehole, *The Life of Captain James Cook* (Stanford, Calif.: Stanford University Press, 1974), 70.

1. Conrad E. Heidenreich, *Explorations and Mapping of Samuel de Champlain, 1603–1632* (Toronto: B. V. Gutsell, 1976), ix.

2. Depending on the purpose of the survey, any number of descriptions, from the high water line to the low water line to a vegetation line, may have been used for the working definition of "shoreline." Researchers must therefore pay particular attention to shoreline definition when comparing positions mapped by different surveyors.

3. In 1807, President Thomas Jefferson established the U.S. Coast Survey (later the U.S. Coast and Geodetic Survey and today the National Oceanic and Atmospheric Administration's National Ocean Survey). He named Ferdinand Hassler the first superintendent. For various reasons, however, including a lack of dedicated funding and the War of 1812, actual survey work did not begin in earnest until 1837. See Florian Cajori, *The Chequered Career of Ferdinand Rudolph Hassler: First Superintendent of the U.S. Coast Survey* (1929; reprint, New York: Arno Press, 1980).

4. Although characterization of this shoreline as "original" is incorrect both scientifically and historically, the term "original shoreline" is frequently encountered as shorthand for that shoreline present at the time of first colonial settlement and before anthropomorphic alteration.

5. Elizabeth H. Boak and Ian Turner, "Shoreline Definition and Detection: A Review," *Journal of Coastal Research* 21 (2005): 688–703; and Robert A. Morton and F. Michael Speed, "Evaluation of Shorelines and Legal Boundaries Controlled by Water Levels on Sandy Beaches," *Journal of Coastal Research* 14 (1998): 1373–84.

6. Stephen T. Mague, "Private Property, Public Rights, and Shifting Sands: The Public Trust Doctrine as a Source of Authority for Coastal Management Decisions" (part 1 of 2), *Surveying and Land Information Systems: American Congress on Surveying and Mapping* 59, no. 2 (1999): 53–68.

7. Donald K. Stauble, "The Use of Shoreline Change Mapping in Coastal Engineering Project Assessment," *Journal of Coastal Research,* special issue 38 (2007): 200: 178–206; and Nicholas C. Kraus and Julie Dean Rosati, "Interpretation of Shoreline-Position Data for Coastal Engineering Analysis," in *Coastal Engineering Technical Note (CETN II-39)* (Vicksburg, Miss.: U.S. Army Engineer Waterways Experiment Station, Coastal and Hydraulics Laboratory, 1997).

8. In its simplest form, "georeferencing" refers to the process by which a map and the features depicted on it are aligned with, or referenced to, their real-world positions on a predefined geographic coordinate system such as a national grid or latitude/longitude.

9. Along these lines, Professor Anne Knowles of Middlebury College has used georeferenced historical maps to re-create accurate landscape maps of the battlefield of Gettysburg to help explore and describe the course of this famous 1863 battle of the American Civil War. Personal communication with the author, June 2006.

10. Nancy S. Seasholes, *Gaining Ground: A History of Landmaking in Boston* (Cambridge: MIT Press, 2003). David W. Trubey, deputy director, Massachusetts Bureau of Underwater Archeology (BUAR), personal communication with the author, November 2007.

11. In Massachusetts the Public Trust Doctrine is codified in chapter 91 of Massachusetts General Law and implemented through the Waterways regulations (310 Code of Massachusetts Regulations [CMR] 9.00) for the purpose of preserving and protecting the public's rights and interests in filled (former) and flowed tidelands. The Waterways regulations are administered by the Waterways Program of the DEP.

12. As set forth in 310 CMR 9.02, "Definitions," the historic high water mark (HHWM) is defined to be "the high water mark that existed prior to human alteration of the shoreline by filling, dredging, excavating, impounding, or other means. In areas where there is evidence of such alteration by fill . . . the historic high water mark is [presumed to be] the farthest landward former shoreline which can be ascertained with reference to topographic or hydrographic surveys, previous license plans, and other historic maps or chart."

13. "Rubbersheeting" refers to a map being stretched and manipulated to fit a contemporary spatial reference system with little or no regard to the mathematical spatial relationships between known positional values of geographic points.

14. Without delving too deeply into the mathematics of plan registration or georeferencing, at its basic level the lower the registration solution or residual, the better the registration or the fit of the plan to points of known geospatial position. Although case-specific and dependent on factors including the number of points used in the registration, the geographic distribution of those points, and the nature of the points, in most cases residuals of 1.0 are assumed to signify an extremely reliable registration solution.

15. Although such assessments do quantify original survey or base map compilation accuracies, they do not address the question of whether the high water line depicted on a plan is in fact correct, and, absent documented procedural standards, there is no practical statistical method that can be used to assess the reliability of depicted shoreline. Consequently, it must be assumed that given the technology, techniques, and standards in use at the time of the survey and any extant instructions, the mapmaker created the most accurate manuscript possible.

16. In surveying, a datum is a predetermined reference system of the earth's surface that is used for making positional measurements. A horizontal datum, such as the North American Datum of 1983 (NAD83), is used for describing the location of a point on the earth's surface in latitude and longitude or another coordinate system. Similarly, vertical datums, such as the North American Datum of 1988 (NAVD88), are used to measure elevations or depths.

17. Seasholes, *Gaining Ground,* 434.

18. Referred to as a national tidal datum epoch (NTDE).

19. *Borax Consolidated, Ltd. v. Los Angeles,* 296 U.S. 10, 26 (1935).

20. A. L. Shalowitz, *Shore and Sea Boundaries* (Washington, D.C.: U.S. Coast and Geodetic Survey, 1964), chapter 4.

21. The U.S. Coast Survey pioneered the use of the plane table and alidade for its land, or topographic, related survey work. Consequently, all fieldwork was compiled on individual T- (or topographic) sheets. H- (or hydrographic) sheets were the bathymetric companion to the T-sheet and depicted soundings data. Compiled at scales from 1:5,000 to 1:20,000, together these field sheets formed the basis for subsequent Coast Survey nautical chart compilation and production.

22. Peter J. Guthorn, *United States Coastal Charts: 1781–1861* (Exton, Pa.: Schiffler Publishing Ltd. 1984), 17–24.

23. DesBarres to Lord Colville, May 1765, Adm. Sec. in Letters (Adm.1-482), in G.N.D. Evans, "Hydrography: A Note on Eighteenth Century Methods," *Mariner's Mirror* 52 (1966): 247–50. Originals at the Public Record Office, London; microfilm copies at the Library of Congress, Washington, D.C.

24. It is interesting to note that DesBarres apparently performed his soundings work relative to an extreme low water datum, a plane of reference representing a worst-

case scenario for mariners compared with contemporary soundings datums of mean low water (MLW) and mean lower low water (MLLW). See Henry Mitchell, "Notes Concerning Alleged Changes in the Relative Elevations of Land and Sea," *Report of the Superintendent of the U.S. Coast Survey Showing the Progress of the Work for the Fiscal Year Ending with June, 1877* (Washington, D.C.: Government Printing Office, 1880), appendix 8.

25. Tidal datums are locally derived base elevations from which to reckon heights or depths and are defined in terms of a certain phase of the tide (for example, mean high water, mean higher high water, mean low water, and so on). See *Tide and Current Glossary* (Washington, D.C.: National Oceanic and Atmospheric Administration, 2000).

26. For example, in areas of extensive tidal flats, with slopes typically in the range of 1 vertical foot in 300 to 400 feet, a difference in height of 0.7 feet would translate into a horizontal displacement of 200 to 300 feet.

27. Chesbrough as quoted in "Report of the City Engineer," in *City Document No. 60: Report of the Joint Standing Committee on Boston Harbor for the Year 1852* (Boston: J. H. Eastburn, City Printer, 1853), 46–47.

28. P. Ruggerio, G. M. Kaminsky, and G. Gelfenbaum, "Linking Proxy-based and Datum-based Shorelines on a High-energy Coastline: Implications for Shoreline Change Analysis," *Journal of Coastal Research,* special issue 38 (2003): 57–82.

29. For early Coast Survey work, it was "not deemed necessary to determine the actual high-water line but rather the outer or seaward edge of the marsh. . . . Therefore, from the topographic survey alone, . . . no conclusion could be drawn as to the exact location of the high-water line, nor as to the condition of the marsh area with reference to the tidal plane of water." Later mapping efforts (after 1870) of the Coast Survey reflected increased attention in terms of procedures and symbology, with the introduction of a representation for marshy areas that were "mostly flooded at high water." Later, with the introduction of aerial mapping technology, the Coast and Geodetic Survey reverted to earlier symbology, and "where the actual shoreline was obscured by marsh grass, . . . the outer edge of vegetation was mapped as an apparent shoreline." See Shalowitz, *Shore and Sea Boundaries,* vol. 2, chapter 4, for an extensive discussion of the shoreline mapping methods and procedures applied by the U.S. Coast Survey and in particular to areas of salt marsh. See also Alfred A. Porro and Lorraine S. Telesky, "Marshland Title Dilemma: A Tidal Phenomenon," *Seton Hall Law Review* 3 (1971): 323, 335–46.

30. E. Robert Thieler, James F. O'Connell, and Courtney A. Schupp, *The Massachusetts Shoreline Change Project: 1800s to 1994.* Technical Report, USGS Administrative Report, generated in collaboration with the Woods Hole Oceanographic Institute (WHOI) Sea Grant Program and Cape Cod Cooperative Extension, to the Massachusetts Office of Coastal Zone Management, September 4, 2001.

31. Seasholes, *Gaining Ground;* Walter Muir Whitehill and Lawrence W. Kennedy, *Boston: A Topographical History,* third edition (Cambridge: Harvard University Press, 2000); and Esther Forbes, *Paul Revere and the World He Lived In* (New York: Houghton Mifflin, 1942).

32. Forbes, *Paul Revere and the World He Lived In,* 45.

33. For a discussion of the natural conditions of the Shawmut Peninsula, see Whitehill and Kennedy, *Boston,* chapter 1.

34. Robert A. Morton and Tara L. Miller, *National Assessment of Shoreline Change: Part 2, Historical Shoreline Changes and Associated Coastal Land Loss Along the U.S. Southeast Atlantic Coast,* Open-File Report 2005-1401 (Washington, D.C: U.S. Geological

Survey, 2005). Also available online at http:/pubs.usgs.gov/of/2005/1401/. See also F. J. Anders and M. R. Byrnes, "Accuracy of Shoreline Change Rates as Determined from Maps and Aerial Photographs," *Shore and Beach* 59 (1991): 17–26; and M. Crowell, S. P. Leatherman, and M. K. Buckley, "Historical Shoreline Change: Error Analysis and Mapping Accuracy," *Journal of Coastal Research* 17 (1991): 839–52; R. C. Daniels and R. H. Huxford, "An Error Assessment of Vector Data Derived from Scanned National Ocean Service Topographic Sheets," *Journal of Coastal Research* 17 (2001): 611–19; and E. R. Thieler and W. W. Danforth, "Historical Shoreline Mapping (1). Improving Techniques and Reducing Positioning Errors," *Journal of Coastal Research* 10 (1994): 549–63.

35. Crowell, Leatherman, and Buckley, "Historical Shoreline Change," 25.

36. Legislation in the form of special wharfing statutes authorized the filling of tidelands and the creation of harborline enactments that created a line beyond which no filling or wharfing could occur are often important instruments for determining the nature and extent of title to waterfront properties.

37. George Cole, *Water Boundaries* (New York: John Wiley & Sons, 1997), 90–91.

38. Ibid., 90.

39. Stephen T. Mague, "Private Property, Public Rights, and Shifting Sands: The Public Trust Doctrine as a Source of Authority for Coastal Management Decisions," part 1 of 2), *Surveying and Land Information Systems: American Congress on Surveying and Mapping* 59, no. 2 (1999): 53–68.

40. Mague, "Private Property, Public Rights, and Shifting Sands," part 2, 95–106.

41. *Austin v. Carter,* 1 Mass. 231 (1804). See also *Adams v. Frothingham,* 3 Mass. 353, 360 (1807), and *Storer v. Freeman,* 6 Mass. 435, 438 (1810); *Commonwealth v. Alger,* 7 Cush. 53, 65 (1861), and *Commonwealth v. Charlestown,* 1 Pick. 180, 182 (1822).

42. Mague, "Private Property, Public Rights, and Shifting Sands," part 2.

43. *Boston Waterfront Development Corp. v. Commonwealth,* 378 Mass. 629, 644 (1979).

44. The Waterways Regulations, 310 CMR 9.02.

45. These ports in descending order of tidelands area include Boston, Salem, New Bedford, Lynn, Fall River, Plymouth, Nantucket, Newburyport, Gloucester, Beverly, Provincetown, Tisbury, and Barnstable. Stephen T. Mague and Robert. W. Foster, "Where's the Shoreline? Sources of Historical High Water Lines in the Context of Massachusetts Coastal Regulations," paper prepared for and presented at the Sixth FIG Regional Conference on "Coastal Areas and Land Administration—Building the Capacity," San Jose, Costa Rica, November 12–15, 2007.

46. The Waterways Regulations, 310 CMR 9.02.

47. This method was developed by David Doyle, chief surveyor, NOAA, National Geodetic Survey. Personal communication with the author, November 2003.

48. Justin Winsor's map entitled "Boston Old and New," included as the frontispiece in volume 1 of his *Memorial History of Boston* (Boston: James R. Osgood & Co. 1881–1883), was also registered. Although not included in this chapter's comparative analysis, it was used as one source for the location of salt marsh surrounding the Shawmut Peninsula.

49. *Report of the Joint Standing Committee on Boston Harbor,* 47.

50. Baldwin was the son of the famous Revolutionary War hero Colonel Loammi Baldwin of Woburn, a proponent of the Middlesex Canal project and also known for the Baldwin apple. His brother, Colonel Loammi Baldwin II, is often considered the father of American civil engineering and was the designer of the first and second graving docks

for the U.S. Navy at the Gosport Navy Yard in Portsmouth, Virginia, and the Charlestown Navy Yard in Boston, Massachusetts.

51. "Report of the City Engineer." See, in particular, "Chesbrough's Report," 46.

52. Interestingly, Captain John Montressor, an engineer in the British Army, was the man who escorted Captain Nathan Hale to the gallows near the East River in New York, where he was executed as a spy.

53. "Report of the City Engineer," 47.

54. Ibid., 40.

55. Seasholes, *Gaining Ground,* 437.

56. E. R. Thieler, E. A. Himmelstross, J. L. Zichichi, and T. L. Miller, *Digital Shoreline Analysis System (DSAS), version 3.0.: An Arcgis Extension for Calculating Shoreline Change.* U.S. Geological Survey Open-File Report 2005-1304, U.S. Geological Survey, Woods Hole Science Center, available online at http://woodshole.er.usgs.gov/project-pages/dsas/version3/index.html, accessed may 26, 2009.

57. In the case of the Baldwin plan, the reconstructed shoreline was provided only for the area known as the Town Dock, on the easterly shore of the peninsula, in the vicinity of Long Wharf. Because of the degree of development depicted on Samuel Clough's 1676 reconstruction, this same area was removed from the Clough reconstruction.

58. The standard deviation is a measure of how widely values are dispersed from the average value (the mean). It is a commonly used measure of the "spread" of data.

6. Remaking Boston, Remaking Massachusetts

1. William Cronon, *Nature's Metropolis: Chicago and the Great West* (New York: W. W. Norton, 1991); Joel A. Tarr, ed., *Devastation and Renewal: An Environmental History of Pittsburgh and Its Region* (Pittsburgh: University of Pittsburgh Press, 2003); and William Deverell and Greg Hise, eds., *Land of Sunshine: An Environmental History of Metropolitan Los Angeles* (Pittsburgh: University of Pittsburgh Press, 2005).

2. Lewis Mumford, *Technics and Civilization* (New York: Harcourt, Brace and Company, 1934); Sam Bass Warner, *Streetcar Suburbs: The Process of Growth in Boston, 1870–1900* (Cambridge: Harvard University Press, 1962); Sam Bass Warner, *The Urban Wilderness: A History of the American City* (New York: Harper and Row, 1972); and John T. Cumbler, *Reasonable Use: The People, the Environment, and the State, New England, 1790–1930* (New York: Oxford University Press, 2001).

3. Martyn Bowden, "Boston: From East Anglian Nuclear Village to English Shiretown to Mercantile Replica of London," essay presented at "Remaking Boston: The City and Environmental Change over the Centuries," Massachusetts Historical Society, Boston, May 2006; and Bernard Bailyn, *The New England Merchants in the Seventeenth Century* (Cambridge: Harvard University Press, 1955).

4. Charles F. Carroll, *The Timber Economy of Puritan New England* (Providence, R.I.: Brown University Press, 1973); and Howard S. Russell, *A Long, Deep, Furrow: Three Centuries of Farming in New England* (Hanover, N.H.: University Press of New England, 1976).

5. Gloria L. Main, *Peoples of a Spacious Land: Families and Cultures in Colonial New England* (Cambridge: Harvard University Press, 2001).

6. Brian Donahue, *The Great Meadow: Farmers and the Land in Colonial Concord* (New Haven, Conn.: Yale University Press, 2004).

7. Ibid.; David Jaffee, *People of the Wachusett: Greater New England in History and Memory, 1630–1860* (Ithaca, N.Y.: Cornell University Press, 1999); and Diana Muir,

Reflections in Bullough's Pond: Economy and Ecosystem in New England (Hanover, N.H.: University Press of New England, 2000).

8. Brooke Hindle, ed., *America's Wooden Age: Aspects of Its Early Technology* (Tarrytown, N.Y.: Sleepy Hollow Restorations, 1975).

9. Robert F. Dalzell Jr., *Enterprising Elite: The Boston Associates and the World They Made* (Cambridge: Harvard University Press, 1987).

10. Michael Rawson, "What Lies Beneath: Science, Nature, and the Making of Boston Harbor," in this volume.

11. Andrew Robichaud, "Ice: A History of the Natural Trade and How It Changed America," senior honor's thesis, Brandeis University, Waltham, Mass., 2004; and Daniel J. Boorstin, *The Americans: The National Experience* (New York: Random House, 1965), 10–16.

12. Morton J. Horwitz, *The Transformation of American Law, 1780–1860* (Cambridge: Harvard University Press, 1977).

13. Nancy S. Seasholes, *Gaining Ground: A History of Landmaking in Boston* (Cambridge: MIT Press, 2003),161–69.

14. Michael Rawson, "The Nature of Water: Reform and the Antebellum Crusade for Municipal Water in Boston," *Environmental History* 9 (2004): 411–35; and Helen Fitch Emery, *The Puritan Village Evolves: A History of Wayland, Massachusetts* (Canaan, N.H.: Phoenix Publishing, 1981).

15. Theodore Steinberg, *Nature Incorporated: Industrialization and the Waters of New England* (New York: Cambridge University Press, 1991).

16. Brian Donahue, "Dammed at Both Ends and Cursed in the Middle: The Flowage of the Concord River Meadows, 1798–1862," *Environmental Review* 13 (1989): 47–67.

17. David R. Foster to author, Harvard Forest, personal conversation, March 25, 2008; Robert E. Shalhope, *Bennington and the Green Mountain Boys: The Emergence of Liberal Democracy in Vermont, 1760–1850* (Baltimore, Md.: Johns Hopkins University Press, 1996), 251–54; and Pauline Swain Merrill, John C. Gowan, and others, *A Small Gore of Land: A History of the Town of New Hampton, New Hampshire* (New Hampton, N.H.: New Hampton Town History Committee, 1976), 76.

18. J. Ritchie Garrison, *Landscape and Material Life in Franklin County, Massachusetts, 1770–1860* (Knoxville: University of Tennessee Press, 1991); Christopher Clark, *The Roots of Rural Capitalism: Western Massachusetts, 1780–1860* (Ithaca, N.Y.: Cornell University Press, 1990); Robert A. Gross, "Culture and Cultivation: Agriculture and Society in Thoreau's Concord," *Journal of American History* 69 (1982): 42–61; and Andrew H. Baker and Holly V. Izard, "New England Farmers and the Marketplace, 1780–1865: A Case Study," *Agricultural History* 65 (1991): 29–52.

19. David R. Foster, *Thoreau's Country: Journey Through a Transformed Landscape* (Cambridge: Harvard University Press, 1999); Brian Donahue, "Another Look from Sanderson's Farm: A Perspective on New England Environmental History and Conservation," *Environmental History* 12 (2007): 9–34; and Richard Judd, *Common Lands, Common People: The Origins of Conservation in Northern New England* (Cambridge: Harvard University Press, 1997).

20. Howard M. Gitelman, *Workingmen of Waltham: Mobility in American Urban Industrial Development, 1850–1890* (Baltimore, Md.: Johns Hopkins University Press, 1974); and Kristen A. Petersen, *Waltham Rediscovered: An Ethnic History of Waltham, Massachusetts* (Portsmouth, N.H.: Peter E. Randall, 1988).

21. Pamela W. Fox, *Farm Town to Suburb: The History and Architecture of Weston, Massachusetts, 1830–1980* (Portsmouth, N.H.: Peter E. Randall, 2002), 475–500.

22. Mumford, *Technics and Civilization,* 154; Warner, *Urban Wilderness*; Joel A. Tarr, *The Search for the Ultimate Sink: Urban Pollution in Historical Perspective* (Akron, Ohio: University of Akron Press, 1996); Martin V. Melosi, *The Sanitary City: Urban Infrastructure in America from Colonial Times to the Present* (Baltimore, Md.: Johns Hopkins University Press, 2000); and Sarah S. Elkind, *Bay Cities and Water Politics: The Battle for Resources in Boston and Oakland* (Lawrence: University Press of Kansas, 1998).

23. James C. O'Connell, "How Metropolitan Parks Shaped Greater Boston, 1893–1945,"in this volume; and Karl Haglund, *Inventing the Charles River* (Cambridge: MIT Press, 2003).

24. Cumbler, *Reasonable Use*; and Elkind, *Bay Cities and Water Politics.*

25. Harold Fisher Wilson, *The Hill Country of Northern New England: Its Social and Economic History, 1790–1930* (New York: Columbia University Press, 1936); John Donald Black, *The Rural Economy of New England: A Regional Study* (Cambridge: Harvard University Press, 1950); Hugh M. Raup, "The View from John Sanderson's Farm: A Perspective for the Use of Land," *Forest History* 10 (1966): 2–11; Michael M. Bell, "Did New England Go Downhill?" *Geographical Review* 79 (1989): 450–66; and Donahue, "Another Look from Sanderson's Farm."

26. Brian Donahue, "Skinning the Land: Economic Growth and the Ecology of Farming in Nineteenth-century Massachusetts," paper presented at the annual meeting of the Social Science History Association, Chicago, November 1988; and David Soll, "Reforestation in Norfolk County, Massachusetts, 1850–1910," in this volume.

27. Robert E. Pike, *Tall Trees, Tough Men* (New York: W. W. Norton, 1967); and Charles H. W. Foster, ed., *Stepping back to Look Forward: A History of the Massachusetts Forest* (Cambridge: Harvard University Press, 1998).

28. O'Connell, "How Metropolitan Parks Shaped Greater Boston"; Haglund, *Inventing the Charles*; Seasholes, *Gaining Ground;* Annalee Saxenian, "Silicon Valley and Route 128: Regional Prototypes or Historic Exceptions?" in Manuel Costells, ed., *High Technology, Space, and Society* (Beverly Hills, Calif.: Sage, 1985), 93–106; and *Urban Affairs Annual Reviews* 28 (1985): 93–101; and Robert W. Preer, *The Emergence of Technopolis: Knowledge-intensive Technologies and Regional Development* (New York: Praeger, 1992).

29. Haglund, *Reinventing the Charles*; and Massachusetts Water Resources Authority, "The Water System," 2008, available online at http://www.mwra.state.ma.us/04water/html/wat.htm.

30. Massachusetts Audubon Society, "Losing Ground: At What Cost?" 2003, available online at http://www.massaudubon.org/news/index.php?id=19&type=news.

31. Mary M. Berlik et al., *The Illusion of Preservation: A Global Environmental Argument for the Local Production of Natural Resources* (Petersham, Mass.: Harvard Forest, 2002).

32. Leo Marx, *The Machine in the Garden: Technology and the Pastoral Ideal in America* (New York: Oxford University Press, 1964); Berlik et al., *Illusion of Preservation*; and David R. Foster et al., *Wildlands and Woodlands: A Vision for the Forests of Massachusetts* (Petersham, Mass.: Harvard Forest, 2005).

33. Mumford, *Technics and Civilization,* 265; and Lewis Mumford, *The Highway and the City* (New York: Harcourt, Brace & World, 1963).

34. Brian Donahue, *Reclaiming the Commons: Community Farms and Forests in a New England Town* (New Haven, Conn.: Yale University Press, 1999); and Foster et al., *Wildlands and Woodlands.*

7. A City (Only Partly) on a Hill: Terrain and Land Use in Pre-twentieth-century Boston

I have derived land elevation values, where not otherwise stated, from the 1903 U.S. Geological Survey, "Boston" 1:62,500 topographic map, supplemented by written sources and by the finer-scale current "Boston North" and "Boston South" 1:25,000 quadrangles. The Sanborn maps referred to in the chapter are the Boston sets of 1867, 1885–88, and 1895–1900 (the last being my chief source for land uses and street addresses at the time of the 1900 Census). In describing the social class of residents in 1900, I have drawn on the three-hundred-plus occupations defined in Bureau of the Census Special Report, *Occupations at the Twelfth Census* (Washington, D.C.: U.S. Government Printing Office, 1904), xxiii–xxv, and the broader categories into which that publication divided them. I have taken as "upper-class" those in the categories "professional service," "bankers and brokers," "merchants and dealers," and "officials of banks and companies."

1. Edward Everett, journal entry, February 7, 1847, *Edward Everett Papers, 1675–1930,* Reel 38, microfilm edition, Massachusetts Historical Society, Boston, 1972.

2. For reviews of studies on the topic, see William B. Meyer, "Bringing Hypsography Back In: Altitude and Residence in American Cities," *Urban Geography* 15 (1994): 505–13; and William B. Meyer, "The Other Burgess Model," *Urban Geography* 21 (2000): 261–70.

3. "The Future of Boston," *Boston Evening Transcript,* December 1, 1853, 1. Similar patterns in premodern American cities are identified by, for example, Carl Abbott, "The Neighborhoods of New York, 1760–1775," *New York History* 55 (1974): 35–54; John K. Alexander, *Render Them Submissive: Responses to Poverty in Philadelphia, 1760–1800* (Amherst: University of Massachusetts Press, 1980), 19–22; Richard M. Bernard, "A Portrait of Baltimore in 1800: Economic and Occupational Patterns in an Early American City," *Maryland Historical Magazine* 69 (1974): 341–60; Stuart M. Blumin, *The Urban Threshold: Growth and Change in a Nineteenth-century American Community* (Chicago: University of Chicago Press, 1976), 109–12; Kenneth T. Jackson, *Crabgrass Frontier: The Suburbanization of the United States* (New York: Oxford University Press, 1985), chapter 1; John Kellogg, "The Formation of Black Residential Areas in Lexington, Kentucky, 1865–1887," *Journal of Southern History* 48 (1982): 21–52; Herbert S. Klein and Edmund P. Willis, "The Distribution of Wealth in Late-eighteenth-century New York City," *Histoire sociale—Social History* 18, no. 36 (1985): 259–83; Mary M. Schweitzer, "The Spatial Organization of Federalist Philadelphia, 1790," *Journal of Interdisciplinary History* 24 (1993): 31–57; Carol Shammas, "The Space Problem in Early United States Cities," *William and Mary Quarterly,* third series, 57 (2000): 505–42; Billy G. Smith, "The Lower Sort": *Philadelphia's Laboring People, 1750–1800* (Ithaca, N.Y.: Cornell University Press, 1990), 163–65; and John Swauger, "Pittsburgh's Residential Pattern in 1815," *Annals of the Association of American Geographers* 68 (1978): 265–77. For a review of earlier studies reaching similar conclusions, see John P. Radford, "The Social Geography of the Nineteenth-century U.S. City," in D. T. Herbert and R. J. Johnston, eds., *Geography and the Urban Environment: Progress in Research and Applications* (Chichester, England: John Wiley & Sons, 1981), vol. 4, 257–93. Some studies suggest that the pattern emerged only in the late eighteenth century with the growth and intensified crowding of cities (for example, see Smith, *"Lower Sort,"* 163–65).

4. On the higher prestige of the center in Boston, see Alan Kulikoff, "The Progress of Inequality in Revolutionary Boston," *William and Mary Quarterly,* third series, 28 (1971): 396–99.

5. Stuart M. Blumin, in *The Emergence of the Middle Class: Social Experience in the American City, 1760–1900* (New York: Cambridge University Press, 1989), 48–50, emphasizes the main street/side street dimension as an important aspect of early residential segregation that is often overlooked when scholars use aggregate figures (for example, for entire wards) to characterize urban neighborhoods. It is also highlighted by, among others, Kellogg, "Formation of Black Residential Areas," 34–35, 38–39; and Smith, *"Lower Sort,"* 163, 165. On State Street, see Kulikoff, "Progress of Inequality," 396–97.

6. On the "steep and tiresome stairs," see George C. Crocker, "The Passenger Traffic of Boston and the Subway," *New England Magazine,* n.s., 19 (1899): 530; William Dean Howells, *A Modern Instance* (1882), in *Novels, 1875–1886* (New York: Library of America, 1982), 305–10; Bainbridge Bunting, *Houses of Boston's Back Bay: An Architectural History, 1840–1917* (Cambridge: Harvard University Press, 1967), 137–39; and Albert Benedict Wolfe, *The Lodging House Problem in Boston* (Boston: Houghton Mifflin, 1906), 60–61.

7. Swauger, "Pittsburgh's Residential Pattern," particularly emphasizes the significance of fringe estates of the independently wealthy, as do Jackson, *Crabgrass Frontier,* 18–19; and Alexander, *Render Them Submissive,* 191–92. On such country estates in the Boston area, see Sam Bass Warner Jr., *Streetcar Suburbs: The Process of Growth in Boston, 1870–1900* (Cambridge: Harvard University Press and MIT Press, 1962), 13–14; and Tamara Plakins Thornton, *Cultivating Gentlemen: The Meaning of Country Life Among the Boston Elite, 1785–1860* (New Haven, Conn.: Yale University Press, 1989).

8. On the characteristic land uses of the premodern fringe in one sector of the Boston area, see Henry C. Binford, *The First Suburbs: Residential Communities on the Boston Periphery, 1815–1860* (Chicago: University of Chicago Press, 1985), 5–6, 39.

9. On the two processes in the Boston area, see Binford, *First Suburbs*; David Ward, "The Emergence of Central Immigrant Ghettoes in American Cities, 1840–1920," *Annals of the Association of American Geographers* 58 (1968): 343–59; and Warner, *Streetcar Suburbs.*

10. Paul L. Knox, "Preface," in Paul L. Knox, ed., *The Restless Urban Landscape* (Englewood Cliffs, N.J.: Prentice Hall, 1993), ix. Notable exceptions today, illustrated by the Beacon Hill neighborhood in Boston, are affluent central-city districts themselves based on amenities: in this case, those of historic architecture and heritage.

11. See Meyer, "Bringing Hypsography Back In" and "Other Burgess Model"; also see Jeff Ueland and Barney Warf, "Racialized Topographies: Race and Altitude in Southern Cities," *Geographical Review* 96 (2006): 50–78.

12. For the concept of the vertical fringe, applied to a case study of Worcester, Massachusetts, see William B. Meyer, "The Poor on the Hilltops? The Vertical Fringe of a Late-nineteenth-century American City," *Annals of the Association of American Geographers* 95 (2005): 773–88.

13. "Architecture in the United States," *American Journal of Science and the Arts* 17 (1830): 250–51.

14. Matthew Klingle, "Changing Spaces: Nature, Property, and Power in Seattle, 1880–1945," *Journal of Urban History* 32 (2006): 197–230 (quotation on 209).

15. On the original area of the peninsula, I have followed Nancy S. Seasholes, *Gaining Ground: A History of Landmaking in Boston* (Cambridge: MIT Press, 2003), 423. On

the hills and their heights in 1630, see Darrett B. Rutman, *Winthrop's Boston: Portrait of a Puritan Town, 1630–1649* (Chapel Hill: University of North Carolina Press, 1965).

16. Walter Muir Whitehill, *Boston: A Topographical History,* second edition (Cambridge: Harvard University Press, 1968), 10. The Hull quotation is from Samuel Eliot Morison, *Builders of the Bay Colony* (Boston: Houghton Mifflin, 1930), 137.

17. On the crucial importance of the hills throughout the Revolutionary siege, see Richard Frothingham, *History of the Siege of Boston* (Boston: C. C. Little and J. Brown, 1851); Charles Francis Adams, "The Battle of Bunker Hill," *American Historical Review* 1 (1896): 401–13; and Allen French, *The First Year of the American Revolution* (Boston: Houghton Mifflin, 1934).

18. For examples of such uses of the hilltops, see *Green & Russell's Boston Post-Boy and Advertiser,* October 22, 1759, 2 (the fall of Quebec); ibid., September 29, 1760, 3 (the fall of Canada); Edmund S. Morgan and Helen M. Morgan, *The Stamp Act Crisis: Prologue to Revolution* (Chapel Hill: University of North Carolina Press, 1953), 124, 125 (the Stamp Act disturbances); Francis G. Walett, "Governor Bernard's Undoing," *New England Quarterly* 38 (1965): 222 (the departure of an unpopular royal governor); Nathaniel Dearborn, *Boston Notions* (Boston: W. D. Ticknor & Co., 1848), 164 (peace with England in 1783); and Whitehill, *Boston,* 49 (the completion of the Charles River Bridge in 1786).

19. David Hackett Fischer, "Boston Common," in William E. Leuchtenberg, ed., *American Places: Encounters with History* (New York: Oxford University Press, 2000), 125–43; Lawrence J. Vale, *From the Puritans to the Projects* (Cambridge: Harvard University Press, 2000), 28–31, 36–38; and Annie Haven Thwing, *The Crooked and Narrow Streets of the Town of Boston, 1630–1822* (Boston: Marshall Jones Co., 1920), 207–8, 221–27.

20. *Report of the Record Commissioners of the City of Boston,* vol. 31 (entry for August 19, 1790) (Boston: City of Boston, 1876–1909), 239.

21. See the texts collected in Daniel A. Cohen, ed., *The Female Mariner and Related Works: Narratives of Cross-Dressing and Urban Vice in America's Early Republic* (Amherst: University of Massachusetts Press, 1997), especially 59, 68, 83, 91–97, 118–27, 137–45, and 170–80; and Whitehill, *Boston,* 70–71.

22. Whitehill, *Boston,* 35–37, 105–7; Allen Chamberlain, *Beacon Hill: Its Ancient Pastures and Early Mansions* (Boston: Houghton Mifflin, 1925); Winthrop S. Scudder, "A History of the Gardiner Greene Estate," *Bostonian Society Publications* 12 (1915): 39–62; and Nathaniel I. Bowditch, "Gleaner" articles in *Report of the Record Commissioners of the City of Boston* (Boston, 1876–1909), vol. 5, especially 54–56, 61–67, 81.

23. Timothy Dwight, *Travels in New England and New York,* ed. Barbara Miller Solomon (Cambridge: Harvard University Press, 1969), vol. 1, 356; and Dearborn, *Boston Notions,* 223.

24. Seasholes, *Gaining Ground,* 135–41.

25. Ibid., 93–96; and Whitehill, *Boston,* 107–9.

26. "Tunnelling the Hill," *Boston Post,* January 29, 1870, 3; Crocker, "Passenger Traffic of Boston and the Subway," 530; "Report of the Hearing Before a Committee of the House of Representatives of Massachusetts on the Occupation and Improvement of the Charlestown Flats in Charles River," *Boston City Documents for 1869,* no. 128, 167–68, 176, 178; and Edward Stanwood, *Boston Illustrated* (Boston: J. R. Osgood & Co., 1872), 23. Christine M. Rosen, *The Limits of Power: Great Fires and the Process of City Growth in America* (New York: Cambridge University Press, 1986), 81, discusses the tendency

of business to avoid high ground and steep slopes whenever possible in the nineteenth-century American city generally and in Boston in particular. Walter Firey's otherwise insightful discussion of values and sentiments as major factors in land use on Beacon Hill and elsewhere gives too little weight to slopes and elevation as a barrier to most forms of nineteenth-century business; see Firey's *Land Use in Central Boston* (Cambridge: Harvard University Press, 1947).

27. Whitehill, *Boston,* 110.

28. Caleb H. Snow, *A Geography of Boston, County of Suffolk, and the Adjacent Towns* (Boston: Carter and Hendee, 1830), 64–65; Adelaide M. Cromwell. "The Black Presence in the West End of Boston, 1800–1864: A Demographic Map," in Donald M. Jacobs, ed., *Courage and Conscience: Black and White Abolitionists in Boston* (Bloomington: Indiana University Press, 1993), 155–67; Josiah Curtis, *Report of the Joint Special Committee on the Census of Boston, May, 1855* (Boston: Moore & Crosby, 1856), 56; "Mortality of the City of Boston in 1870," in *Second Annual Report of the State Board of Health of Massachusetts* (Boston: Wright & Potter, 1871), 350; "City Improvements," *The Commercial Bulletin,* October 27, 1866, 1; "City Government," *Boston Evening Transcript,* May 24, 1872, 4; "How the Poor Live: Pictures from Boston's Tenement Houses," *Boston Courier,* June 8, 1879, 1; "How the Poor Live: Degradation and Crime on Beacon Hill," *Boston Courier,* March 14, 1880, 1; "How the Poor Live: More Beacon Hill Tenements Brought to Light," *Boston Courier,* March 21, 1880, 1; "Part of the 'Other Half'," *Boston Evening Transcript,* January 7, 1893, 6; "The Shady Side of Beacon Hill," *Boston Evening Transcript,* March 11, 1893, 13; and "To Improve Beacon Hill," *Boston Evening Transcript,* March 23, 1893, 5.

29. Even after the construction of the reservoir, buildings high on the Hill, and especially their upper floors, often suffered from low pressure and unreliable service during times of high demand, until the building of an improved "high-service" system after the Civil War.

30. "The New Court House," *Boston Daily Advertiser,* November 11, 1871, 2; and "The New Court House," *Boston Herald,* November 21, 1871, 1.

31. For early complaints about the gasworks nuisance, see, for example, "Boston Gas Works," *Boston Evening Transcript,* September 15, 1832, 2. On the choice of a site for the smallpox hospital (which was never opened, the epidemic having waned in time), see "The Board of Health," *Boston Daily Advertiser,* January 20, 1873, 1; "About-Home Matters: The Board of Health," *Boston Post,* January 20, 1873, 3; and "Small-pox," *Boston Post,* February 1, 1873, 3. A permanent mainland isolation hospital was eventually constructed on the city's horizontal and vertical fringe: a reporter judged that from a distance, it looked like a gentleman's country estate: "At the Smallpox Hospital," *Boston Evening Transcript,* November 29, 1881, 4.

32. Paula J. Todisco, *Boston's First Neighborhood: The North End* (Boston: Boston Public Library, 1976), 27.

33. Jacqueline Barbara Carr, *After the Siege: A Social History of Boston, 1775–1800* (Boston: Northeastern University Press, 2005).

34. Oscar Handlin, *Boston's Immigrants: A Study in Acculturation* (Cambridge: Harvard University Press, 1959), 90, 93–94, 97; *Report of the Commissioners on the Hours of Labor* (Boston: Wright & Potter, 1867), 59; and "Assault upon an Officer," *Boston Evening Transcript,* March 30, 1857, 2.

35. "The Removal of Fort Hill," *Boston Saturday Evening Gazette,* February 18, 1865, 2; and David Ward, "The Emergence of Central Immigrant Ghettoes in American Cities, 1840–1920," *Annals of the Association of American Geographers* 58 (1968): 352.

36. "Destructive Conflagration on Fort Hill," *Boston Herald,* July 12, 1852, 2; "Fort Hill Reclaimed for Business Purposes," *Boston Herald,* July 14, 4; and Seasholes, *Gaining Ground,* 61. On the results expected from the project, see "New Mercantile Palaces," *Boston Evening Transcript,* March 3, 1854, 2.

37. Quotations are from "The Fort Hill Improvement," *Boston Post,* April 3, 1865, 1; "The Removal of Fort Hill," *Saturday Evening Gazette,* February 18, 1865, 2; and an untitled editorial, *Daily Evening Traveller,* March 22, 1865, 2. For similar views, see also "Fort Hill," *Daily Evening Traveller,* May 31, 1865; 2; "Fort Hill," *Daily Evening Traveller,* September 10, 1866, 2; "City Improvements: The Removal of Fort Hill," *Daily Evening Traveller,* October 1, 1866, 2; and "The Church Street District," *Boston Evening Transcript,* June 4, 1867, 2.

38. "The Fort Hill Movement," *Boston Post,* September 6, 1866, 4; "A Fatal Accident," *Boston Post,* October 16, 1866, 4; "The Fort Hill Improvement," *Boston Herald,* October 31, 1866, 4; *Annual Report of the Superintendent of Streets,* Boston City Document no. 6, 1867, 14–16; "High Prices—No. 5," *Boston Pilot,* February 27, 1869, 4; "City and Vicinity," *Boston Pilot,* October 16, 1869, 5; and "The Poor White Man," *Boston Pilot,* May 7, 1870, 1.

39. "Improvements on Fort Hill," *Boston Post,* May 12, 1870, 3. For overviews of the project, see Seasholes, *Gaining Ground,* 61–70; and Charles Phillips Huse, *The Financial History of Boston from May 1, 1822, to January 30, 1909* (Cambridge: Harvard University Press, 1916), 127–28.

40. J. Anthony Lukas, *Common Ground: A Turbulent Decade in the Lives of Three American Families* (New York: Random House, 1985), 78.

41. Ibid.; Snow, *Geography of Boston,* 136; Robert H. Lord, John E. Sexton, and Edward T. Harrington, *History of the Archdiocese of Boston in the Various Stages of Its Development, 1604 to 1943,* vol. 2 (New York: Sheed & Ward, 1944), 206–7; Seasholes, *Gaining Ground,* 405–6; "Report of the Commissioners on Grading Bunker Hill," *Charlestown Advertiser,* September 9, 1871, 2; and Cynthia Zaitzevsky, *Frederick Law Olmsted and the Boston Park System* (Cambridge: Harvard University Press, 1982), 101–2, 104.

42. 1900 Manuscript Census, Boston, Ward 4, Enumeration District 1198, 1–8, and Enumeration District 1200, 13; and John S. Billings, *Vital Statistics of Boston and Philadelphia, Covering a Period of Six Years Ending May 31, 1890* (Washington, D.C.: U.S. Government Printing Office, 1895), 51, 87–88. The population of the hilltop also included the resident staff of the church and Catholic school.

43. On the original elevation of the Heights, see "Sketches of Home, and History of Mattapannock. No. 1," *South Boston Gazette,* January 5, 1850, 2; Charles Dickens, *American Notes and Pictures from Italy* (Oxford: Oxford University Press, 1987), 30; *Thirty-ninth Annual Report of the Trustees of the Perkins Institution and Massachusetts Asylum for the Blind* (Boston: Wright & Potter, 1871), 7; *Thirty-seventh Annual Report* (1869), 25–29; *Thirty-fifth Annual Report* (1867), 6; *Twenty-sixth Annual Report* (1858), 5; *Sixteenth Annual Report* (1848), 18–19; and Charles Sumner, *The True Grandeur of Nations* (Boston: J. H. Eastburn, 1845), 80.

44. 1900 Manuscript Census, Boston, Ward 15, Enumeration District 1378, 22–23; Enumeration District 1379, 16–18; Enumeration District 1380, 8–15, 17–19, 21; and Enumeration District 1381, 1–4, 13–14. See also the untitled editorial, *South Boston Bulletin,* July 23, 1892, 8.

45. William H. Sumner, *A History of East Boston* (Boston: J. E. Tilton, 1858), 546, 589; and 1900 Manuscript Census, Boston, Ward 1, Enumeration District 1168, 30–34.

46. Charles W. Eliot, *Charles Eliot, Landscape Architect* (Boston: Houghton Mifflin, 1902), 403.

47. Michael P. Conzen and George K. Lewis, *Boston: A Geographical Portrait* (Cambridge, Mass.: Ballinger Publishing Company, 1976), 7–9; and Derwent Whittlesey, "Chorography of the Boston (Mass.) Locality: A Report of Reconnaissance," *Annals of the Association of American Geographers* 20 (1930): 48–49.

48. George H. M. Rowe, "Historical Description of the Buildings and Grounds of the Boston City Hospital," in David W. Cheever et al., eds., *A History of the Boston City Hospital from Its Foundation Until Today* (Boston: Municipal Printing Office, 1904), 2. For objections to the Winthrop site, see "Letter from George B. Emerson to Samuel C. Cobb," Boston City Document no. 113, December 9, 1869, 3–5; "The City Buildings," *Boston Daily Advertiser*, October 5, 1871, 2; an untitled editorial, *Boston Daily Advertiser*, October 13, 2; and "Who Shall Rule?," *Boston Daily Advertiser*, October 30, 1871, 1.

49. Eliot, *Charles Eliot*, 351, 562; and Karl Haglund, *Inventing the Charles River* (Cambridge: MIT Press, 2003), especially 107–14.

50. As with the hills, certain specialized uses took advantage of the islands' site features. They too were favored sites for fortifications to defend the city from attack. Castle Island was the early site of Fort Independence; Governor's and George's held large defensive works in later times. The Beacon Hill signal mast had its counterpart in the shipping beacon on the islet of Nix's Mate and the main harbor lighthouses on Long and Great Brewster islands. On the harbor islands generally, see Nathaniel B. Shurtleff, *A Topographical and Historical Description of Boston* (Boston: Noyes, Holmes and Co., 1872); James H. Stark, *Illustrated History of Boston Harbor* (Boston: Photo-Electrotype Company, 1879); M. F. Sweetser, *King's Handbook of Boston Harbor* (Cambridge: M. King, 1882); and Edward Rowe Snow, *The Islands of Boston Harbor, 1630–1971* (New York: Dodd, Mead & Company, 1971).

51. Dorothea L. Dix, "Location of a City Hospital for the Insane Poor of Boston," *Boston Evening Transcript*, October 28, 1; "Final Report of the Grand Jury," *Boston Evening Transcript*, June 24, 1857, 2; "Deer Island During the Storm," *Boston Herald*, April 18, 1851, 2; "The Horrors of the Passage to Deer and Rainsford Islands," *Boston Evening Transcript*, January 19, 1855, 2; and "Report of the Committee on Health of the Common Council on the Orders for the Erection of a Small-Pox Hospital at Gallup's Island," Boston City Document no. 91, September 12, 1872, 6.

52. 1900 Manuscript Census, Boston, Ward 2, Enumeration Districts 1809, 1827, 1830, and 1831.

8. Reforestation in Norfolk County, Massachusetts, 1850–1910

1. "The Farmer's Outlook," *New England Farmer* 38, n.s. (January 6, 1883): 1.

2. A. W. Cheever, "The Use of Commercial Fertilizers in New England," *New England Farmer* 38, n.s. (March 24, 1883): 1. Cheever's January speech was republished in the *New England Farmer*'s March issue.

3. W. D. Philbrick, "Ensilage at Florence Farm," *New England Farmer* 38, n.s. (February 3, 1883): 1.

4. W. D. Philbrick, "Another Silo," *New England Farmer* 38, n.s. (March 10, 1883): 1.

5. "Modern Agriculture: Farming in Norfolk County," *New England Farmer* 38, n.s. (April 7, 1883): 1.

6. "Running in Debt for Farms," *Massachusetts Ploughman and New England Journal of Agriculture* 58 (February 18, 1899): 1.

7. David Foster and John O' Keefe, *New England Forests Through Time: Insights from the Harvard Forest Dioramas* (Petersham, Mass.: Harvard Forest, 2000), 10.

8. Studies that emphasize the steep and rapid decline of agriculture in New England in the late nineteenth century include John D. Black, *The Rural Economy of New England* (Cambridge: Harvard University Press, 1950); Harold Fisher Wilson, *The Hill Country of Northern New England* (New York: Columbia University Press, 1936); Clarence Danhoff, *Change in Agriculture* (Cambridge: Harvard University Press, 1969); and Robert Eisenmenger, *The Dynamics of Growth in New England's Economy, 1870–1964* (Middletown, Conn.: Wesleyan University Press, 1967).

9. Michael M. Bell, "Did New England Go Downhill?" *Geographical Review* 79 (October 1989), 450–66.

10. "Abandoned Farms," *Boston Daily Advertiser,* November 2, 1887, 4.

11. Massachusetts Bureau of Statistics of Labor, *Census of the Commonwealth of Massachusetts, 1875,* vol. 3, *Agricultural Products and Property* (Boston: Wright & Potter Printing Co., 1878); and Massachusetts Bureau of Statistics of Labor, *Census of the Commonwealth of Massachusetts, 1895,* vol. 6, *The Fisheries, Commerce and Agriculture* (Boston: Wright & Potter Printing Co., 1899).

12. William Cronon, *Nature's Metropolis: Chicago and the Great West* (New York: W. W. Norton, 1991).

13. For background on New England farming practices in the first half of the nineteenth century, see Howard S. Russell, *A Long, Deep Furrow: Three Centuries of Farming in New England* (Hanover, N.H.: University Press of New England, 1982).

14. Emerson as quoted in Carl Reidel et al., *The Yankee Forest. A Prospectus: A Report of the Yale School of Forestry and Environmental Studies Fifth Forest Project* (New Haven, Conn.: Yale University, 1978), 11.

15. *First Annual Report of the Secretary of the Massachusetts State Board of Agriculture* (Boston: William White, Printer to the State, 1854), 70.

16. *First Annual Report,* 68–69.

17. Brian Donahue's seminal study of farming in Concord, Massachusetts, suggests that the switch to extensive farming was a response to the market revolution. In embracing a more intensive mode of agriculture, Bay State farmers were returning to the kind of farming practiced by their great-grandparents. See Brian Donahue, *The Great Meadow: Farmers and the Land in Colonial Concord* (New Haven, Conn.: Yale University Press, 2004).

18. Porter as quoted in *First Annual Report,* 333.

19. T. Whitaker, "Improving a Farm," *New England Farmer* 28, n.s. (May 24, 1873): 1.

20. Henry Colman, *Fourth Report of the Agriculture of Massachusetts* (Boston: Massachusetts Agricultural Survey, 1841), 264.

21. Lawson as quoted in *First Annual Report,* 345–46.

22. Fred Bateman, "Improvement in American Dairy Farming, 1850–1910: A Quantitative Analysis," *Journal of Economic History* 28 (1968): 255–73; and Frank Smith, *Dover Farms* (Dover, Mass.: Historical and Natural History Society, 1914), 114.

23. *Resident and Business Directory of Milton, Massachusetts, 1885–1886* (Needham, Mass.: Local Directory and Publishing Co., 1886), 22.

24. E. Melanie DuPuis, *Nature's Perfect Food: How Milk Became America's Drink* (New York: New York University Press, 2002), 37.

25. George Whitaker, *The Milk Supply of Boston and Other New England Cities,* USDA Bureau of Animal Industry, Bulletin No. 20. (Washington, D.C.: U.S. Department of Agriculture, 1898), 27.

26. "The Low Price of Milk," *New England Farmer* 38, n.s. (March 17, 1883): 1.

27. Whitaker, *Milk Supply of Boston,* 15.

28. Ibid., 10.

29. John G. Palfrey, *Statistics of the Condition and Products of Certain Branches of Industry in Massachusetts, for the Year Ending April 1, 1845* (Boston: Dutton and Wentworth, 1846), 247.

30. *Annual Report of the Receipts and Expenditures for the Year Ending February 1, 1870* (Medfield, Mass., 1870), 18.

31. Oliver Warner, secretary of the Commonwealth of Massachusetts, *Statistical Information Relating to Certain Branches of Industry for the Year Ending May 1, 1865* (Boston: Wright and Potter Printing Co., 1866); *Census of the Commonwealth of Massachusetts,* 1875, vol. 3, *Milk Production Data.*

32. Whitaker, *Milk Supply of Boston,* 12–22.

33. George Whitaker, *The Milk Supply of Boston, New York, and Philadelphia,* USDA, Bureau of Animal Industry, Bulletin No. 81 (Washington, D.C.: U.S. Government Printing Office, 1905), 24.

34. Massachusetts Bureau of Statistics of Labor, *Census of the Commonwealth of Massachusetts, 1885,* vol. 3, *Agricultural Products and Property* (Boston: Wright and Potter Printing Co., 1887).

35. Department of Agriculture, *Report of the Commissioner of Agriculture for the Years 1881 and 1882* (Washington, D.C.: U.S. Government Printing Office, 1882).

36. Massachusetts State Board of Agriculture, 1881 *Annual Report* (Boston: William White, 1882), 112.

37. *Massachusetts Register and Business Directory of 1867* (Boston: George Adams, 1867); *Massachusetts Register and Business Directory of 1878* (Boston: Sampson, Davenport & Co., 1878); *Massachusetts Year Book and Business Directory of Every Town and City in the State* (Worcester, Mass.: F. S. Blanchard & Co., 1899); and Gail Lembo, ed., *The Diaries of George M. Wadsworth, 1857–1893* (Franklin, Mass: Gail Lembo, 1998).

38. "Concentrated Farming," *New England Farmer* 34, n.s. (May 24, 1879): 1.

39. *Census of the Commonwealth of Massachusetts,* 1875, vol. 3, xix.

40. Brian Donahue, "Skinning the Land: Economic Growth and the Ecology of Farming in Nineteenth-century Massachusetts," paper presented at the annual meeting of the Social Science History Association, Chicago, November 1988, p. 28.

41. Individual level census returns, *1870 and 1880 Federal Censuses,* town of Medfield, viewed at the National Archives (NARA), Waltham, Mass.

42. Norfolk Agricultural Society, *1865 Transactions of the Norfolk Agricultural Society* (Dedham, Mass.: Norfolk Agricultural Society, 1865), 24–32.

43. Ibid., 24–32.

44. Ibid., 32.

45. Ibid., 29.

46. Edward Pierce Hamilton, *A History of Milton* (Milton, Mass.: Milton Historical Society, 1957), 49–51; and *1895 Massachusetts State Census,* vol. 6, *Value of Agricultural Property and Assessed Real Estate* (Boston: Wright and Potter Printing Co., 1899), 341.

47. *Census of the Commonwealth of Massachusetts,* 1895, vol. 6, *Milk Production Data; and Census of the Commonwealth of Massachusetts,* 1905, vol. 4, *Agriculture, Fisheries, and Commerce* (Boston: Wright and Potter Printing Co., 1909).

48. "A Bit of Experience," *Massachusetts Ploughman and New England Journal of Agriculture* 58 (October 22, 1898), 8.

49. Albert K. Teele, *The History of Milton, Massachusetts, 1640 to 1887* (Milton, Mass.: Press of Rockwell and Churchill, 1887), 81.

50. "Modern Agriculture: Farming in Norfolk County," *New England Farmer* 38, n.s. (April 7, 1883), 1.

51. *Census of the Commonwealth of Massachusetts,* 1895, vol. 6; and *Census of the Commonwealth of Massachusetts,* 1905, vol. 4, *Milk Production and Wooden Product Data.*

52. Lucien Gove, "The Eastern Farmer's Handicap: An Object Lesson," *Massachusetts Ploughman and New England Journal of Agriculture* 65 (April 28, 1906): 1.

53. Oliver Warner, secretary of the Commonwealth of Massachusetts, *Statistical Information Relating to Certain Branches of Industry for the Year Ending May 1, 1865; 1895 and 1905 Census of the Commonwealth of Massachusetts, Agricultural Volumes.*

54. Mollie Beattie, Charles Thompson, and Lynn Levine, *Working with Your Woodland: A Landowner's Guide,* revised edition (Hanover, N.H.: University Press of New England, 1993), 12.

55. John Philip Wentling, *Woods Used for Packing Boxes in New England,* Forest Service Circular 78 (Washington, D.C.: United States Department of Agriculture, 1907), 2.

56. Ibid., 3.

57. Ibid., 3.

58. Hu Maxwell, *A Study of the Massachusetts Wood-using Industries* (Boston: Wright & Potter, 1910), 10.

59. Wentling, *Woods Used for Packing Boxes,* 4.

60. *The Natural Replacement of White Pine on Old Fields in New England,* USDA, Forest Service Bulletin No. 63 (Washington, D.C.: U.S. Government Printing Office, 1905), 21.

61. Allen Chamberlain, "Possibilities for Farm Forestry in Massachusetts," in *Massachusetts State Board of Agriculture, 1900 Annual Report* (Boston: Wright & Potter, 1901), 295.

62. "A Prominent Milk Farmer," *Massachusetts Ploughman and New England Journal of Agriculture* 64 (January 14, 1905): 4.

63. "Farm Land Is Good Property," *Massachusetts Ploughman and New England Journal of Agriculture* 65 (June 9, 1906): 4.

9. How Metropolitan Parks Shaped Greater Boston, 1893–1945

1. Robert H. Wiebe, *The Search for Order, 1877–1920* (New York: Hill and Wang, 1967), 166. Wiebe argued that Progressivism originated in Eastern cities seeking solutions to the social problems stemming from rampant urban and industrial growth.

2. Jon A. Peterson, *The Birth of City Planning in the United States, 1840–1917* (Baltimore, Md.: Johns Hopkins University Press, 2003), 98.

3. Karl Haglund, *Inventing the Charles River* (Cambridge: MIT Press, 2002), 50–51; and "The Park Question," *Boston Daily Advertiser,* December 2, 1869. Ernest W. Bowditch, an engineer who had worked with Copeland, published a comparable "Metropolitan Park Plan" of his own in 1874.

4. Landscape architect Walter L. Creese argued that "the Metropolitan Park System represented its [the Emerald Necklace's] evolutionary glory." See Walter L. Creese, *The Crowning of the American Landscape: Eight Great Spaces and Their Buildings* (Princeton, N.J.: Princeton University Press, 1985), 183.

5. "Parks for Growing Cities," *Garden and Forest,* February 10, 1892, 61.

6. Eliot as quoted in ibid., 655.

7. Galen Cranz, *The Politics of Park Design: A History of Urban Parks in America* (Cambridge: MIT Press, 1982), 5, 69. Cranz called Olmsted's theory of parks "popularized transcendentalism."

8. Center for Public Service, Tufts University, *The Greening of Boston: An Action Agenda: A Report from the Boston Foundation* (Boston: Boston Foundation, 1987), 44.

9. Baxter as cited in ibid., 48.

10. Quotation as cited in Landscape Research and Office of Planning and Community Development, City of Somerville, *Beyond the Neck: The Architecture and Development of Somerville* (Somerville, Mass.: City of Somerville, 1982), 60.

11. Charles W. Eliot Sr., *Charles Eliot, Landscape Architect* (Amherst: University of Massachusetts Press, 1999), 318.

12. Eliot, *Charles Eliot,* 608; and Geoffrey Blodgett, *The Gentle Reformers: Massachusetts Democrats in the Cleveland Era* (Cambridge: Harvard University Press, 1966), 125–26.

13. Edward C. Kirkland, *Charles Francis Adams, Jr., 1835–1915: The Patrician at Bay* (Cambridge: Harvard University Press, 1965), 187.

14. Board of Metropolitan Park Commissioners, *Report of the Metropolitan Park Commissioners* (Boston: Commonwealth of Massachusetts, 1893).

15. Eliot, *Charles Eliot,* 402.

16. Ibid., 351.

17. Kirkland, *Charles Francis Adams, Jr.,* 188.

18. "Landscape-art in Public Parks," *Garden and Forest,* May 3, 1893, 191.

19. Ibid., 381.

20. "The Park System of Greater Boston," *Garden and Forest,* June 23, 1897, 241.

21. Eliot, *Charles Eliot,* 395.

22. Ibid., 401.

23. Leah A. Schmidt, *Revere Beach* (Charleston, S.C.: Arcadia Publishing, 2002), 9; and Sylvester Baxter, "Seaside Pleasure-Grounds for Cities," *Scribner's Magazine* 5, no. 23 (1898): 681.

24. Eliot, *Charles Eliot,* 411.

25. Lena Lencek and Gideon Bosker, *The Beach: The History of Paradise on Earth* (New York: Viking, 1998), 183, 187.

26. Sylvester Baxter, "Boston's New Metropolitan Parks," *Garden and Forest,* January 17, 1894, 22.

27. Commonwealth of Massachusetts, *A Description of the Topographical Model of Metropolitan Boston* (Boston: Wright & Potter Printing Co., 1900), 4. The model included a scale-model State House, Boston Common with the Frog Pond and each pathway, the Harbor Islands, each church spire, 250 miles of railroads, 157,000 dwellings, and 200,000 trees.

28. Sylvester Baxter, "Twenty-five Years of the Premier Park System of the World—in Boston," *Boston Transcript,* July 13, 1918.

29. Peterson, *Birth of City Planning,* 335.

30. William Wilson, *City Beautiful Movement* (Baltimore, Md.: Johns Hopkins University Press, 1995), 79. Peterson, *Birth of City Planning,* 162. Peterson added: "The Boston area had pioneered this conception [creating regional park systems] in the 1890s. By 1902–1905 many other cities had commissioned similar plans, among them Baltimore, Philadelphia, Providence, and Washington in the East; Chicago, Minneapo-

lis, and St. Paul in the Midwest; and Portland and Seattle in the Far West. . . . Of all the cities to emulate Boston, in 1902–1905, none exceeded Chicago for sheer audacity" (in Peterson, *Birth of City Planning,* 163–64).

31. Peterson, *Birth of City Planning,* 221.

32. Mel Scott, *American City Planning Since 1890* (Berkeley: University of California Press, 1969), 116–17. The historian Jon Peterson stated of the Progressive Era: "No other reform period in American history has so fruitfully addressed the city as a setting for life, especially its public realm" (in Peterson, *Birth of City Planning,* 393).

33. Massachusetts Green Ribbon Commission, *Enhancing the Future of the Metropolitan Park System: Final Report and Recommendations of the Green Ribbon Commission* (Boston: Commonwealth of Massachusetts, 1996), 18.

34. Cranz, *Politics of Park Design,* 62.

35. "Parks for the People," *Christian Science Monitor,* July 10, 1912.

36. "Public Slow to Make Full Use of Metropolitan Parks," *Boston Globe,* October 8, 1916.

37. Metropolitan District Commission, *Annual Report of the Metropolitan District Commission for the Year 1923* (Boston: Metropolitan District Commission, 1924), 9.

38. Board of Metropolitan Park Commissioners, *Report of the Board of the Metropolitan Park Commissioners, December 1915* (Boston: Metropolitan Park Commission, 1916), 17, 180. Until after World War II the MDC bathhouses rented bathing suits for two hours at a time. As late as 1951, the MDC had thirty thousand bathing suits on hand; but the "antideluvian" women's bathing suits, with white-tiered skirts, had few takers. Swimmers started wearing their own, more daring bathing suits, and the MDC got out of the swimsuit rental business. "MDC Parkmen Ready Swim Spots for Hot Weather," June 16, 1951, *Boston Herald,* Clippings File, Boston University Beebe Communications Library.

39. Board of Metropolitan Park Commissioners, *Report of the Board of the Metropolitan Park Commissioners, December 1914* (Boston: Metropolitan Park Commission, 1915), 30–31.

40. Cranz called the 1930–65 period the "Recreation Facility Era" because of its emphasis on building facilities (see Cranz, *Politics of Park Design,* 103).

41. Metropolitan District Commission, *Annual Report of the Metropolitan District Commission for the Year 1938* (Boston: Metropolitan District Commission, 1939), 5.

42. Ibid., 21.

43. *Annual Report of the Metropolitan District Commission for the Year 1939* (Boston: Metropolitan District Commission, 1940), 64.

44. Eliot, *Charles Eliot,* 604. Massachusetts Department of Conservation and Recreation Division of Planning and Engineering, "Historic Parkway Preservation Treatment Guidelines," May 24, 2006. Planners have made a distinction between the avenue or boulevard and the parkway. The avenue or boulevard traversed urbanized areas and created appealing addresses for affluent people to reside. The verdant parkway either linked the city with outlying parks, linked parks, or created its own linear park. City planner Charles Mulford Robinson explained the distinctions: "Considered closely, however, the parkway may have been a development that belongs to neither boulevard nor avenue and that justifies its separate discussion. In speaking of the former thoroughfares, it was noted that the first requirement was that they should afford ease of communication and that the second was that they should have a certain 'dignified and stately' beauty. When we come to parkways, there is no restriction as to the kind of

beauty that may be given. It may be as picturesque, gentle, and softly winning as we please." See Charles Mulford Robinson, *Modern Civic Art, or, The City Made Beautiful* (New York: G. P. Putnam's Sons, 1904), 308.

45. Robinson, *Modern Civic Art,* 209.

46. Eliot, *Charles Eliot,* 537–38.

47. Frederick Law Olmsted recognized the urban-planning aspect of parkways in an 1886 letter to the Minneapolis Park Commissioners, in which he wrote that parkways "are likely to become the stems of systems of streets which will be the framework of the permanent residence quarters of our cities of the future." Olmsted, as cited in August Heckscher, *Open Spaces: The Life of American Cities* (New York: Harper & Row, Publishers, 1977), 195.

48. Eliot, *Charles Eliot,* 606.

49. Ibid., 404.

50. "Country's Finest Pleasure Drive Is Now Complete," *Boston Globe,* July 9, 1904. Even in the face of growing automobile use of parkways to reach reservation, the MPC sought to limit automobiles in favor of horses, bicycles, and pedestrians. As late as 1913, the MPC reaffirmed its rule against automobiles using Revere Beach Parkway to travel to the beach on Saturdays, Sundays, and holidays. See "Rule Against Autos Stands," *Boston Globe,* July 3, 1913.

51. Clay McShane, *Down the Asphalt Path: The Automobile and the American City* (New York: Columbia University Press, 1994), 191.

52. Baxter, as cited in Haglund, *Inventing the Charles River,* 222.

53. "Destructive to Roads," *Boston Globe,* August 31, 1906. "Maintenance of Parks," *Boston Globe,* March 22, 1916. Sylvester Baxter, "A Vast Metropolitan Province," *Boston Evening Transcript,* April 5, 1913.

54. Wilson, *City Beautiful Movement,* 10.

55. Richard England, *Report of the Board of the Metropolitan Park Commissioners, December, 1911* (Boston: Metropolitan Planning Commission, 1912), 23. In 1957 the *Christian Science Monitor* recognized that "without this step [the establishment of the metropolitan park system in 1893], Boston and its immediate suburbs today might be a completely unrelieved congestion of wood, brick and cement." See "Suburbs Sprawl Outward in Bay State," *Christian Science Monitor,* May 18, 1957.

56. Chairman de la Casas, as quoted in "Is Greater Boston Getting Its Money's Worth from the Metropolitan Parks?" *Boston Globe,* August 16, 1908.

57. Sam Bass Warner Jr., *Streetcar Suburbs: The Process of Growth in Boston 1870–1900* (Cambridge: Harvard University Press, 1962), 26.

58. *A Handbook of New England* (Boston: Porter E. Sargent, 1917), 404.

59. Ibid., 506.

60. "Great Parkway to the Fells," *Boston Globe,* December 3, 1896.

61. Landscape Research and Office of Planning and Community Development, City of Somerville, *Beyond the Neck,* 58.

62. Ibid., 58–60.

63. Advertisement cited in *Northwest Arlington, Massachusetts: An Architectural and Historical Study,* second edition (Arlington, Mass.: Arlington Historic Commission, 1995), 26.

64. Boston Chamber of Commerce, Committee on Municipal and Metropolitan Affairs; Committee on Public Utilities, *Metropolitan Planning and Development in Boston and Its Environs* (Boston: Boston Chamber of Commerce, 1922), 18.

65. Heckscher, *Open Spaces,* 24.

66. Metropolitan District Commission, *Annual Report of the Metropolitan District Commission for the Year 1926* (Boston: Metropolitan District Commission, 1927), 3.

67. Board of Metropolitan Park Commissioners, *Report of the Board of the Metropolitan Park Commissioners, December 1915* (Boston: Metropolitan Park Commission, 1916), 16.

68. Trustees of Public Reservations of Massachusetts, *The Bay Circuit* (Boston: Trustees of Public Reservations of Massachusetts, 1937), 3.

69. Metropolitan District Commission, *Annual Report of the Metropolitan District Commission for the Year 1936* (Boston: Metropolitan District Commission, 1937), 22.

70. John Nolen and Henry V. Hubbard, *Parkways and Land Values* (Cambridge: Harvard University Press, 1937), 8, 128.

71. "Blue Hills Provide Bit of Wild and Rugged Beauty at One's Door," *Christian Science Monitor,* August 2, 1927.

72. "Swampscott-to-Quincy Belt Nears Completion," August 28, 1958, unidentified Boston newspaper, *Boston Herald,* Clippings File, Beebe Communications Library, Boston University.

73. Metropolitan District Commission, Parks Division, *Study and Recommended Programs of the Development of Parks and Reservations and Recreational Facilities of the Metropolitan Parks District, 1956* (Boston: Metropolitan District Commission, 1956), 4.

74. Metropolitan Green Ribbon Commission, *Enhancing the Future,* 19.

75. Norman T. Newton, *Design on the Land: The Development of Landscape Architecture* (Cambridge: Harvard University Press, 1971), 333.

76. The MDC observed in its 1956 recreation plan that "glowing pictures have been presented of the tax advantages and other economic benefits to local communities of the transfer of large and apparently idle public areas to private ownership for industrial or residential development." See Metropolitan District Commission, *Study,* 1956, iii.

77. Ibid., 34, 80.

78. "Met. Dist. Area Extension Hit," *Boston Herald,* March 26, 1941.

79. Metropolitan Area Planning Council, *The 1976 Regional Open Space Plan,* vol. 1, *Open Space and Recreation Programs for Metropolitan Boston* (Boston: Metropolitan Area Planning Council, 1976), 15.

80. Parks historian Galen Cranz explained the role of contemporary parks in sustaining communities: "It also reflects a new place that parks have assumed in popular consciousness as a sort of urban asset or resource, a space permanently full of potential, whether it is put to use or not, for contributions to urban well-being" (see Cranz, *Politics of Park Design,* 222).

10. Reclaiming the Middle Charles River Reservation

1. "MDC Begins Picking Land for Acquisition," *Boston Globe,* May 25, 1988, 21.

2. Walter L. Creese, *The Crowning of the American Landscape: Eight Great Spaces and Their Buildings* (Princeton, N.J.: Princeton University Press, 1985), 168, 183; Commonwealth of Massachusetts, Metropolitan District Commission, *Study and Recommended Program of Development of Park and Reservation and Recreational Facilities of the Metropolitan Parks District,* Edwards, Kelcey and Beck, consultants (Boston: Metropolitan District Commission, 1956).

3. Commonwealth of Massachusetts, *Enhancing the Future of the Metropolitan Park System: Final Report and Recommendations of the Green Ribbon Commission* (Boston: Green Ribbon Commission, 1996), 25, 43.

4. Julius Gy Fabos and Jack Ahern, eds., *Greenways: The Beginning of an International Movement* (New York: Elsevier, 1996), 415–57; and Charles D. Little, *Greenways for America* (Baltimore, Md.: Johns Hopkins University Press, 1990), 179–98.

5. On the construction of the dam and the creation of the Charles River Basin as a public space, see Karl Haglund, *Inventing the Charles River* (Cambridge: MIT Press, 2003).

6. Massachusetts Metropolitan Area Planning Council, *Open Space and Recreation Program for Metropolitan Boston,* vol. 3 (Boston: Metropolitan Area Planning Council, 1969), 5.

7. In *Greenways for America,* 23–24, Little attributes the first use of the word "greenway" to an unreferenced study by Edmund Bacon cited in William H. Whyte, *Securing Open Space for Urban America: Conservation Easements,* Urban Land Institute Technical Bulletin no. 36 (1959): 23.

8. Massachusetts Metropolitan Area Planning Council, *Open Space and Recreation Program,* vol. 3, 32.

9. Julia B. O'Brien, interview with Karl Haglund, July 10, 2008.

10. Boston Conservation Commission, *Charles-to-Charles: A Conservation and Recreation Corridor for Boston, Brookline, and Newton* (Boston: Boston Conservation Commission, 1972); William Giezentanner, *Charles River Pathway Plan* (Newton, Mass.: Newton Conservation Commission, 1975); John Wacker & Associates, *The Charles River, Waltham* (Waltham, Mass.: Waltham Conservation Commission, 1974); and John Wacker & Associates, *The Charles River, Watertown* (Watertown, Mass.: Watertown Conservation Commission, 1974).

11. O'Brien interview.

12. Marty Carlock, "Hopes Rise for Charles River Pathway," October 4, 1992, *Boston Globe,* West Weekly, 18.

13. Haglund, *Inventing the Charles River,* 422.

14. Carlock, "Hope Rises for Charles River Pathway," 18.

15. Susan Hershey, "New River Walkway Dedicated," *Watertown Sun,* June 10, 1992, 1A.

16. Driscoll, "Corridor of Life," 11.

17. Dave Denison, "MDC Seeks to Reclaim Land for Walkway Along Charles," *Boston Globe,* June 4, 1992.

18. Driscoll, "Corridor of Life," 11.

19. Denison, "MDC Seeks to Reclaim Land."

20. Ibid.

21. Robin Estrin, "Bicycle Path Is Nearly Done, but Some See Hazards," *News Tribune,* October 16, 1992, 1, 6.

22. Denison, "MDC Seeks to Reclaim Land."

23. Jennifer R. Yelin, *Realizing the Vision: Reclaiming Public Open Space in the Upper Charles River Reservation* (Boston: Metropolitan District Commission, 1997), A-93.

24. Denison, "MDC Seeks to Reclaim Land."

25. Carlock, "Hope Rises for Charles River Pathway," 18.

26. Kevin Hollenbeck, July 17, 1997, interview in Yelin, *Realizing the Vision,* A-15.

27. Driscoll, "Corridor of Life," 15.

28. Yelin, *Realizing the Vision,* 35–38 and appendixes G, H, I.

29. Richard Higgins, "Gradually, MDC Gets Back Land Along Charles," *Boston Globe,* August 10, 1997, 11.

30. "Upper Charles River Restoration Program," October 8, 1997, Department of Conservation and Recreation Archives, Boston.

31. Higgins, "Gradually, MDC Gets Back Land Along Charles," 11.

32. Amy Mednick, "MDC Moves to Reclaim a Bank of the Charles," *Boston Globe,* April 11, 1993, 5.

33. "Minutes," Citizens Advisory Committee, Upper Charles River Reservation Restoration Master Plan, November 15, 1994.

34. Julia O'Brien, July 9, 1997, interview, in Yelin, *Realizing the Vision,* A-21, A-22.

35. For a sense of the opportunities and challenges from a national perspective at the time this project began, see Robert S. Patten with Amy Derry, Hal Heimstra, and Marianne Fowler, "ISTEA and Trails: Merging Transportation Needs and Recreation Values," available online at http://ntl.bts.gov/DOCS/mtn.html.

36. Commonwealth of Massachusetts, *Upper Charles River Reservation Master Plan* (Boston: Metropolitan District Commission, 1998), 7, 10.

37. Ralph Ranalli, "New Bridge Lovely, but Needed Repairs," *Boston Globe,* December 17, 2006, quoted in Driscoll, "Corridor of Life," 34.

38. Christopher Rogers, "MDC Is Taking It Back," *Watertown Sun,* October 9, 1996, 18.

39. Nick Cartier, "Riverwalk Continuing This Summer," *Newton Tab,* January 8, 1998, 16.

40. Elizabeth Harris, "Neighbors Warm up to Pleasant Street Greenway," *Watertown TAB and Press,* May 5, 1997, 30.

41. Harrison Sheppard, "Down by the River," *News Tribune,* February 25, 1998, 1, 24.

42. Harris, "Neighbors Warm up to Pleasant Street Greenway," 30.

43. Joel S. Finkelstein to David Balfour, August 11, 1997, Department of Conservation and Recreation Archives, Boston.

44. Gil Wooley, "Favorite Places: The Middle Charles," *Massachusetts Sierran* (Summer 2000): 7.

45. Emily Sweeney, "Charles River Walkway Takes a Big Step Forward," *Boston Globe,* Globe West, August 15, 2002, W1, 6.

46. Daniel Wolfman, "Upper Charles River Walk Sees Path to Completion Clearing," *Watertown TAB,* August 16, 2002, 3.

47. Bill Thompson, "Banking on a River: A New Greenway Reconnects Boston Area Residents with the Forgotten Charles," *Landscape Architecture Magazine* (September 1998): 50, 52–55.

48. Louise Bowditch, "A Walk Along the Charles," *Boston Globe,* January 13, 2005.

11. Boston's Weather and Climate Histories

1. Gregory A. Zielinski and Barry D. Keim, *New England Weather, New England Climate* (Hanover, N.H.: University Press of New England, 2003).

2. On the meteorology of the sea breeze, see James P. Barbato, "The Sea Breeze of the Boston Area and Its Effect on the Urban Atmosphere," Ph.D. dissertation, Boston University, 1975.

3. Ralph Waldo Emerson, "Boston," *Atlantic Monthly* 69 (1892): 27.

4. David M. Ludlum, *Early American Winters* (Boston: American Meteorological Society, 1966–68), and Ludlum, *The Country Journal New England Weather Book* (Boston: Houghton Mifflin, 1976), contain much material relating to Boston. For an example of meteorological history of a more recent vintage, see Paul J. Kocin, Alan D. Weiss, and Joseph J. Wagner, "The Great Arctic Outbreak and East Coast Blizzard of February 1899," *Weather Forecasting* 3 (1988): 305–18.

5. William R. Baron, "Eighteenth-century New England Climate Variation and Its Suggested Impact on Society," *Maine Historical Society Quarterly* 21 (1982): 201–18; William R. Baron et al., "Frost-Free Season Record Reconstruction for Eastern Massachusetts," *Journal of Climate and Applied Meteorology* 23 (1984): 317–19; and William R. Baron, "Historical Climates of the Northeastern United States: Seventeenth to Nineteenth Centuries," in George P. Nicholas, ed., *Holocene Human Ecology in Northeastern North America* (New York: Plenum Press, 1988), 29–46.

6. For an application of this approach to the United States, see William B. Meyer, *Americans and Their Weather: A History* (New York: Oxford University Press, 2000).

7. William B. Meyer, "Edward Bellamy and the Weather of Utopia," *Geographical Review* 94 (2004): 43–54; and C. P. Yaglou and Anne Mesmer, "The Importance of Clothing in Air Conditioning," *Journal of the American Medical Association* 117 (1941): 1261–62.

8. Anne Rowe Cunningham, ed., *Letters and Diary of John Rowe* (1903; reprint, New York: New York Times, 1969), 100, 138; and David Grayson Allen et al., eds., *The Diary of John Quincy Adams,* vol. 2 (Cambridge: Harvard University Press, 1981), 143.

9. See, for example, Tim Armstrong, "'A Good Word for Winter': The Poetics of a Season," *New England Quarterly* 60 (1987): 568–83; and Adam W. Sweeting, *Beneath the Second Sun: A Cultural History of Indian Summer* (Hanover, N.H.: University Press of New England, 2003). See Lauri Bauer Coleman, "'Rain Down Righteousness': Interpretations of Natural Events in Mid-eighteenth-century Boston" in this volume.

10. Lewis Mumford, *Technics and Civilization* (New York: Harcourt, Brace and Company, 1934). See Brian Donahue, "Remaking Boston, Remaking Massachusetts" in this volume.

11. David Cressy, "The Vast and Furious Ocean: The Passage to Puritan New England," *New England Quarterly* 57 (1984): 520–21.

12. William Wood, *New England's Prospect,* ed. Alden T. Vaughan (Amherst: University of Massachusetts Press, 1977), 26–27.

13. Fitz-Henry Smith, "Some Old-fashioned Winters in Boston," *Proceedings of the Massachusetts Historical Society* 65 (1932–36): 270, 305.

14. Baron, "Eighteenth-century New England Climate Variation," 213–14.

15. Karen Ordahl Kupperman, "The Puzzle of the American Climate in the Early Colonial Period," *American Historical Review* 87 (1982): 1262–89.

16. Sewall, as cited in Karen Ordahl Kupperman, "Climate and Mastery of the Wilderness in Seventeenth-Century New England," in David D. Hall and David Grayson Allen, eds., *Seventeenth-century New England* (Boston: Colonial Society of Massachusetts, 1984), 3–37 (quotation on 37).

17. Wood, *New England's Prospect,* 112; and James Axtell, *The Invasion Within: The Contest of Cultures in Colonial North America* (New York: Oxford University Press, 1985), 141.

18. William B. Meyer, "Harvard and the Heating Revolution," *New England Quarterly* 77 (2004): 588–94; and Cunningham, *Letters and Diary of John Rowe,* 119 (also 223, 225, 239).

19. Karen Ordahl Kupperman, "Fear of Hot Climates in the Anglo-American Colonial Experience," *William and Mary Quarterly,* third series, 41 (1984): 213–40.

20. John B. Blake, *Public Health in the Town of Boston, 1630–1822* (Cambridge: Harvard University Press, 1959).

21. Samuel Brown, *A Treatise on the Nature, Origin, and Progress of the Yellow Fever* (Boston: Manning & Loring, 1800), 22–23, 25. For similar observations, see W. Gamage, "Medical: For the Centinel," *Columbian Centinel,* May 29, 1799, 1; and "To the Board of Health," *Columbian Centinel,* June 26, 1802, 1. On discussions of the same phenomenon elsewhere in the same period, see William B. Meyer, "Urban Heat Island and Urban Health: Early American Perspectives," *The Professional Geographer* 43 (1991): 38–48.

22. Eric G. Nellis, "The Working Poor of Pre-Revolutionary Boston," *Historical Journal of Massachusetts* 17 (1989): 137–59; Richard B. Morris and Jonathan Grossman, "The Regulation of Wages in Early Massachusetts," *New England Quarterly* 11 (1938): 470–500; *Report of the Record Commissioners of the City of Boston,* vol. 31 (Boston: City Printers, 1876–1909), 239 (entry for August 19, 1790); and David Cressy, "The Seasonality of Marriage in Old and New England," *Journal of Interdisciplinary History* 16 (1985): 1–21.

23. Charles Francis Adams, *Three Episodes of Massachusetts History,* vol. 2 (Boston: Houghton Mifflin, 1892), 676; Ludlum, *Country Journal New England Weather Book,* 129; and "The Boston Traveller: Number 1," *The American Herald,* August 23, 1784, 2.

24. "Railway," *Boston Patriot & Mercantile Advertiser,* February 2, 1827, 2.

25. Clay McShane, "Gelded Age Boston," *New England Quarterly* 74 (2001): 275, 276.

26. Samuel Eliot Morison, *One Boy's Boston, 1887–1901* (Boston: Houghton Mifflin, 1962), 18.

27. Joseph Hatton, *Henry Irving's Impressions of America* (Boston: James R. Osgood, 1884), 237; Winslow Homer, "Sleighing in Haymarket Square, Boston" and "Sleighing on the Road, Brighton, Near Boston," *Ballou's Pictorial,* January 29, 1859; William Dean Howells, *Novels, 1875–1886* (New York: Library of America, 1982), 890–92; and McShane, "Gelded Age Boston," 282. For general background, see Blake McKelvey, *Snow in the Cities: A History of America's Urban Response* (Rochester, N.Y.: University of Rochester Press, 1995).

28. W. H. Bunting, *Portrait of a Port: Boston, 1852–1914* (Cambridge: Harvard University Press, 1971); and Robert G. Albion, William A. Baker, and Benjamin W. Labaree, *New England and the Sea* (Middletown, Conn.: Wesleyan University Press, 1972), 181–88, 191.

29. Meyer, "Harvard and the Heating Revolution," 594–99.

30. For examples, see Frank Draper, "Ventilation and Warming," in *Six Lectures on School Hygiene* (Boston: Ginn & Company, 1886), 33–64; "The Superheated American House," *Boston Medical and Surgical Journal* 163 (1910): 777; Robert DeC. Ward, "The Relative Humidity of Our Houses in Winter," *Boston Medical and Surgical Journal* 142 (1900): 217–19; and Meyer, "Harvard and the Heating Revolution."

31. Meyer, *Americans and Their Weather,* 128–29, 145. Such interruptions, of course, would not end with the neotechnic change from domestic coal to imported oil for heating.

32. *Twelfth Census of the United States, vol. 9: Manufactures, Part 3: Special Reports on Selected Industries* (Washington, D.C.: U.S. Census Office, 1902), 686; Richard O. Cummings, *The American Ice Harvests: A Historical Study in Technology, 1800–1918* (Berkeley: University of California Press, 1949); and Oscar Edward Anderson Jr., *Refrigeration in*

America: A History of a New Technology and Its Impact (Princeton, N.J.: Princeton University Press, 1953).

33. Howard S. Russell, *A Long, Deep Furrow: Three Centuries of Farming in New England* (Hanover, N.H.: University Press of New England, 1976), 451–53, 478, 505–07; and "New Winter Gardening Methods," *Boston Globe,* March 11, 1900, 29.

34. Alexander Keyssar, *Out of Work: The First Century of Unemployment in Massachusetts* (New York: Cambridge University Press, 1986), 62–68, 159–60; Bunting, *Portrait of a Port,* 224, 254; "Winter on the Wharves," *Boston Globe,* January 14, 1894, 27; McShane, "Gelded Age Boston," 286; *Boston Evening Transcript,* September 2, 1856, 2; William I. Cole, "Boston's Pauper Institutions," *New England Magazine* 24 (1898): 238; and *Second Annual Report of the Pauper Institutions Trustees of the City of Boston, for the Year Ending January 31, 1899* (Boston: Municipal Printing Office, 1899), 7–8.

35. John S. Billings, *Vital Statistics of Boston and Philadelphia, Covering a Period of Six Years Ending May 31, 1890* (Washington, D.C.: U.S. Government Printing Office, 1895), 21–22, 33; Paul Beaven, ed., "A History of the Boston Floating Hospital," *Pediatrics* 19 (1957): 629–38; and Josephine Halberstadt, "The Boston Floating Hospital, Season of 1906," *American Journal of Nursing* 7 (1907): 368.

36. William B. Meyer, "The Worst Weather Disaster in New England History," *Yankee* 61, no. 1 (1997): 48–54; and G. M. Shattuck and M. M. Hilferty, "Causes of Death from Heat in Massachusetts," *New England Journal of Medicine* 209 (1933): 319–29.

37. "Report of the Hearing Before a Committee of the House of Representatives of Massachusetts on the Occupation and Improvement of the Charlestown Flats in Charles River," *Boston City Documents for 1869,* no. 128, 46–47, 75, 81, 283–85, 289–90; and *Charles River Dam: Evidence and Arguments Before the Board of Harbor and Land Commissioners and Report Thereon* (Boston: Wright & Potter, 1895), 110, 160–61, 176.

38. Emerson, "Boston," 27; and William A. Koelsch, "Robert DeCourcy Ward, 1867–1931," in T. W. Freeman, ed., *Geographers: Biobibliographical Studies,* vol. 7 (1983), 145–51.

39. Compare, for example, Graham Storey, ed., *The Letters of Charles Dickens,* vol. 11, *1865–1867* (Oxford: Clarendon Press, 1999), 496, and Thomas Colley Grattan, *Civilized America,* vol. 1 (London: Bradbury and Evans, 1859), 105–6, with Curtis Guild, *Britons and Muscovites; or, Travels in Two Empires* (Boston: Lee and Shepard, 1888), 14–15, 167–17, 19–23, 32–34; and Margaret Williamson, *John and Betty's English History Visit* (Boston: Lothrop, Lee & Shepard, 1910), 18. On American/English differences, see Meyer, *Americans and Their Weather,* 74, 78–79, 127, and William B. Meyer, "Why Indoor Climates Change: A Case Study," *Climatic Change* 55 (2002): 395–407.

40. Martyn J. Bowden et al., "The Effect of Climate Fluctuations on Human Populations: Two Hypotheses," in T.M.L. Wigley, M. J. Ingram, and G. Farmer, eds., *Climate and History* (Cambridge: Cambridge University Press, 1981), 479–513.

41. "No Longer Isolated," *Daily Evening Item,* February 16, 1899, 7; and "The Power of the Elements," *New York Tribune,* November 30, 1898, 6.

42. Thomas Harrison Eames, "The Wreck of the Steamer 'Portland'," *New England Quarterly* 13 (1940): 191–206.

43. Edward L. Ullman, "Amenities as a Factor in Regional Growth," *Geographical Review* 44 (1954): 122–23.

44. William B. Meyer, "Vulnerability to Severe Storms Along the Massachusetts Coast: A Historical Perspective," paper presented at annual meeting, New England–

St. Lawrence Valley Regional Division of the Association of American Geographers, Boston, November 1996.

45. Sebastian Junger, *The Perfect Storm: A True Story of Men Against the Sea* (New York: Norton, 1997).

46. R. I. Glass, P. O'Hare, and J. L. Conrad, "Health Consequences of the Snow Disaster in Massachusetts, February 6, 1978," *American Journal of Public Health* 69 (1979): 1047–49.

47. Pablo Suarez, William Anderson, Vijay Mahal, and T. R. Lakshmanan, "Impacts of Flooding and Climate Change on Urban Transportation," *Transportation Research, Part D: Transport and Environment* 10 (2005): 231–44.

48. Paul H. Kirshen and N. M. Fennessey, "Possible Climate-Change Impacts on Water Supply of Metropolitan Boston," *Journal of Water Resources Planning and Management—ASCE* 121, no. 1 (1995): 61–70.

49. Rutherford H. Platt, "The 2020 Water Supply Study for Metropolitan Boston," *Journal of the American Planning Association* 61 (1995): 185–99; Sarah Asefa et al., "Damp Northeast Still Not Immune to Drought," *Salem News*, May 19, 2006, B7; and Troy Hill and Colin Polsky, "Adaptation to Drought in the Context of Suburban Sprawl and Abundant Rainfall," *Geographical Bulletin* 47, no. 2 (2006): 85–100.

50. Russ Lopez and H. Patricia Hynes, "Sprawl in the 1990s: Measurement, Distribution, and Trends," *Urban Affairs Review* 38 (2003): 325–55.

51. James Bishop Peabody, ed., *The Holmes-Einstein Letters: Correspondence of Mr. Justice Holmes and Lewis Einstein, 1903–1935* (New York: St. Martin's Press, 1964), 32.

52. With at least one (unsurprising) exception: Henry David Thoreau, *Walden*, ed. J. Lyndon Shanley (Princeton, N.J.: Princeton University Press, 1989), 131.

12. "Rain Down Righteousness": Interpretations of Natural Events in Mid-eighteenth-century Boston

Epigraph: Boston Gazette, November 24, 1755, 1.

1. J. E. Ebel, "A New Look at the 1755 Cape Ann, Massachusetts Earthquake," *Seismological Research Letters* 73 (2002): 418–19. According to Ebel, seismologists now measure the magnitude of the seismic moment rather than relying on older scales such as the one Richter developed. The 1755 earthquake also measured VIII on the Modified Mercalli Intensity scale, meaning that fright and panic were general but structural damage was slight. This scale is a quantitative measure, based on the reports of those who actually experienced an earthquake. For further discussion of this material, see Weston Geophysical Research, "The Historical Seismicity of New England," U.S. Nuclear Regulatory Commission submittal docket No. 50–471 (DES67601) (1976); and John E. Ebel, "Major Historical Earthquakes in Northeastern North America and Their Effects in Boston," paper presented at the Massachusetts Historical Society's Boston Environmental History Seminar, November 12, 2002.

2. Thomas Prince, *Earthquakes the Work of God, and Tokens of His Just Displeasure* (Boston: D. and Z. Fowle, 1755); Thomas Prince, *An Improvement of the Doctrine of Earthquakes Being the Works of God, and Tokens of His Just Displeasure* (Boston: D. and Z. Fowle, 1755); and Charles Chauncy, *Earthquakes as a Token of God's Anger* (Boston: Edes and Gill, 1755).

3. William G. McLoughlin, ed., *The Diary of Isaac Backus*, vol. 1 (Providence, R.I.: Brown University Press, 1979), 392.

4. Experience Wight Richardson, "Diary of Experience Wight Richardson, 1728–1782," transcribed and compiled by Ellen Richardson Glueck and Thelma Smith Ernst, typescript (1978), Massachusetts Historical Society, Boston, p. 35.

5. John Winthrop, *Lecture on Earthquakes* (Boston: Edes and Gill, 1755).

6. The term "supernatural rationalism" was first used by A. C. McGiffert in 1911 to describe the religious platform embraced by some members of the Church of England in reaction to the continuous religious unrest of the seventeenth century. For McGiffert, supernatural rationalism was natural religion and pragmatic Christian tolerance combined. A. C. McGiffert, *Protestant Thought Before Kant* (New York: Harper and Brothers, 1961), 186–254. The historian Conrad Wright later appropriated the term to define the religious beliefs of a broad spectrum of New England intellectuals in the eighteenth century. In this later context Wright identified a similar desire to align scripture and nature despite larger doctrinal differences. See Wright, *The Beginnings of Unitarianism in America* (Boston: Starr King Press, 1955); and Wright, *The Liberal Christians: Essays on American Unitarian History* (Boston: Beacon Press, 1970), 1–21. Although he labeled it "rational religion" rather than "supernatural rationalism," the historian Leslee Koch Gilbert has agreed with Wright that this set of beliefs was nearly universal in eighteenth-century learned circles. See Gilbert, "The American Enlightenment: Science, Politics, and the Religion of Reason," Ph.D. dissertation, University of Kentucky, 1998. The historian James Delbourgo has used the term "physico-theology" to refer to the set of beliefs in seventeenth- and eighteenth-century America in which "reason and piety were linked." See Delbourgo, *A Most Amazing Scene of Wonders: Electricity and Enlightenment in Early America* (Cambridge: Harvard University Press, 2006), especially 63–72.

7. John Winthrop, *A Letter to the Publishers of the Boston Gazette Containing an Answer to Rev. Mr. Prince's Letter* (Boston: Edes and Gill, 1756), 2.

8. Their acceptance of Christian Revelation is what distinguished supernatural rationalists from deists who believed that God's will could be discerned though human reason alone. See Wright, *Liberal Christians,* 5–8.

9. Newton's findings led him to believe that the system of the universe would, over time, deteriorate without a source of regeneration. He identified this source as God, who both designed and actively administered his creation. See Steven Shapin, *The Scientific Revolution* (Chicago: University of Chicago Press, 1996), 152–53.

10. These ideas have been addressed by a number of scholars. See, for example, Shapin, *Scientific Revolution,* 123–26; Gilbert, "American Enlightenment," 20–99; Michael Heyd, *"Be Sober and Reasonable": The Critique of Enthusiasm in the Seventeenth and Early Eighteenth Centuries* (Leiden: E. J. Brill, 1995), 3–8, 158–63; Michael P. Winship, *Puritan Providentialism in the Restoration and Early Enlightenment* (Baltimore, Md.: John Hopkins University Press, 1996), 32–44; and Lorraine Daston and Katharine Park, *Wonders and the Order of Nature: 1150–1750* (New York: Zone Books, 1998), 329–35.

11. As early as 1680, the sermons of Archbishop Tillotson were required reading at Harvard College. Tillotson summarized and popularized the tenets of natural religion. Sara Errington, "Wonders and the Creation of Evangelical Culture in New England, 1720–1820," Ph.D. dissertation, Brown University, 2000, 24.

12. For the history of the Royal Society and the development of their transatlantic communications, see Andrea Rushnock, "Correspondence Networks and the Royal Society, 1700–1750," *British Journal for the History of Science* 32 (1999): 155–69; and Richard Sorrenson, "Towards a History of the Royal Society in the Eighteenth Century," *Notes and Records of the Royal Society of London* 50 (1996): 29–46.

13. Thomas Prince, *The Agency of God in Droughts and Rains* (Boston: Kneeland and Green, 1749).

14. Prince, *Earthquakes the Work of God,* title page.

15. Harry Stout, *The New England Soul* (New York: Oxford University Press, 1986), 23–28.

16. Quoted in Eliphalet Williams, *The Duty of a People, Under Dark Providences, or Symptoms of Approaching Evil, to Prepare to Meet Their God* (New London, Conn.: Timothy and John Green, 1756), 43–44.

17. John Jerman, *The American Almanack for 1752* (Philadelphia: B. Franklin and D. Hall, 1751). This quotation comes from "A Hymn to the Creator and Preserver of All Things, the Great Being of All Beings," 4–5.

18. Wright, *Liberal Christians,* 16.

19. See Christopher Grasso, *A Speaking Aristocracy: Transforming Public Discourse in Eighteenth-century Connecticut* (Chapel Hill: University of North Carolina Press, 1999), 74, especially n65.

20. For a discussion of Bacon's plan to reform natural knowledge, see Shapin, *Scientific Revolution,* 139–50; and Daston and Park, *Wonders and the Order of Nature,* 231–52.

21. The relationship between God's direct intervention and his use of second causes is addressed in Shapin, *Scientific Revolution,* 142–50; and Gilbert, "American Enlightenment," 150–66.

22. Grasso has considered the fate of covenant language and corporate identity in eighteenth-century New England in *A Speaking Aristocracy,* 39–48; Wright has described the development of a benevolent image of God among eighteenth-century intellectuals, especially the Reverend Charles Chauncy; see Wright, *Beginnings of Unitarianism,* 161–83.

23. Nathaniel Ames, *An Astronomical Diary or Almanack for 1758* (Portsmouth, N.H.: Daniel Fowle, 1757), 15; and Nathaniel Ames, *An Astronomical Diary or Almanack for 1757* (Boston: J. Draper, 1756), title page.

24. Frank Lambert, *Inventing the Great Awakening* (Princeton, N.J.: Princeton University Press, 1999). Grasso has also argued that the Great Awakening was an important interpretive event in which public figures sought to establish their own explanations as the most valid; see Grasso, *Speaking Aristocracy,* 89.

25. For more on the learned reaction against enthusiasm, see Heyd, *"Be Sober and Reasonable."* For this topic in the context of eighteenth-century New England, see Errington, "Wonders and the Creation of Evangelical Culture," chapter 3.

26. In making this contention, I argue against those scholars who posit either the seventeenth-century scientific revolution or the religious upheavals in England and America as the catalyst for a linear move away from belief in the supernatural powers of God. Winship (*Puritan Providentialism*), Errington ("Wonders and the Creation of Evangelical Culture"), and Daston and Park (*Wonders and the Order of Nature*) all make this argument in some form. Instead, I agree with another group of historians of religion and science who argue for the complexity of the interpretive environment of early America, which embraced both rational and supernatural explanations of natural phenomena. These historians include, among others, Shapin (*Scientific Revolution*), Grasso (*Speaking Aristocracy*), Delbourgo (*Most Amazing Scene of Wonders*), and Nina Reid-Maroney (*Philadelphia's Enlightenment, 1740–1800: Kingdom of Christ, Empire of Reason* [Westport, Conn.: Greenwood Press, 2001]).

27. The war lasted from 1744 to 1748. William DeLoss Love, *The Fast and Thanksgiving Days of New England* (Cambridge, Mass.: Riverside Press, 1895), 319, see chapter 22, "Spells of Weather," for a descriptive account of the 1749 drought, 314–27.

28. *Boston Weekly News-letter,* June 22, 1749, 2; and *Boston Gazette,* June 20, 1749, 2.

29. Richardson, "Diary of Experience Wight Richardson," 10.

30. Francis G. Walett, ed., *The Diary of Ebenezer Parkman,* vol. 1 (New Castle, Del.: Oak Knoll Press, 1974), 201.

31. Love, *Fast and Thanksgiving Days,* 319. In addition to 1748 and 1749, two other years in the mid-eighteenth century were considered drought years: 1757 and 1762.

32. *Boston Gazette,* June 6, 1749, 2.

33. Love, *Fast and Thanksgiving Days,* 178, 319. According to Love, of all the weather-related calamities, fast days were held most often on account of drought.

34. Aaron Smith, *Some Temporal Advantages of Keeping Covenant with God in Two Discourses* (Boston: Kneeland, 1749).

35. Ibid., 16–17.

36. Walett, *Diary of Ebenezer Parkman,* 197.

37. Smith, *Some Temporal Advantages,* 13.

38. Thomas Prince, *The Agency of God in Droughts and Rains* (Boston: Kneeland and Green, 1749). Prince wrote this sermon on the occasion of the provincewide Thanksgiving day on August 24, 1749.

39. Prince dedicated the sermon to the Society for "clear[ing] the way for our admiring views and adorations of the sovereign creator and actual ruler of the universe"; see Prince, *God in Droughts and Rains,* dedication.

40. Ibid., 4–5.

41. Ibid., 25.

42. Ibid., 5.

43. McLoughlin, *Diary of Isaac Backus,* 65.

44. *Boston Evening Post,* July 24, 1749, 1.

45. Ibid., 2.

46. *Boston Gazette,* August 14, 1749, 1.

47. Fred Anderson, *Crucible of War: The Seven Years War and the Fate of Empire in British North America, 1754–1766* (New York: Random House, 2000). For information on the disastrous defeat of General Braddock at Fort Duquesne, see 94–107.

48. Williams, *Duty of a People,* 39, 49.

49. Charles Chauncy, *The Earth Delivered from the Curse to Which It Is, at Present, Subjected* (Boston: Edes and Gill, 1756), 23.

50. Jonathan Mayhew, *The Expected Dissolution of All Things* (Boston: Edes, Gill, and Draper, 1755). Excerpt reprinted in the *Boston Gazette,* September 13, 1756, 1.

51. Thomas Foxcroft, *The Earthquake: A Divine Visitation* (Boston: S. Kneeland, 1756), 45, 46.

52. The passage is from Psalms 18:7 and was cited by Foxcroft, *Earthquake,* 5; Chauncy, *Earth Delivered,* 24; Williams, *Duty of a People,* 44–45; and Prince, *Earthquakes the Work of God,* 3.

53. Timothy Harrington, *Prevailing Wickedness and Distressing Judgments: Ill-boding Symptoms on a Stupid People* (Boston: Edes and Gill, 1756), 13.

54. Mather Byles Jr., *Divine Power and Anger Displayed in Earthquakes* (Boston: S. Kneeland, 1755), 26.

55. Mather Byles, *The God of Tempest and Earthquake* (Boston: D. and Z. Fowle, 1755), 7–8.

56. Williams, *Duty of a People,* 47.

57. Byles, *Divine Power and Anger Displayed,* 22.

58. Eric Seeman, *Pious Persuasions: Laity and Clergy in Eighteenth-century New England* (Baltimore, Md.: Johns Hopkins University Press, 1999), 149–51.

59. Chauncy, *Earthquakes as a Token,* 27–28; and Thomas Prentice, *Observations Moral and Religious, on the Late Terrible Night of the Earthquake* (Boston: S. Kneeland, 1756), 3.

60. Foxcroft, *Earthquake,* 5.

61. Byles, *Divine Power and Anger Displayed,* 7–8.

62. Chauncy, *Earthquakes as a Token,* 8.

63. Ibid., 8.

64. Gilbert also makes this argument, in his "American Enlightenment," 105. On this point I disagree with Errington, whose analysis of the 1755 earthquake sermons leads her to believe that, between 1727 and 1755, ministers had become wary of claiming God's immediate intervention and assigning supernatural causes to earthquakes because of the effects of enthusiasm related to the Great Awakening. See Errington, "Wonders and the Creation of Evangelical Culture," 149–61.

65. Byles, *Divine Power and Anger Displayed,* 1.

66. Williams, *Duty of a People,* 12; and Prentice, *Observations Moral and Religious,* 5.

67. Foxcroft, *Earthquake,* 27.

68. Prentice, *Observations Moral and Religious,* 5.

69. Williams, *Duty of a People,* 43, 46.

70. Foxcroft, *Earthquake,* 30.

71. Chauncy, *Earthquakes as a Token,* 9–10.

72. Prince, *Earthquakes the Work of God,* 9, 4–9, and 16–18; Prince, *Improvement,* 4–5, 14; and Prince, *Earthquakes the Work of God,* 15 and 12.

73. Prince, *Earthquakes the Work of God,* 20–23.

74. The Royal Society of London published a special edition of its *Philosophical Transactions* following two earthquakes in London in 1750. In this edition the Reverends Stephen Hales and William Stuckely argued that electricity was the cause of these earthquakes, which occurred exactly one month apart. See *Philosophical Transactions* 46 (1749–1750).

75. William Stuckely, "The Reverend William Stuckely to the President, on the Causes of Earthquakes," *Philosophical Transactions* 46 (1749–1750): 642, 645.

76. Winthrop, *Lecture on Earthquakes,* 29.

77. John Winthrop, "An Account of the Earthquake Felt in New England, and the Neighboring Parts of America, on the 18th of November, 1755," *Philosophical Transactions* 50 (1757): 1–18.

78. Winthrop, *Lecture on Earthquakes,* 7–8.

79. Prince, *Earthquakes the Work of God,* 12.

80. Winthrop, *Lecture on Earthquakes,* 27.

81. Ibid., 29–31.

82. Ibid., 29.

83. Ibid., appendixes 32–38.

84. Ibid., 36.

85. Winthrop, *Letter to the Publishers of the Boston Gazette,* 7.

86. According to Daston and Park, certain elements of society in the seventeenth and eighteenth centuries were more likely to be influenced by their passions: "because a self fortified by reason and will could resist the onslaughts of the unholy trinity of enthusiasm, superstition, and imagination, not everyone was believed to be equally at risk. Particularly susceptible were women, the very young, the very old, primitive people, and the uneducated masses, a motley group collectively designated as 'the vulgar'" (see Daston and Park, *Wonders and the Order of Nature,* 343).

87. Charles Chauncy, *Enthusiasm Described and Cautioned Against* (Boston: Draper for Eliot and Blanchard, 1742), 26.

88. Williams, *Duty of a People,* 36.

89. T. D. Kendrick, *The Lisbon Earthquake* (London: Methuen, 1956), 11–14.

90. For a discussion of the panicked response to the quake, see Kendrick, *Lisbon Earthquake,* 62–63 and 88–90 specifically.

91. *Two Very Circumstantial Accounts of the Late Dreadful Earthquake at Lisbon* (Boston: Reprinted by D. and Z. Fowle, 1756), 6.

92. To feed the popular demand for earthquake literature, Boston printers published a number of reprints from the London press, some of them several times. These included *Two Very Circumstantial Accounts: An Account of the Late Dreadful Earthquake of Lisbon* (Boston: Green and Russell, 1756); Thomas Sherlock, *Letter from the Lord Bishop of London to the Clergy and People of London on Occasion of the Late Earthquakes* (Boston: Reprinted by John Draper, 1750); and *Lines Made after the Great Earthquake* (Boston: 1755).

93. John Winthrop, *Two Lectures on Comets* (Boston: Green and Russell, 1759), appendix, i.

94. *Boston Evening-Post,* May 7, 1759, 2.

95. Winthrop, *Two Lectures on Comets,* 39–41.

96. Ibid., 41–42. The natural philosopher Thomas Clap also acknowledged the salutary effect of comets on the atmosphere in his publication on meteors, in his *Terrestrial Comets: Conjectures upon the Nature and Motion of Meteors* (Norwich, Conn.: John Trumbull, 1781).

97. Winthrop, *Two Lectures on Comets,* 40, 42.

98. Ibid., 42; Clap, *Terrestrial Comets,* 12, 13; and Benjamin Franklin, *Poor Richard Improved, Being an Almanack and Ephemeris for 1757* (Philadelphia: B. Franklin and D. Hall, 1756), 29–30. The theory that the flood was caused by a comet, published by Cambridge professor and clergyman William Whiston, was widely referred to in both popular and more learned publications of the 1750s.

99. Winthrop, *Two Lectures on Comets,* 42–43.

100. Ibid., 8.

101. *Blazing Stars Messengers of God's Wrath: In a Few Serious and Solemn Meditations upon the Wonderful Comet* (Boston: R. Draper and Fowle and Draper, 1759).

102. *Boston Evening-Post,* May 21, 1759, 2.

103. Winthrop, *Two Lectures on Comets,* 44.

104. Clap used the phrase "omnipotent creator" in *Conjectures upon the Nature and Motion of Meteors,* 11. Winthrop referred to God's "consummate wisdom" in *Two Lectures on Comets,* 43.

105. *Conjectures upon the Nature and Motion of Meteors* was published posthumously in 1781 but was written by Clap in 1755.

106. Almanac writers capitalized on the predicted arrival of the comet, devoting significant space in their annual publications to articles on the nature and function of

comets. Obviously, earthquakes did not garner the same sort of attention in almanacs because they were unpredictable events more suitably addressed by newspaper reports and occasional sermons.

107. Marion Barber Stowell, *Early American Almanacs: The Colonial Weekday Bible* (New York: B. Franklin, 1977), x.

108. Elizabeth Carroll Reilly, "Common and Learned Readers: Shared and Separate Spheres in Mid-eighteenth-century New England," Ph.D. dissertation, Boston University, 1994, 11; see 408–15 for further discussion.

109. George Wheten, *An Astronomical Diary or an Almanack for 1753* (Boston: D. Fowle), 19–23.

110. David Sewall, *An Astronomical Diary or Almanack for 1758* (Portsmouth, N.H.: D. Fowle), 15.

111. James Davis, *An Astronomical Diary or an Almanack for 1758* (Boston: Edes and Gill), 15.

112. Sewall, *Astronomical Diary,* title page.

113. John Jerman, *The American Almanack for 1752* (Philadelphia: B. Franklin and D. Hall, 1751). This quotation comes from "A Hymn to the Creator and Preserver of All Things, the Great Being of All Beings," 4–5.

114. Ames, *Astronomical Diary or Almanack for 1759,* 19.

115. The almanac writer Jesse Parsons made the same reference to comets as former planets punished by God, *The American Ephemeris and Almanack for 1757* (New York: J. Parker and W. Weyman), 22. The almanac writer John Tobler discussed second causes, in *The Pennsylvania Town and Country-man's Almanack for 1758* (Philadelphia: Soloman Fussel, 1757), 32.

116. Franklin, *Poor Richard Improved,* 30–31.

117. Parsons used the phrase "instruments of divine vengeance," in *American Ephemeris,* 20. The comet as the cause of the flood is also mentioned by Parsons, in ibid., 21, and Sewall, *Astronomical Diary,* 16.

118. Tobler, *Pennsylvania Town and Country-man's Almanack,* 2–3.

119. Ames, *Astronomical Diary or Almanack for 1757,* title page.

120. Winthrop, *Two Lectures on Comets,* 44.

121. Winthrop, *Lecture on Earthquakes,* 31.

122. Roger Sherman, *An Astronomical Diary* (New Haven, Conn.: James Parker and Company, 1756) , 3.

123. Chauncy, *Earthquakes as a Token,* 15, 7.

124. Franklin, *Poor Richard Improved,* 29–30.

125. Winthrop, *Lecture on Earthquakes,* 31.

126. Winthrop, *Two Lectures on Comets,* 8; and Winthrop, *Letter to the Publishers of the Boston Gazette,* 7.

13. Biological Responses to Climate Change in Boston

1. M. N. Jensen, "Climate Warming Shakes up Species," *BioScience* 54 (2004): 722–29; C. Parmesan, "Ecological and Evolutionary Responses to Recent Climate Change," *Annual Review of Ecology, Evolution, and Systematics* 37 (2006): 637–69; and J. A. Pounds, M. R. Bustamante, L. A. Coloma, J. A. Consuegra, M.P.L. Fogden, P. N. Foster, E. La Marca, K. L. Masters, A. Merino-Vinteri, R. Puschendorf, S. R. Ron, G. A. Sánchez-

Azofeifa, C. J. Still, and B. E. Young, "Widespread Amphibian Extinctions from Epidemic Disease Driven by Global Warming," *Nature* 439 (2006): 161–67.

2. C. Parmesan and G. Yohe, "A Globally Coherent Fingerprint of Climate Change Impacts Across Natural Systems," *Nature* 421 (2003): 37–42; and T. L. Root, J. T. Price, K. R. Hall, S. H. Schneider, C. Rosenzweig, and J. A. Pounds, "Fingerprints of Global Warming on Wild Animals and Plants," *Nature* 421 (2003): 57–60.

3. A. J. Miller-Rushing and R. B. Primack, "Climate Change and Plant Conservation: Plant Conservation Strategies Need to Anticipate Climate Change," *Plant Talk* 35 (2004): 34–38; and T. Sparks, "Lateral Thinking on Data to Identify Climate Impacts," *Trends in Ecology and Evolution* 22 (2007): 169–71.

4. D. Primack, C. Imbres, R. B. Primack, A. J. Miller-Rushing, and P. Del Tredici, "Herbarium Specimens Demonstrate Earlier Flowering Time in Response to Warming in Boston," *American Journal of Botany* 91 (2004): 1260–64.

5. A. Ledneva, A. J. Miller-Rushing, R. B. Primack, and C. Imbres, "Climate Change as Reflected in a Naturalist's Diary, Middleborough, Massachusetts," *Wilson Bulletin* 116 (2004): 224–31.

6. A. J. Miller-Rushing, R. B. Primack, D. Primack, and S. Mukunda, "Photographs and Herbarium Specimens as Tools to Document Phenological Changes in Response to Global Warming," *American Journal of Botany* 93 (2006): 1667–74.

7. Sarah Chapin, *Concord Flora 1834–1836 Observed by Edward Jarvis* (Concord, Mass.: Sarah Chapin, 1994); Henry David Thoreau, *The Journal of Henry D. Thoreau* (Boston: Houghton Mifflin, 1906); A. J. Miller-Rushing and R. B. Primack, "Global Warming and Flowering Times in Thoreau's Concord: A Community Perspective," *Ecology* 89 (2008): 332–41; Alfred W. Hosmer Botanical Manuscripts, 1878–1903, William Munroe Special Collections, Concord Free Public Library, Concord, Mass.; and R. J. Eaton, *A Flora of Concord* (Cambridge: Museum of Comparative Zoology, Harvard University, 1974).

8. J. H. Conover, "Highlights of the History of the Blue Hill Observatory and the Early Days of the American Meteorological Society," *Bulletin of the American Meteorological Society* 66 (1985): 30–37.

9. G. A. Hodgkins and I. C. James II, "Historical Ice-Out Dates for 29 Lakes in New England: U.S. Geological Survey Open-File Report 02-34," 2002.

10. E. A. Holyoke, "A Meteorological Journal for the Year 1786 to the Year 1822, Inclusive," *Memoirs of the American Academy of Arts and Sciences* 1 (1833): 107–215; J. P. Hall, "Register of the Thermometer for 36 Years, from 1821 to 1856, to Which Is Added the Quantity of Rain Falling in Boston, Mass., for 34 years, from 1823 to 1856," *Memoirs of the American Academy of Arts and Sciences* 4 (1857): 1–2, Parts I and II; U.S. Patent Office, *Results of Meteorological Observations: Made under the Direction of the United States Patent Office and the Smithsonian Institution from the Year 1854 to 1859* (Washington, D.C.: U.S. Government Printing Office, 1864); and L. Hill, *Meteorological and Chronological Register: Comprising a Record of the Weather . . . from a Personal Diary of the Author from 1806 to 1869* (Plymouth, Mass.: Moses Bates, 1869).

11. Intergovernmental Panel on Climate Change (IPCC), *Climate Change 2007, The Physical Science Basis: Contribution of Working Group 1 to the Fourth Assessment Report of the Intergovernmental Panel on Climate Change,* ed. S. Solomon, D. Qin, M. Manning, Z. Chen, M. Marquis, K. B. Averyt, M. Tignor, and H. L. Miller (Cambridge: Cambridge University Press, 2007).

12. New England Regional Assessment Group, *New England Regional Assessment* (Durham, N.H.: University of New Hampshire, Institute for the Study of Earth, Oceans, and Space, 2001).

13. Primack et al., "Herbarium Specimens Demonstrate Earlier Flowering Time," 1260–64.

14. A. H. Fitter, R.S.R. Fitter, I.T.B. Harris, and M. H. Williamson, "Relationships Between First Flowering Date and Temperature in the Flora of a Locality in Central England," *Functional Ecology* 9 (1995): 55–60; T. H. Sparks and P. D. Carey, "The Responses of Species to Climate over Two Centuries: An Analysis of the Marsham Phenological Record, 1736–1947," *Journal of Ecology* 83 (1995): 321–29; T. H. Sparks, E. P. Jeffree, and C. E. Jeffree, "An Examination of the Relationship Between Flowering Times and Temperature at the National Scale Using Long-Term Phenological Records from the UK," *International Journal of Biometeorology* 44 (2000): 82–87; and D. R. Cayan, S. A. Kammerdiener, M. D. Dettinger, J. M. Caprio, and D. H. Peterson, "Changes in the Onset of Spring in the Western United States," *Bulletin of the American Meteorological Society* 82 (2001): 399–415.

15. Ledneva et al., "Climate Change as Reflected in a Naturalist's Diary," 224–31.

16. Fitter et al., "Relationships Between First Flowering Date and Temperature in the Flora," 55–60; Sparks and Carey, "Responses of species to Climate over Two Centuries," 321–29; N. L. Bradley, A. C. Leopold, J. Ross, and W. Huffaker, "Phenological Changes Reflect Climate Change in Wisconsin," *Proceedings of the National Academy of Sciences, USA* 96 (1999): 9701–04; Sparks, Jeffree, and Jeffree, "Examination of the Relationship," 82–87; and D. R. Cayan, S. A. Kammerdiener, M. D. Dettinger, J. M. Caprio, and D. H. Peterson, "Changes in the Onset of Spring in the Western United States," *Bulletin of the American Meteorological Society* 82 (2001): 399–415.

17. Hosmer Botanical Manuscripts; and Miller-Rushing and Primack, "Global Warming and Flowering Times," 332–41.

18. Root et al., "Fingerprints of Global Warming on Wild Animals and Plants," 57–60; and C. Parmesan, "Influences of Species, Latitudes, and Methodologies on Estimates of Phenological Response in Global Warming," *Global Change Biology* 13 (2007): 1860–72.

19. D. W. Inouye, B. Barr, K. B. Armitage, and B. D. Inouye. "Climate Change Is Affecting Altitudinal Migrants and Hibernating Species," *Proceedings of the National Academy of Sciences, USA* 97 (2000): 1630–33; and C. Both, S. Bouwhuis, C. M. Lessells, and M. E. Visser, "Climate Change and Population Declines in a Long-Distance Migratory Bird," *Nature* 441 (2006): 81–83.

CONTRIBUTORS

Lauri Bauer Coleman is a Ph.D. candidate at the College of William and Mary and an adjunct professor at Moravian College.

Brian Donahue is associate professor of American environmental studies on the Jack Meyerhoff Foundation at Brandeis University.

Daniel Driscoll is the director of recreation facilities planning at the Massachusetts Department of Conservation and Recreation.

Duncan M. FitzGerald is professor of earth sciences at Boston University.

Karl Haglund for the past twenty years has been the project manager at the Massachusetts Department of Conservation and Recreation for the New Charles River Basin, where the "lost half mile" of the river meets Boston Harbor.

Stephen T. Mague is a senior project manager with Durand & Anastas Environmental Strategies, Inc., an environmental consulting firm in Boston.

William B. Meyer is a lecturer in geography at Colgate University.

Abraham J. Miller-Rushing worked on his contribution as a doctoral student at Boston University. He is now the coordinator of the Wildlife Phenology Program of the USA National Phenology Network and The Wildlife Society.

James C. O'Connell is a planner and historian with the National Park Service's Northeast Region Office in Boston.

Anthony N. Penna is professor emeritus of history at Northeastern University.

Richard B. Primack is a professor of biology at Boston University.

Michael Rawson is an assistant professor of history at Brooklyn College (CUNY).

Peter S. Rosen is an associate professor of geology at Northeastern University.

Steven M. Rudnick retired as the director of the environmental studies program at the University of Massachusetts Boston.

David Soll is a Mellon Post-Doctoral Fellow in Environmental Studies at Lafayette College.

Conrad Edick Wright is the Ford Editor of Publications and director of research at the Massachusetts Historical Society.

INDEX

Note: Page numbers in italic type indicate the presence of figures, illustrations, or tables.